OCR AS/A level Chemistry A

Second Edition

1

Sam Holyman
David Scott
Victoria Stutt

ALWAYS LEARNING

PEARSON

Published by Pearson Education Limited, 80 Strand, London, WC2R 0RL.

www.pearsonschoolsandfecolleges.co.uk

Edited by Tracey Cowell, Kate Redmond and Tony Clappison
Designed by Elizabeth Arnoux for Pearson Education Limited
Typeset by Tech-Set Ltd, Gateshead

Illustrated by Tech-Set Ltd, Gateshead and Peter Bull Art Studio
Cover design by Juice Creative
Picture research by Alison Prior

First edition published 2008

This edition published 2015

18 17 16 15
10 9 8 7 6 5

British Library Cataloguing in Publication Data
A catalogue record for this book is available from the British Library

ISBN 978 1 447 99078 9

Websites
Pearson Education Limited is not responsible for the content of any external internet sites. It is essential for tutors to preview each website before using it in class so as to ensure that the URL is still accurate, relevant and appropriate. We suggest that tutors bookmark useful websites and consider enabling students to access them through the school/college intranet.

Printed in Italy by Lego S.p.A

This resource is endorsed by OCR for use with specification OCR Level 3 Advanced Subsidiary GCE in Chemistry A (H032) and OCR Level 3 Advanced GCE in Chemistry A (H432). In order to gain OCR endorsement this resource has undergone an independent quality check. OCR has not paid for the production of this resource, nor does OCR receive any royalties from its sale. For more information about the endorsement process please visit the OCR website www.ocr.org.uk

Acknowledgements

The publisher would like to thank the following for their kind permission to reproduce their photographs:

(Key: b-bottom; c-centre; l-left; r-right; t-top)

Alamy Images: Jonathan Eastland 123t, Justing Kase z12z 179tl, Libby Welch 101b, Lowe Photo 123c, Peter Titmuss 189t, Science Photos 213tr, Steve Bloom Images 191c; **Brown University:** Wang lab 100; **Dave Gent:** Dave Gent 206br; **DK Images:** Steve Gorton 190c; **Fotolia.com:** Philip Kinsey 128br, Thierry Milherou 99r; **Getty Images:** Jeff Foot / Discovery Channel Images 128l; **Martyn F. Chillmaid:** 21l, 21c, 22l, 61c, 183c, 210b, 213cl, Martyn Chillmaid 11r; **NASA:** NASA / ESA / S.Beckwith (STScl) and the HUDF Team 71bl; **Pearson Education Ltd:** Trevor Clifford 27b, 50tr, 178l, 203l, Gareth Boden 61t, Oxford Designers & Illustrators Ltd 119tr, 119cl, Coleman Yuen 190b; **Rob Ritchie:** Rob Ritchie 42l, 49tl, 49tc, 49cl; **Science Photo Library Ltd:** 10l, 18br, 107bl, 178br, 196-197, Andrew Lambert Photography 10r, 12br, 14bl, 14br, 15bl, 18cl, 23br, 35tr, 46, 69cl, 91l, 116l, 118l, 120c, 125bl, 127tl, 138l, 201tr, Charles Clarivan 208r, Charles D Winters 42r, 63t, 139tl, 139tc, 153tl, Geoff Tompkinson 38l, Gusto Images 13cr, Kenneth Libbrecht 74-75, Martyn F Chillamid 20bl, Martyn F Chillmaid 25r, 62cl, 65tl, 143r, 148bl, 178tr, Mikkel Juul Jensen 12c, Monty Rakusen 15tr, NASA / Goddard Space Flight Centre 209bl, P J Stewart 104-105, Paul Rapson 208bl, Phil Hill 30-31, Prof K Seddon & J van den Berg / Queen's University, Belfast 162-163, Russell Knightley 12cl, 12bl, Victor De Schwannberg 200c; **SciLabware Ltd:** 210t; **Shutterstock.com:** alterfalter 189b, Andrey Prokhorov 132-133, Huguette Roe 188tr, Leungchopan 188br, MarcelClemens 193tl, Mopic 119bl, Rido 189c, Tischenko Irina 8-9

Cover images: *Front:* **Science Photo Library Ltd:** Steve Gschmeissner

All other images © Pearson Education

We are grateful to the following for permission to reproduce copyright material:

Figures
Figure on page 193 adapted from 'Catalysts for a green industry', reproduced by permission of the Royal Society of Chemistry from http://www.rsc.org/education/eic/issues/2009July/catalyst-green-chemistry-research-industry.asp; Figure on page 227 adapted from 'Five rings good, four rings bad', reproduced by permission of the Royal Society of Chemistry from http://www.rsc.org/education/eic/issues/2010Mar/FiveRingsGoodFourRingsBad.asp

Text
Article on page 70 adapted from 'Stars to Stalagmites: How Everything Connects', Wspc (Paul S Braterman); Article on page 100 from 'First boron buckyballs roll out of the lab', New Scientist (Jacob Aron), © 2014 Reed Business Information - UK. All rights reserved. Distributed by Tribune Content Agency; Article on page 128 adapted from 'Some of our Selenium is missing', New Scientist Magazine issue 2004 (http://www.newscientist.com/article/mg14820044.600-some-of-our-selenium-is-missing.html), © Copyright Reed Business Information Ltd, © 1995 Reed Business Information - UK. All rights reserved. Distributed by Tribune Content Agency; Article on page 158 adapted from 'Organic Chemists Contribute to Renewable Energy', http://www.rsc.org/Membership/Networking/InterestGroups/OrganicDivision/organic-chemistry-case-studies/organic-chemistry-biofuels.asp, © Royal Society of Chemistry 2014, reproduced by permission of the Royal Society of Chemistry from http://www.rsc.org/Membership/Networking/InterestGroups/OrganicDivision/organic-chemistry-case-studies/organic-chemistry-biofuels.asp; Article on page 192 adapted from 'Catalysts for a green industry', reproduced by permission of the Royal Society of Chemistry from http://www.rsc.org/education/eic/issues/2009July/catalyst-green-chemistry-research-industry.asp; Article on page 226 adapted from 'Five rings good, four rings bad', reproduced by permission of the Royal Society of Chemistry from http://www.rsc.org/education/eic/issues/2010Mar/FiveRingsGoodFourRingsBad.asp

Every effort has been made to contact copyright holders of material reproduced in this book. Any omissions will be rectified in subsequent printings if notice is given to the publishers.

Contents

Module 3
The periodic table and energy

Module 4
Core organic chemistry

How to use this book

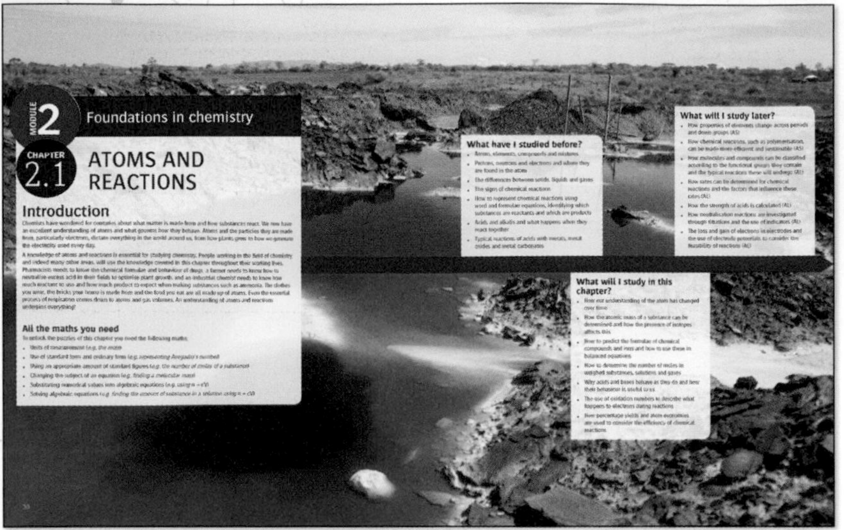

Welcome to your OCR AS/A level Chemistry A student book. In this book you will find a number of features designed to support your learning.

Chapter openers

Each chapter starts by setting the context for that chapter's learning:

- Links to other areas of Chemistry are shown, including previous knowledge that is built on in the chapter and future learning that you will cover later in your course.
- The **All the maths I need** checklist helps you to know what maths skills will be required.

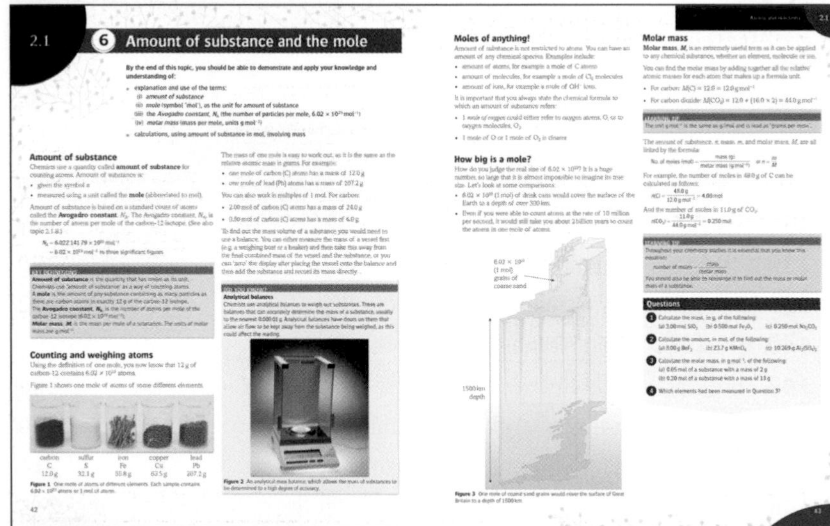

Main content

The main part of the chapter covers all of the points from the specification you need to learn. The text is supported by diagrams and photos that will help you understand the concepts.

Within each topic, you will find the following features:

- **Learning objectives** at the beginning of each topic highlight what you need to know and understand.
- **Key terms** are shown in bold and defined within the relevant topic for easy reference.
- **Worked examples** show you how to work through questions, and how your calculations should be set out.
- **Investigations** provide a summary of practical experiments that explore key concepts.
- **Learning tips** help you focus your learning and avoid common errors.
- **Did you know?** boxes feature interesting facts to help you remember the key concepts.

At the end of each topic, you will find **questions** that cover what you have just learned. You can use these questions to help you check whether you have understood what you have just read, and to identify anything that you need to look at again.

2.1 7 Types of formulae

What is an empirical formula?

Calculating empirical formulae

What is a molecular formula?

Calculating molecular formulae

Questions

Thinking Bigger

At the end of each chapter there is an opportunity to read and work with real-life research and writing about science. These sections will help you to expand your knowledge and develop your own research and writing techniques. The questions and tasks will help you to apply your knowledge to new contexts and to bring together different aspects of your learning from across the whole course. The timeline at the bottom of the spread highlights which other chapters of your book the material relates to.

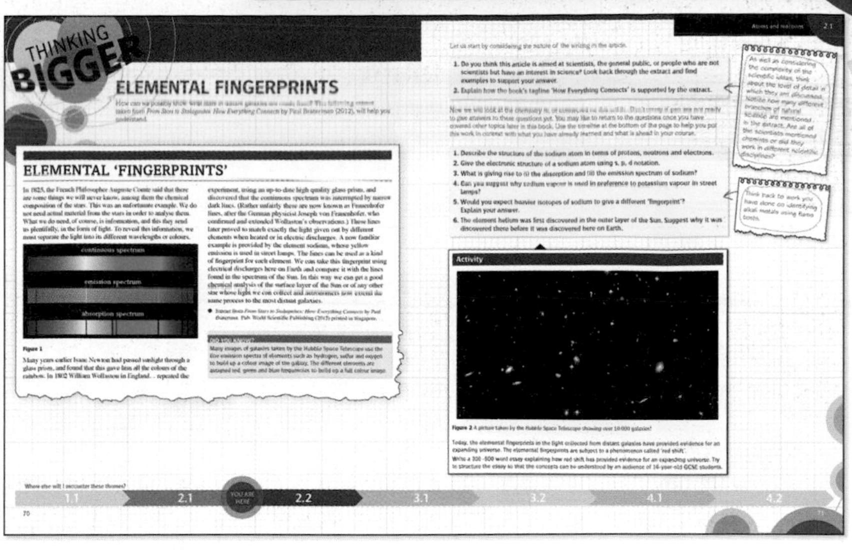

These spreads will give you opportunities to:

- read real-life material that's relevant to your course
- analyse how scientists write
- think critically and consider relevant issues
- develop your own writing
- understand how different aspects of your learning piece together.

Practice questions

At the end of each chapter, there are **practice questions** to test how fully you have understood the learning.

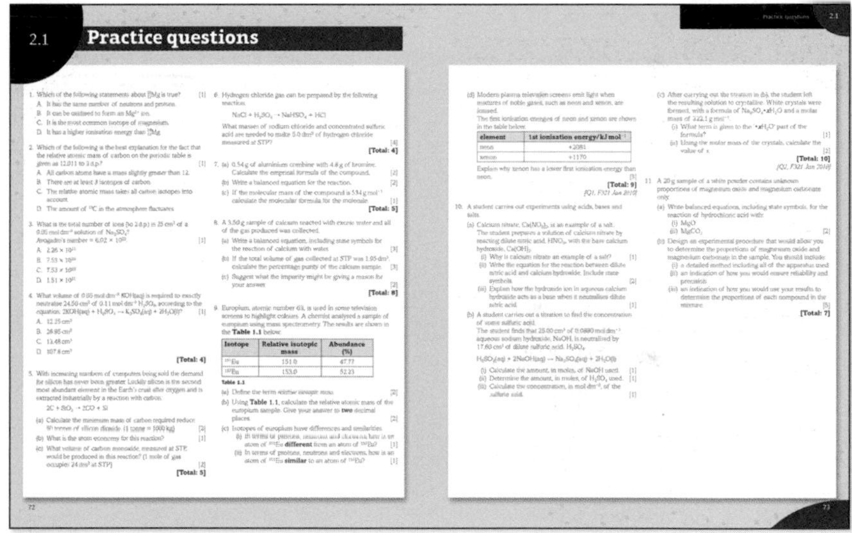

Getting the most from your ActiveBook

Your ActiveBook is the perfect way to personalise your learning as you progress through your OCR AS/A level Chemistry A course. You can:

- access your content online, anytime, anywhere
- use the inbuilt highlighting and annotation tools to personalise the content and make it really relevant to you
- search the content quickly.

Highlight tool

Use this to pick out key terms or topics so you are ready and prepared for revision.

Annotations tool

Use this to add your own notes, for example, links to your wider reading, such as websites or other files. Or make a note to remind yourself about work that you need to do.

1 Practical skills in chemistry

CHAPTER 1.1

PRACTICAL SKILLS ASSESSED IN A WRITTEN EXAMINATION

Introduction

The philosophers in Ancient Greece would ponder the world around them and generate ideas to explain their observations. Some were correct, like the idea of the atom, and others were not, like the hypothesis that the world was flat. It wasn't until the age of the alchemists that logical planned practical investigations into the material world were made. These forerunners of modern day chemists would mix chemicals, heat, cool and observe – all in their practical search for gold. This experimentation gave a lot of advances to the sciences and medicine, including discovering many elements, such as phosphorus. So, experiments are an important part of science.

Practical science allows researchers to test theories, explore the world and make new discoveries. To make sure that scientists don't miss anything a clear repeatable approach is used in practical work. This involves setting an aim, which is then used to plan and carry out an investigation. Having the skills to logically plan out an experiment and interpret the results while modifying techniques underpins our understanding of many areas of chemistry.

All the maths you need

To unlock the puzzles of this chapter you need the following maths:

- Use the correct units to quantify a variable
- Display data in different formats such as tables, charts and graphs
- Be able to calculate averages (mean)
- Be able to calculate the gradient of a trend line on a graph
- Identify anomalies in a data set
- Be able to represent numbers in standard form
- Be able to represent numbers in appropriate significant figures

What have I studied before?

- Definitions of dependent, independent and control variables
- Followed a method to obtain results
- Recorded practical results in a table
- Generated line graphs with lines of best fit
- Draw simple conclusions consistent with the results
- Evaluated simple methods

What will I study later?

- How experiments can be used to test a hypothesis
- How experiments can be used to expand our knowledge
- How experiments can give information that may be interpreted incorrectly

What will I study in this chapter?

- How to write an experimental aim and use it to determine if an investigation is valid
- The difference between a method and an outline
- How to display results in the most appropriate way
- How to mathematically manipulate results
- How to fully evaluate a method

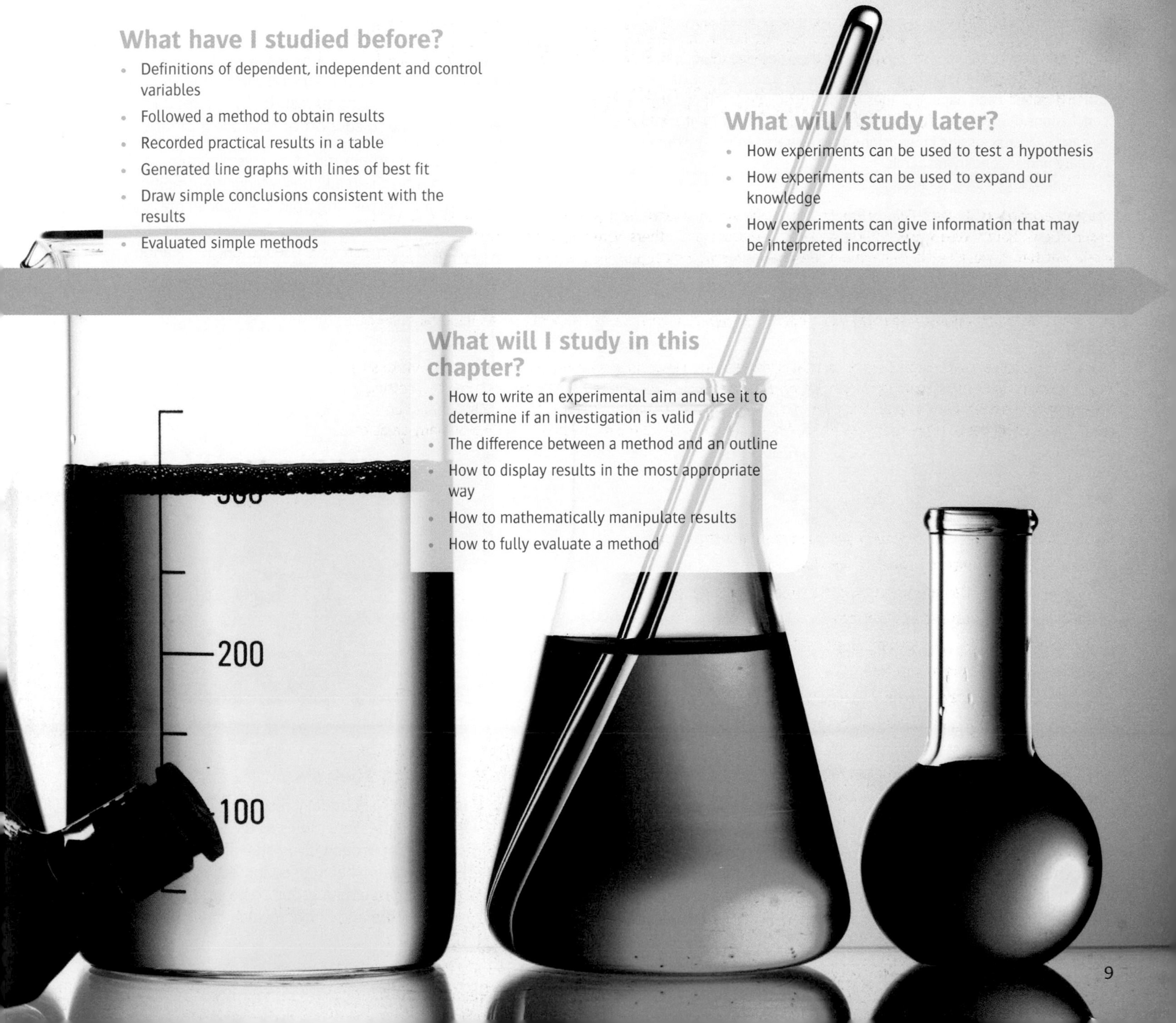

1.1

Experimental design

By the end of this topic, you should be able to demonstrate and apply your knowledge and understanding of:

- experimental design, including to solve problems set in a practical context

Types of research

Scientists investigate using two techniques:

- Primary research – new data is collected and conclusions are then drawn.
- Secondary research – data from other studies is used in different ways to draw conclusions.

DID YOU KNOW?

Scientists share their ideas by publishing their findings in periodicals called journals. Their investigations are written up into a document called a paper. Each paper outlines the scientist's research and their conclusions. Papers are often used by other scientists in their secondary research. Most scientific conclusions are generated from a mixture of primary and secondary research.

Figure 1 Scientists create an aim for their research and can then apply for funding.

Before a scientist starts a piece of research, they set out their aim. This is either a question that they want to answer or a hypothesis that they want to test. When they are planning their research, they must ensure that the method generates results that can achieve the aim. Suitable results that can test the aim are described as **valid**.

Once the aim has been agreed, a scientist will create a **hypothesis**. This is a prediction using their scientific knowledge and trusted information from text books, colleagues and papers. The scientist then decides on their method for testing their ideas.

Surveys

A survey is a type of primary research. It sets out limits to observe something that is already happening.

For example, if you took a daily photograph of a piece of copper attached to the roof of a building for one year, this would be a survey of how the metal tarnished. In A level Chemistry we rarely use surveys.

Experiments

An **experiment** is a type of primary research. It is an ordered set of practical steps that are used to test the hypothesis. The results can be used as evidence to support or disprove a theory.

An example of an experiment might be investigating how the chain length of straight-chain alkanes affects their melting point by directly measuring the melting points. A special apparatus can be used to accurately measure the melting points of organic chemicals. In A level Chemistry we use experiments a lot.

Figure 2 Melting point apparatus.

Meta study or meta analysis

A meta study is a type of secondary research. This uses the raw data from a variety of different studies to try to answer a new aim. This is a mathematical approach using statistics and is often used by social scientists. It is also being used increasingly in medical research, as it allows a large data group to be studied. However, the data may be unreliable and biased, which can cause errors in the conclusions.

In A level Chemistry we sometimes use data from a variety of sources to draw conclusions. For example, periodicity data (like melting points of period 3 elements) could come from a variety of sources.

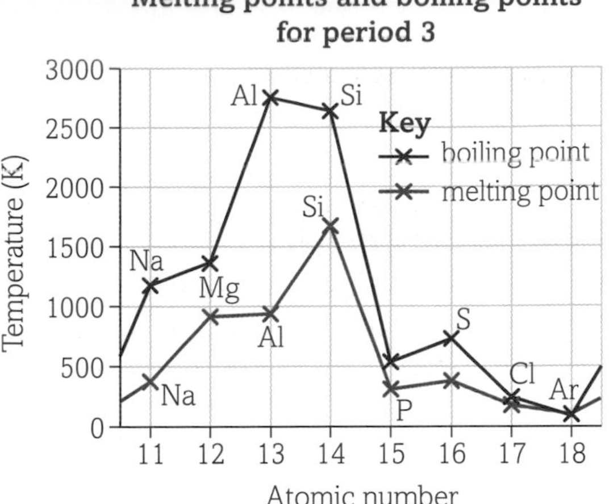

Figure 3 Melting point and boiling point line graphs of period 3 elements are the result of a meta study.

KEY DEFINITIONS

A **valid** experiment provides information to test the aim of the experiment.
A **hypothesis** is a prediction and explanation of the chemistry behind the prediction.
An **experiment** is an ordered set of practical steps, which are used to test the hypothesis.
Quantitative data is a quantity (number) of what is being observed.
Qualitative data is a description of what is being observed.
Resolution is the smallest change in the quantity being measured that can be observed.
Accuracy is how close to the true value a measurement is.

DID YOU KNOW?

The use of information technology is making it quicker to gather large amounts of data. As databases become easier to search, valid meta studies can be completed quickly, rather than many researchers repeating identical studies and generating the same conclusions.

Types of data

Results can be:

- **quantitative** – have a numerical value and require a measuring instrument in order to be observed
- **qualitative** – descriptions of what is observed.

Most experiments in A level Chemistry give quantitative data. The data can be manipulated using mathematical techniques to help see patterns and draw conclusions.

To observe quantitatively, you use measuring equipment like thermometers, burettes and top pan balances. These pieces of equipment allow for measurements to be made as they contain a scale, calibration mark or a digital display. Some equipment, like a thermometer, can be used to measure a range of different values. Others, like a bulb pipette, can only measure one quantity.

For qualitative data scientists need to use their senses. Researchers carry out the experiment and note down what they observe. This is usually what they see occurring, but observations can also involve smell, sound and, in a suitable environment, even taste.

Qualitative research is prone to bias, as the researcher often unconsciously screens what they think is important and notes those observations, rather than everything that occurs. This observational bias can be reduced by using modern technology, such as filming experiments. Then a group of researchers can review the clip, discuss their observations and draw a common conclusion.

LEARNING TIP

Surveys are used more in social sciences like psychology. In these areas, lots of factors are likely to affect an outcome. It is therefore difficult to select just one dependent variable to monitor and so design a valid experiment.

Choosing equipment

Quantitative data requires a measuring instrument to give a quantity to a variable. When choosing equipment, you should consider two factors:

- **Resolution** – this is the smallest change in the quantity being measured that can be observed. If you are measuring liquids, a measuring jug has less resolution than a measuring cylinder.
- **Accuracy** – this is how close to the true value a measurement is. When measuring the volume of a liquid, a bulb pipette will be the most accurate way of measuring a standard volume.

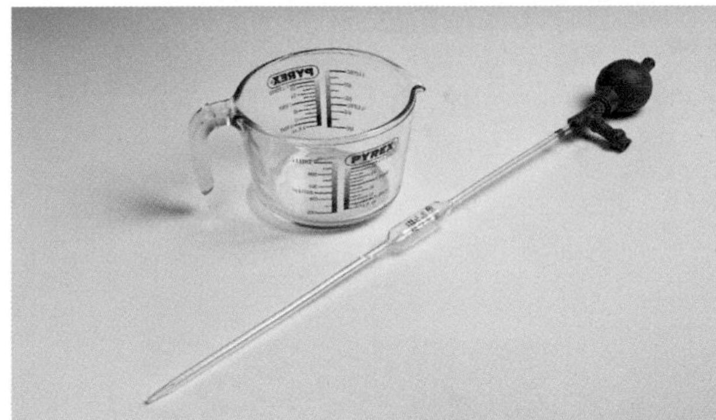

Figure 4 A bulb pipette has less resolution than a measuring jug but is more accurate at measuring 25 cm^3 of liquid at 25 °C.

Questions

1. What are the features of a valid aim?
2. Describe the similarities and differences between meta studies and experiments.
3. In an experiment, a student wanted to measure the mass of water to 1 g. Suggest how this can be done without using a balance.

1.1

2 Types of variable

By the end of this topic, you should be able to demonstrate and apply your knowledge and understanding of:

- **identification of variables that must be controlled, where appropriate**
- **appropriate units for measurements**

What are variables?

Variables are the factors that can affect the outcome of an experiment. In an aim, a scientist will often suggest a connection between two variables. All other variables must then be kept the same in order to discover whether these two variables are related.

Independent variable

An **independent variable** is the factor that you are interested in changing to see the effect it has on another factor. Independent variables are chosen before an experiment starts. They are listed in the first column of a results table and can be filled in before the experiment begins.

DID YOU KNOW?

An experiment is undertaken to discover how chemical structure affects melting point. The independent variable is the structure, as you choose the structures that you will investigate.

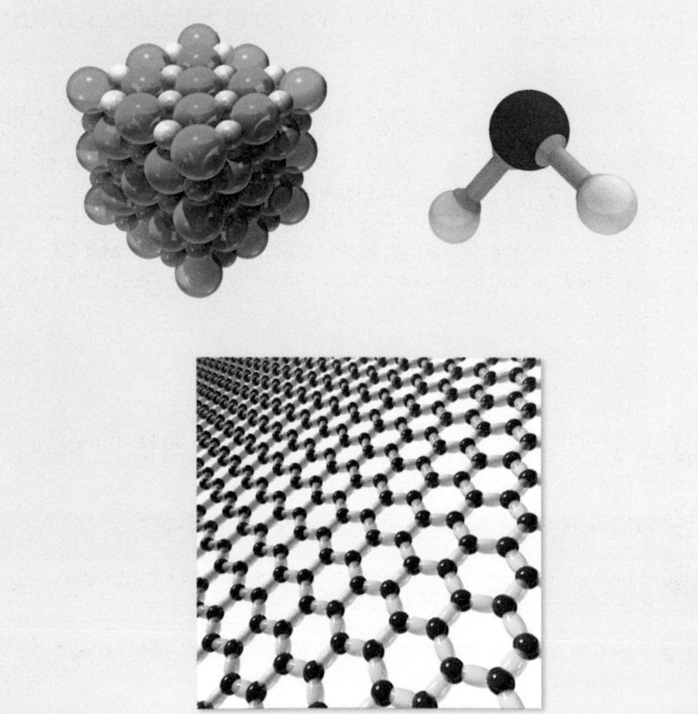

Figure 1 Find out more about how structure affects melting point in topic 3.1.7.

Dependent variable

A **dependent variable** is the factor that you measure or observe in an experiment. This data is added to the results table as the experiment progresses.

INVESTIGATION

In an experiment to measure the enthalpy change of a reaction, the dependent variable is the temperature change of the solution. This is monitored during the experiment and recorded.

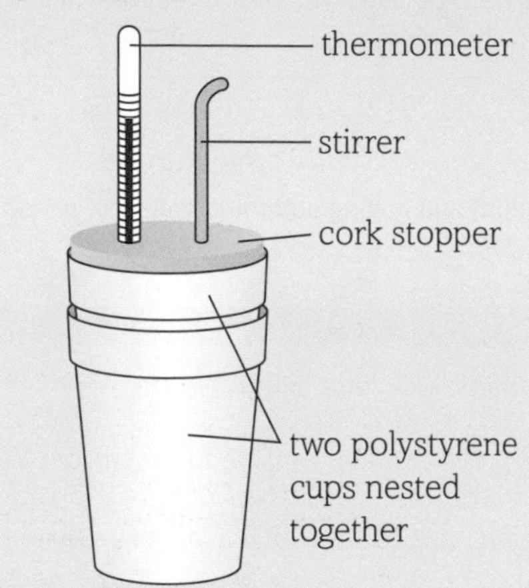

Figure 2 Find out more about coffee cup calorimetry in topic 3.2.4.

Control variable

A **control variable** is a factor that you must keep constant between each run of the experiment, so that results can be compared. Control variables must be considered before an experiment starts and measured each time a run happens. They are not recorded in a results table.

INVESTIGATION

In a titration experiment to measure the concentration of a solution of sodium hydroxide, the control variables include the concentration and type of acid. This allows concordant results to be generated.

Figure 3 Find out more about titrations in topic 2.1.17.

Extraneous variable

An **extraneous variable** is a factor that may affect the experiment but that you have not measured or controlled. You may not be able to control this variable (for example, atmospheric conditions) or it may be something like an accident. Extraneous variables can cause error, but may not affect the outcome. Always note any extraneous variables during an experiment and bear them in mind when interpreting the results.

INVESTIGATION

In a calorimetry experiment to measure the enthalpy of combustion of a fuel, there are many extraneous variables. Examples include the atmospheric conditions and the height of the flame compared to the calorimeter. These variables may impact on the dependent variable, but their effect should be small and similar in each run. So, valid conclusions may still be drawn.

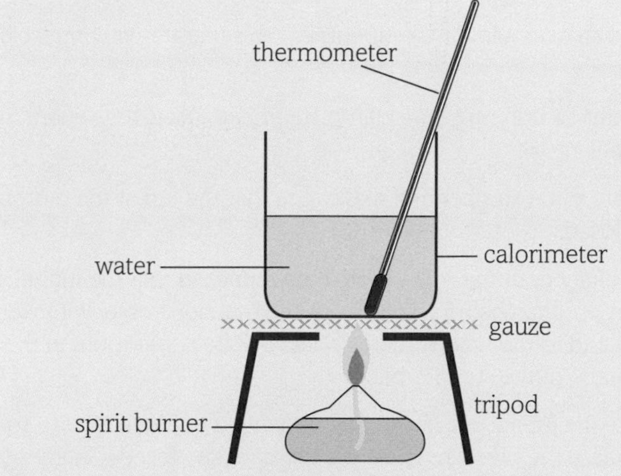

Figure 4 Find out more about copper calorimetry in topic 3.2.5.

KEY DEFINITIONS

Variables are the factors that can affect the outcome of an experiment.
An **independent variable** is the factor that you are interested in changing to see the effect it has on one other factor.
A **dependent variable** is the factor that you observe in an experiment.
A **control variable** is a factor that you must keep constant between experimental runs so that you can compare results.
An **extraneous variable** is a factor that is not controlled or measured in an experiment but may introduce error into the results.

Units of measurement

The International System of units (SI) states the unit that quantities should be measured in. For example, distance is measured in metres (m), time in seconds (s) and mass in kilograms (kg).

Sometimes these units are not appropriate for use in an experiment. This could be due to:

- the measuring equipment: the resolution may not allow you to measure in SI units, or the type of measuring equipment may not use SI units
- the scale of the experiment: the diameter of an atom is about 0.000 000 000 1 m, so a standard form prefix such as 1.0×10^{-10} m could be used, or a different unit like 1 Å.

Figure 5 The SI unit for temperature is kelvin, although in Europe temperature is often measured in Celsius and in the US Fahrenheit is used.

Some quantities are measured with more than one unit. For example, the concentration of a solution can be thought of as the amount of chemical per unit volume. Therefore, the unit is mol dm^{-3}, where mol measures the amount of substance and dm^{-3} is per volume.

LEARNING TIP

When considering time, the SI unit is the second (s). However, in many practical experiments, time is measured by a stopwatch in minutes and seconds. Often students incorrectly record time in a results table. Remember that if a stopwatch shows 1:30 it means 1 minute and 30 seconds. This is actually 1.5 minutes, not 1.3 minutes (as 30 seconds is half a minute).

Questions

1. What is the difference between control variables and extraneous variables?
2. Why is an independent variable important in an experiment?
3. What are all the different measuring units for temperature and how do you interconvert between them?

Writing a plan

By the end of this topic, you should be able to demonstrate and apply your knowledge and understanding of:

- evaluation that an experimental method is appropriate to meet the expected outcomes
- how to use a wide range of practical apparatus and techniques correctly

What to include in a plan

When designing an investigation, it is important to focus on the agreed aim:

- Use the aim to determine the independent and dependent variables.
- Then choose the range (smallest and largest quantity) for each variable. This could be limited by the measuring instruments or time.
- Next, consider all the control variables and choose sensible values. Think about the limits of the equipment and any safety considerations.

At this point you may wish to write an outline **plan**. This is a summary of the experiment that you wish to complete. Here is an example.

INVESTIGATION

Aim: to find out which halide is present in a variety of samples.
Independent variable: samples of chemicals from the range of fluorides, chlorides, bromides and iodides (the range).
Dependent variable: presence and colour of precipitate from the range of no precipitate, white, cream or yellow precipitate.
Control variables: same concentration and volume of samples and reagents.
Outline method: add a few drops of acidified silver nitrate to a sample. Observe colour of precipitate to determine the halide. White is chloride, cream is bromide and yellow is iodide.

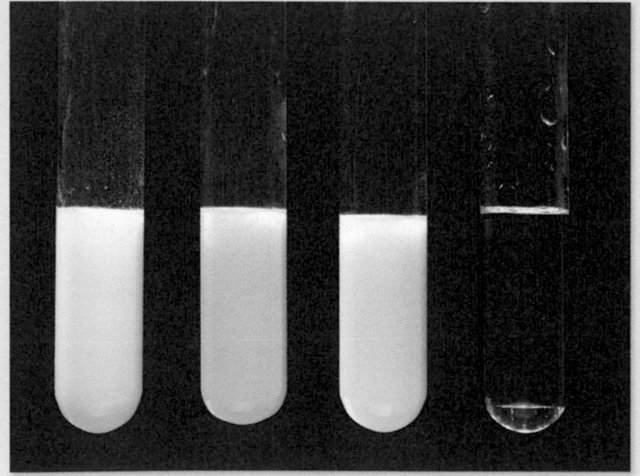

Figure 1 The results for halide ion testing: iodide has a yellow precipitate, bromide a cream precipitate, chloride a white precipitate and fluoride no precipitate. Find out more about halide ion testing in topic 3.1.10.

The method

The **method** is a step-by-step detailed explanation of how to complete an experiment. It should be written in such a way that anyone can follow it and complete the experiment in the same way. The equipment and reagents needed are listed before the method or detailed in a fully labelled diagram.

Here is an example for an experiment to separate two immiscible liquids.

1. Mount an iron ring on a clamp stand and insert the separating funnel into it.
2. Remove the stopper and make sure that the tap at the bottom is closed.
3. Carefully pour the mixture into the funnel so that the funnel is no more than half full. Wash out the reaction vessel with water and add to the funnel; there should still be some room in the funnel. Put the stopper back.
4. Take the funnel out of the ring and invert it. Open the tap to equalise the pressure. Turn the tap back to closed. Gently shake the mixture in the funnel and equalise the pressure. Repeat until you no longer hear a whistle.
5. Replace the funnel in the iron ring and leave the mixture to separate into layers. Remove the stopper, place a beaker under the spout and open the tap. Collect the first, water layer in a beaker. Turn the tap off. As the organic product is in the second layer, this aqueous layer can be discarded.
6. Using a clean, dry beaker, open the tap and collect the desired organic product.

Figure 2 A chemist using a separating funnel in an organic synthesis. Find out more about organic techniques in topic 4.2.7.

Evaluating the method

Once a method has been written, it is important to evaluate it. You should consider whether the results will allow you to achieve the aim of the experiment.

Sometimes, the method may not actually allow you to answer the question the aim posed. In that case:

- the aim could be modified; for example, it may currently be impossible to investigate the aim with available technology
- the method could be adapted to ensure that the data it generates allows the aim to be achieved.

DID YOU KNOW?

Sometimes when a procedure is started, it becomes clear that some of the planning choices are not ideal. You may then wish to modify the procedure.
It is important to try out the method and make changes after the first run-through. This is known as preliminary experimentation. All the changes should be clearly noted on the method, and you should make a note of why these adaptations have been made.

INVESTIGATION

A titration experiment was planned to accurately find out the concentration of a 0.1 $mol\,dm^{-3}$ alkali to three significant figures. In the initial method a concentration of 10 $mol\,dm^{-3}$ of hydrochloric acid was used. However, this produced a very small titre, which gave a high experimental error. The method was therefore modified to use a much lower concentration of acid. This was also a safer option.

Figure 3 Find out more about titration in topic 2.1.17.

DID YOU KNOW?

Procedures are scientific methods that have been generally agreed. This means that the results are reproducible and allows comparison of results generated in different laboratories.
Procedures are often used in forensic science laboratories and in pathology laboratories.

Figure 4 Forensic scientists use procedures, which are precise methods agreed across the industry.

KEY DEFINITIONS

A **plan** is a summary of an experiment that you wish to complete.
A **method** is a step-by-step detailed explanation of how to complete an experiment.

LEARNING TIP

Draft an outline plan before you try to write a full method. This will help you to clarify the stages of procedure.

Questions

1. What is the difference between a method and an outline plan?
2. Write an outline plan for distillation to separate ethanol and water.
3. Write a method for distillation to separate ethanol and water.
4. Explain how you can use a meta study to generate a valid method.

4 Planning an investigation

By the end of this topic, you should be able to demonstrate and apply your knowledge and understanding of:

- experimental design, including an experiment to solve problems set in a practical context
- identification of variables that must be controlled, where appropriate
- evaluation that an experimental method is appropriate to meet expected outcomes

Designing an experiment: first steps

In this topic we are focusing on the combustion of alcohols as an example of how an experiment can be designed, planned and carried out.

Identifying the aim

Primary alcohols have the –OH functional group on the end of the carbon chain. The aim of this investigation is to discover whether the length of the carbon chain has an effect on the enthalpy of combustion.

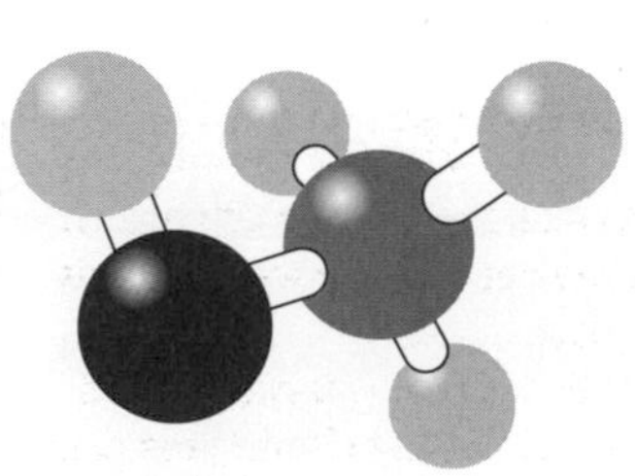

methanol alcohol molecule

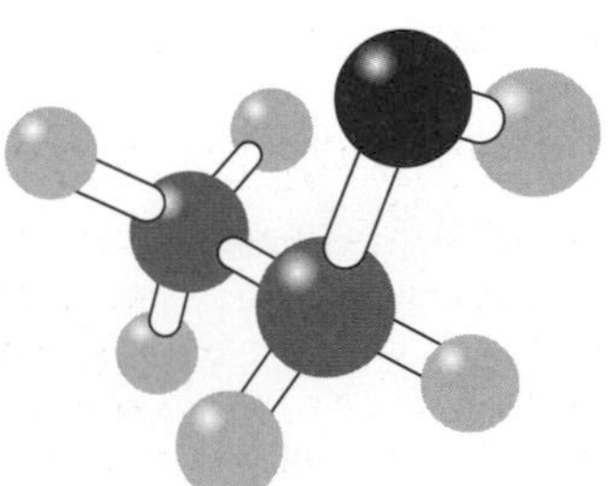

ethanol molecule

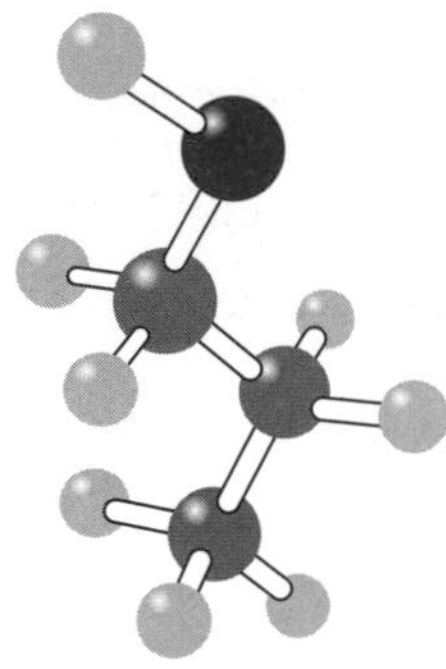

propan-1-ol molecule

Figure 1 Methanol, ethanol and propan-1-ol are all examples of primary alcohols. Find out more about alcohols in topic 4.2.1.

Initial planning

If you have been given the aim, as is the case here, the first task is to determine the two main variables. In this investigation:

- the independent variable is the type of alcohol
- the dependent variable is the temperature change.

Next, consider the outline plan of the experiment:

- Using copper calorimetry and spirit burners, measure the temperature change as a known mass of each liquid alcohol is burnt.
- The results then need to be mathematically processed to generate a value for enthalpy of combustion.

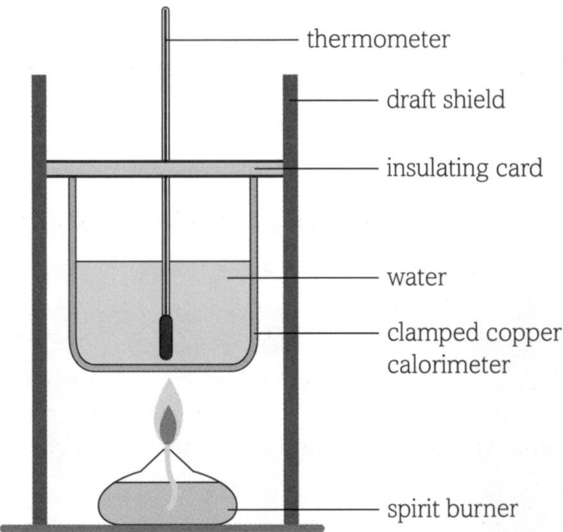

Figure 2 Copper calorimetry would be used. Find out more about this technique in topic 3.2.5.

Is the plan valid?

Once you have drafted a plan, review it to consider whether it is valid. Will it generate results able to answer the question posed in the aim?

By recording the temperature of the water in the calorimeter at the start and end of the experiment, the following equation can be used:

$q = mc\Delta T$, where q = energy, m = mass of substance being heated, c = specific heat capacity and ΔT = temperature change.

This gives the energy released in the observed reaction.

By taking the mass of the fuel before and after combustion, the mass of

the alcohol used can be calculated. Then, using $n = \frac{m}{M_r}$, where n = moles,

m = mass and M_r = relative molecular mass, the amount of alcohol used can be calculated.

Using these two values it is possible to calculate the enthalpy of combustion per mole. This can then be compared between the different alcohols. Therefore, the experiment is valid.

Adding the detail

Choose the range of your independent variable. In this experiment, methanol, ethanol and propan-1-ol will be used, as they are the first three primary alcohols and are readily available in schools and colleges.

Consider the range of the dependent variable. Tap water is about 19 °C and pure water boils at 100 °C. Therefore, use thermometers with a range of –10 °C to 110 °C.

Control variables

In this experiment, one control variable is the mass of water to be heated. It needs to fit into the copper calorimeter and be enough that an appreciable amount will not evaporate. The suggested mass is 100 g.

Another control variable is the atmospheric conditions – ensure that there is a draught shield around the apparatus to minimise effects of drafts.

Extraneous variables

Some of the possible extraneous variables for this experiment include:

- The starting temperature of the water – this will not affect the calculations, as it is the temperature *change* that is needed.
- The time for which the fuel is combusted – this will not affect the calculations, as the mass of the fuel before and after combustion is used to generate the enthalpy of combustion.
- The height of the flame from the calorimeter – use a ruler to try to ensure the same distance between the flame and the calorimeter.

Hypothesis

As we explained in topic 1.1.1, a hypothesis is a prediction using scientific ideas. In this experiment, the hypothesis is that the enthalpy of combustion for primary, aliphatic, straight-chain alcohols increases as the carbon chain length increases. We can make this prediction based on the fact that there are more bonds present in longer molecules and so more stored chemical energy.

LEARNING TIP

Hypothesis is a prediction with an explanation using science ideas.

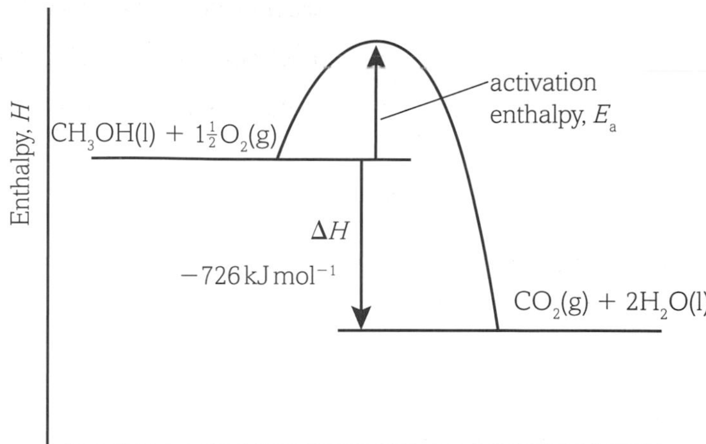

Figure 3 Enthalpy diagram to show the complete combustion of methanol. Find out more about enthalpy profile diagrams in topic 3.2.2.

To find out more about alcohols go to topic 4.2.1 and to find out more about bond energy go to topic 3.2.5.

DID YOU KNOW?

It is not always necessary to collect primary data in order to achieve an investigation aim. For example, the experimental aim considered in this topic could be answered using secondary data.
You could use a printed data book or an online digital database to find the enthalpy of combustion of the primary alcohols. By plotting the data (relative molecular mass against enthalpy of combustion), you could then identify a trend in the data and answer the aim.

Questions

1. Write a method for the experiment in this topic.
2. Design a results table for the experiment in this topic.
3. Outline how you could answer the aim using secondary research.

Recording data

By the end of this topic, you should be able to demonstrate and apply your knowledge and understanding of:

- appropriate units for measurement
- presenting observations and data in an appropriate format
- processing, analysing and interpreting qualitative and quantitative experimental results

Observing variables

In an experiment, the variables can be categorised in two ways:

1. **Qualitative** – a description of the variable; for example, the presence or absence of a silver mirror using Tollens' reagent allows us to differentiate between aldehydes and ketones
2. **Quantitative** – a numerical measure of a variable; for example, the temperature change caused by a reaction.

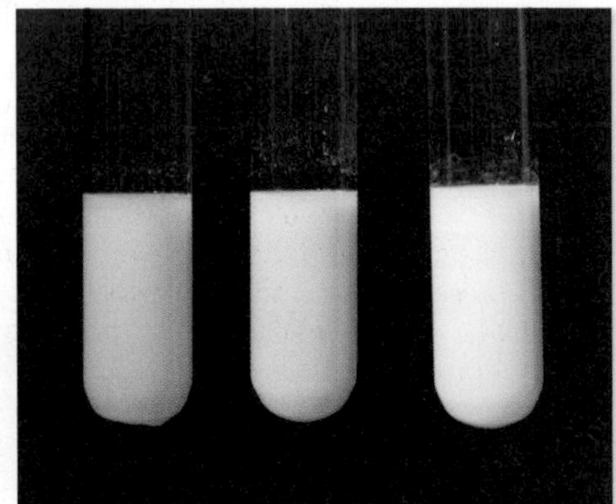

Figure 1 A qualitative result could be the colour of a precipitate indicating that a halide is present: chloride is white, bromide is cream and iodide is yellow.

KEY DEFINITIONS

Qualitative data is a description of the variable.
Quantitative data is a numerical measure of a variable.
Continuous variables are a measured value that could be any number.
Discrete variables can only be particular defined numbers.
Categoric variables mean a qualitative description of a variable.

Presenting observations and data

Tables

During an experiment all results should be noted in a table. A table is a clear and structured way of recording information about an experiment.

- The first column should be for the independent variable and this can be completed before the experiment begins.
- The other columns should contain the dependent variable.
- Each column heading must be labelled with the variable and its unit of measurement. The entries in the table should be without units, as they are recorded in the headings.

Tables may include variables that are generated by the observed data. For example, in a titration, the start and end volumes on the burette are measured and used to calculate the titre. The titre will be recorded in the results table, as this is data that is useful for interpretation.

	Rough	Run 1	Run 2
Final volume (cm^3)	0.00	0.00	0.10
Initial volume (cm^3)	12.50	12.00	12.00
Titre (cm^3)	12.50	12.00	11.90
Mean titre (cm^3)		11.95	

Table 1 A results table for a titration. Find out more about titrations in topic 2.1.17.

Data loggers

Data loggers are measuring instruments attached to a method of recording the data. There are many different types of sensor, from pH probes to temperature probes. The data can be displayed in a table or in a chart.

The advantage of using a data logger is that you do not need to be able to interpret a scale to obtain a measurement. The pictorial display of the data allows real-time analysis and trends can often be seen quickly.

Figure 2 A data logger being used to follow a neutralisation reaction.

Graphs and charts

Often, patterns in data are more easily seen in graphs or charts. Different types of variable are suitable for different types of display. Variables can be:

- **continuous** – a measured value that could be any number, for example temperature
- **discrete** – values can only be defined numbers, for example atomic number
- **categoric** – a qualitative description, for example colour of a precipitate.

When drawing a graph or chart, the independent variable must be on the horizontal x-axis and the dependent variable on the y-axis. The graphs and charts you may use in A level Chemistry include:

- a scatter graph with a line of best fit – both variables must be continuous
- a line graph – the independent variable can be discrete or categoric, the dependent variable can be continuous or discrete
- a bar chart – the independent variable can be discrete or categoric, the dependent variable can be continuous or discrete.

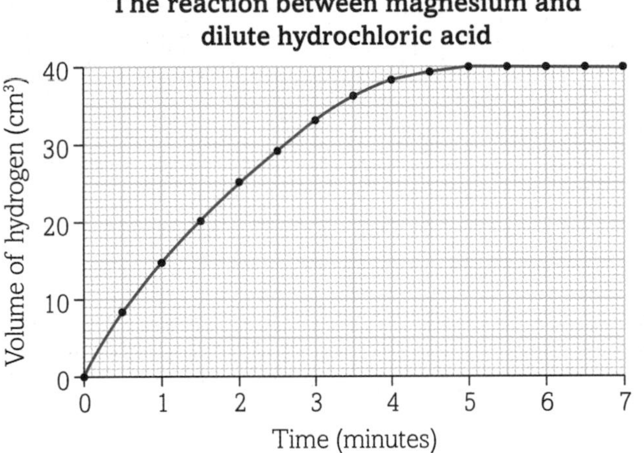

Figure 3 A scatter graph with line of best fit showing the rate of reaction of magnesium with hydrochloric acid. Find out more about rate of reaction in topic 3.2.7.

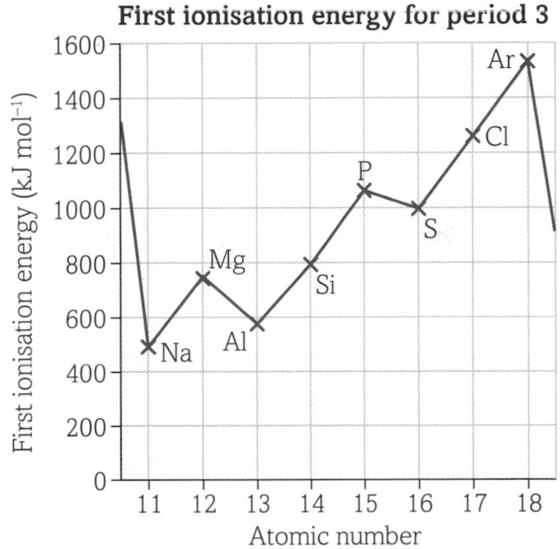

Figure 4 A line graph of the first ionisation energy of period 3. Find out more about periodicity of period 3 in topic 3.1.2.

LEARNING TIP

All data from an experiment should be recorded in a table before it is displayed in other ways.

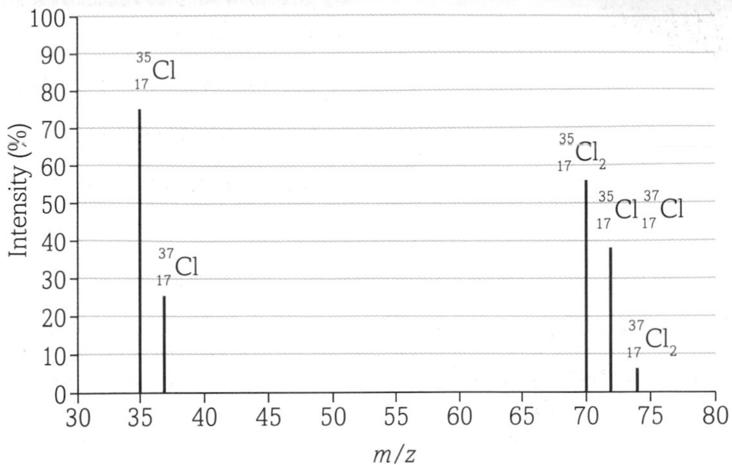

Figure 5 A mass spectrum is a complex bar chart. The two bars on the left represent isotopes of individual atoms. The three bars on the right represent isotopic versions of a chlorine molecule. Find out more about mass spectrometry in topic 4.2.10.

DID YOU KNOW?

When data has been plotted on a scatter graph, it is important to see if there is a correlation and determine whether the two variables are related. If there is no pattern, no line of best fit should be drawn and there is no relationship between the variables. This means that the independent variable does not have an effect on the variable that you thought was a dependent variable.

Questions

1. Draw a suitable results table to record the temperature change in a coffee cup calorimeter.
2. Give an example from A level Chemistry of a quantitative discrete variable.
3. Explain how you would display melting point data for period 3 elements.

Manipulating data

By the end of this topic, you should be able to demonstrate and apply your knowledge and understanding of:

- **processing, analysing and interpreting qualitative and quantitative experimental results**
- **use of appropriate mathematical skills for analysis of quantitative data**
- **appropriate use of significant figures**
- **plotting and interpreting suitable graphs from experimental results, including:**
  (i) selection and labelling of axes with appropriate scales, quantities and units
  (ii) measurement of gradients

Standard form

Standard form is a method for writing very small or very large numbers. Standard form is always written as $A \times 10^n$, where A is the first significant figure and n is a description of the direction and number of places that the decimal point has moved.

- For a large number like atmospheric pressure, 100 000 Pa becomes 1.0×10^5 Pa, as the decimal place has moved five places to the left.
- For a very small number like the diameter of a hydrogen molecule, 0.000 000 000 74 m becomes 7.4×10^{-10} m, as the decimal place has moved ten places to the right.

Significant figures

When we record or manipulate quantitative data, it is important to think about the meaning of the numerical value. We use significant figures to represent a quantity that has meaning. This also relates to the confidence that you have in the degree of accuracy.

Example 1

A top pan balance has an accuracy of +/− 0.01 g. Therefore, if an object has a mass of 1.00 g as measured on the top pan balance, its accurate mass could be as small as 0.99 g or as large as 1.01 g. So, the mass should be reported to two decimal places.

Figure 1 A top pan balance has an accuracy of +/− 0.01 g, so the mass should be reported to two decimal places.

Here are the rules for significant figures:

- Zeros after the first non-zero digit are significant. This is because zeros before the first non-zero digit are not.
- You may need to round numbers up or down. Count the number of significant figures desired and look at the next value along. If that number is five or more, round up. If it is four or less, round down.

Example 2

Calculate the number of moles in 1.1 cm^3 of 0.6 $mol\,dm^{-3}$ solution of sodium hydroxide.

Use:

$$\text{moles (mol)} = \text{concentration (mol dm}^{-3}\text{)} \times \text{volume (dm}^3\text{)}$$

$$\text{moles} = 1.1/1000 \times 0.6 = 0.00066 \text{ moles}$$

- In standard form, this is 6.6×10^{-4} moles, reported to two significant figures.
- It would become 7×10^{-4} moles for one significant figure.
- It would become 6.60×10^{-4} moles for three significant figures.

As the least accurate piece of data is concentration, with just one significant figure, the answer should be reported to one significant figure, 7×10^{-4} moles.

Older balances may have an analogue scale. This equipment will have a needle which moves around a scale when masses are put on the plate. It is important to calibrate the balance before it is used. Take known masses, place them on the top pan and use the calibration wheel to move the needle so that it is in line with the accurate mass. Make sure that you look at the scale and needle directly to reduce the chance of parallax error. Read the value to the smallest graduation, then estimate one digit beyond those shown directly by the measurement scale. So, if the needle is half way between 5.1 g and 5.2 g graduations, the reported mass would be 5.15 g.

Burettes are always an analogue scale. Again they should be read with the eye line directly in line with the meniscus to reduce the change of parallax error. The scale is to the nearest 0.1 cm^3, and the values should be reported to 2 decimal places with the last digit estimated.

LEARNING TIP

The general rule for multiplication or division is that your answer should have no more significant figures than the least accurate number.
For addition or subtraction, the result should be reported to the same number of decimal places as the number with the fewest decimal places.
Examination questions often state how many significant figures your answer should include.

Scatter graphs

When constructing a scatter graph, it is important to maximise the amount of graph paper used for plotting the data so that patterns and trends are easy to observe. Choose a suitable scale, which may mean not starting the graph at the origin (0,0).

The independent variable must be continuous and noted on the *x*-axis. The dependent variable must also be continuous but displayed on the *y*-axis.

Line of best fit

The points should be plotted with x, not ●, as this gives a more accurate plot. Hold the plotted graph at arm's length and see the shape in the data.

- If there is no pattern, do not draw a line of best fit.
- If a data point doesn't fit the pattern, circle it and ignore the anomalous result.
- If the pattern is a straight line, use a ruler and a sharp pencil to draw the line. Do not connect the dots; draw the pattern the data shows.
- If the pattern is a curve, hold a pencil firmly on the desk and move the paper to make a smooth curve.

(a) 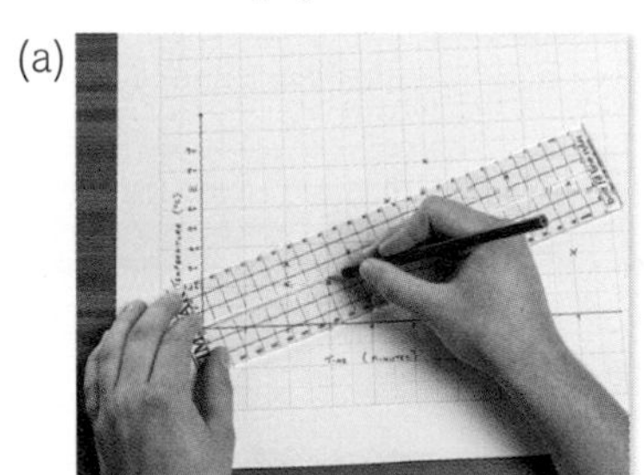

(b) 

Figure 2 (a) Use a ruler to draw a straight line of best fit. (b) Draw a smooth curve freehand.

Gradient

The gradient is the measure of the amount of slope in a graph.

For a straight line, choose two points on the graph. Note the change in the *y*-axis and divide this by the change in the *x*-axis.

$$\text{gradient} = \frac{\Delta y}{\Delta x}$$

For a curve, draw a tangent to the curve. This is a straight line that just touches the curve at the point of interest. Calculate the gradient of the straight tangent line, as described above. The units of the gradient will be the dependent variable units divided by the independent variable units.

Gradient = 0.16 − 0.08/10 − 35 = −0.0032 mol dm^{-3} s^{-1}

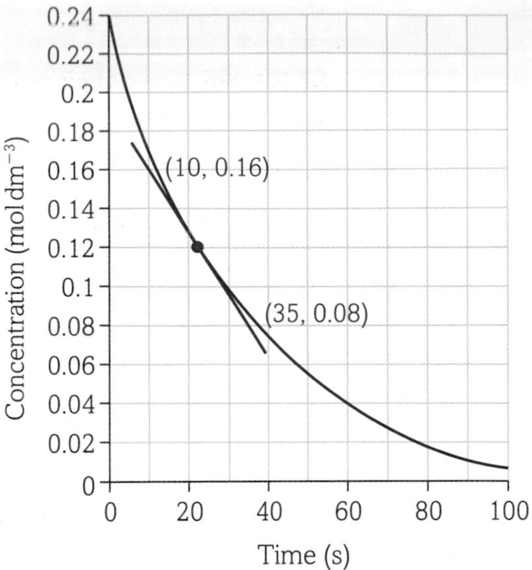

Figure 3 Rate of reaction graph showing how to calculate the gradient of a curve. Find out more about rate of reaction in topic 3.2.7.

DID YOU KNOW?

Computer graphic packages can be very useful for adding trend lines. However, these may not be a line of best fit, as anomalous results are not discarded by the computer. Therefore, you need to remove any anomalous points before asking the computer to draw a trend line.
You can also use a computer package to calculate the gradient of the trend line at any point.

Averages

Concordant results are values that are close to each other and therefore represent reliable data. For example, titre values should be within 0.1 cm^3 and are then described as concordant. Results that are not concordant are **anomalous**.

Only concordant results are used to calculate an average.
In A level Chemistry we often use the mean, which is a calculated central value generated by adding up the concordant results, then dividing by the number of concordant results.

KEY DEFINITIONS

Concordant results are values that are close to each other and therefore represent reliable quantitative data.
Anomalous results do not follow the general pattern of the data.

Questions

1. Write (a) 0.000 003 and (b) 5 400 000 in standard form.
2. Write 0.003 4567 to one, two, three, four, five and six significant figures.
3. Choose two concordant results from this list and calculate an average: 0.23, 0.24, 0.26.

Evaluating results and drawing conclusions

By the end of this topic, you should be able to demonstrate and apply your knowledge and understanding of:

- how to evaluate results and draw conclusions
- the limitations in experimental procedures

Evaluating results

It is important to consider whether the results of an experiment are reliable. There are two things to consider:

- Are the results repeatable? Each time the experiment is repeated, are the results similar or concordant?
- Are the results reproducible? When other people complete the experiment, do they get similar results?

Some results can be repeatable but not reproducible.

Example 1

If there is a systematic error, such as you incorrectly reading a burette, this will have the same effect on each result. The results will therefore be repeatable and concordant. However, other people using the burette correctly will get different results, so the results will not be reproducible.

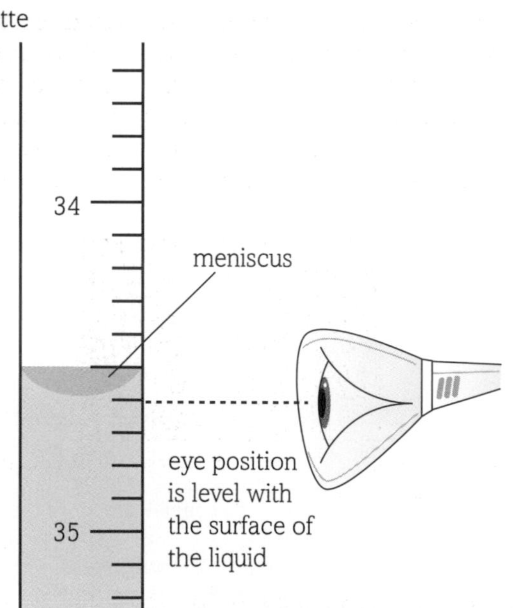

Figure 1 When using a burette, always take the reading from the bottom of the meniscus. Find out more about titrations in topic 2.1.17.

Results that are close to the true value are described as **accurate**. True values can be found from trusted sources, for example, a teacher's value, or from a text book or data book. Some results may be accurate but if they are not repeated then we cannot comment on how **reliable** they are.

Example 2

When copper calorimetry is used to investigate the enthalpy of combustion, usually the fuel is only combusted once by each student. If the enthalpy of combustion is close to the teacher's value, it is likely to be accurate. However, as the experiment has not been repeated we cannot comment on repeatability. If the whole class compare their results, we can comment on reproducibility.

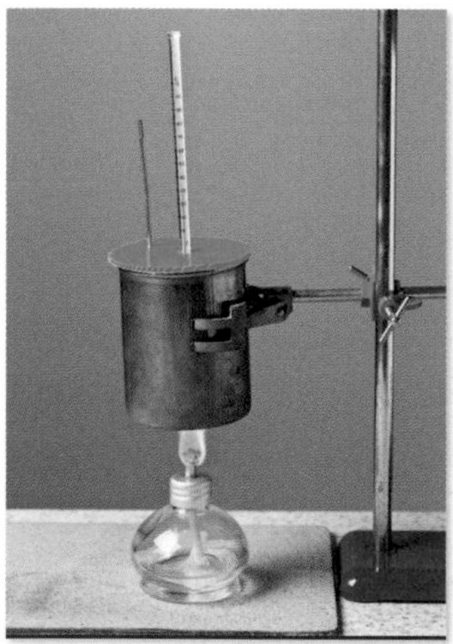

Figure 2 Using copper calorimetry. Find out more about copper calorimetry in topic 3.2.5.

KEY DEFINITIONS

Accurate results are close to the true value.
Reliable results are similar when they are repeated.
A **false positive** in a chemical test is when a positive result is produced but not due to the desired product being formed.

Drawing conclusions consistent with results

Once the data from an experiment has been collected and displayed, the next step is to draw conclusions. Look back at the aim of the experiment and try to use the data to answer the question that was posed.

Lines of best fit

To draw a line of best fit there must be at least five data points. Ideally, these should be distributed so that the trend in the data can be clearly seen. Sometimes the range of the data is not large enough, so a conclusion can be drawn that fits the results but is not the whole story.

Example

In an experiment to study the rate of reaction between magnesium ribbon and hydrochloric acid, the hydrogen gas produced was collected. The rate of reaction was very fast at the start, slowed down and then stopped. When the experiment was observed frequently and to completion, the graph of the results allowed an accurate conclusion to be drawn.

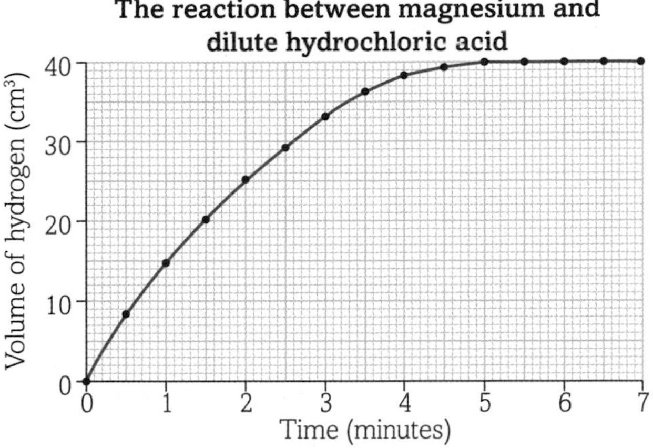

Figure 3 Rate of reaction graph for the whole reaction.

If you only took data for the first 20 seconds of the experiment, the line of best fit would be a straight line going through the origin, showing a directly proportional correlation.

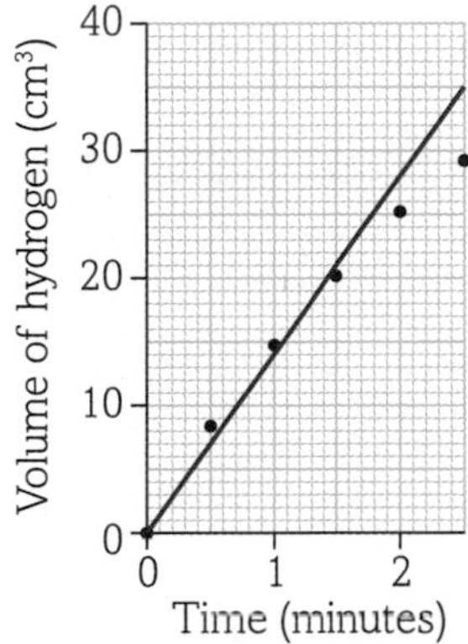

Figure 4 Rate of reaction graph for the first part of the reaction only.

If you only collected data after five minutes, the line of best fit would be a horizontal straight line, showing no change in the rate of reaction.

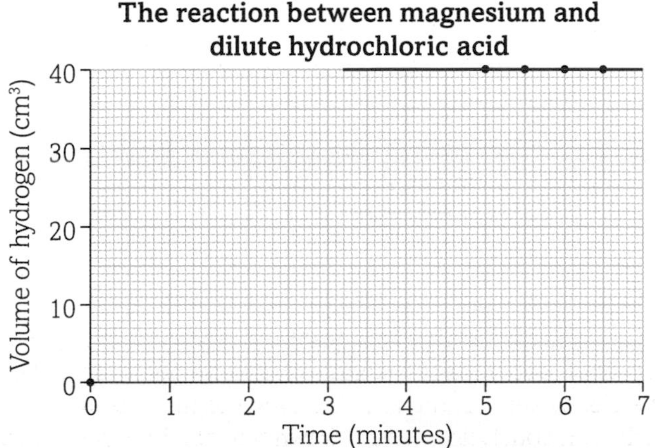

Figure 5 Rate of reaction graph for the end part of the reaction.

DID YOU KNOW?

If the results you obtain are vastly different from what you were expecting, consider whether it is the equipment, method or prediction that is incorrect. Calibrate all the equipment, review the method and look up the scientific knowledge that underpins your prediction.
Next, repeat the experiment and see whether the original results were anomalous or actually just unexpected. A different explanation may be needed to describe the relationship between the variables.

LEARNING TIP

To draw a line of best fit hold the plotted graph at arm's length to see the pattern. Some graphs need more points to be confident with the conclusion.

Limitations in experimental procedures

The limitations of an experiment relate to how accurate the generated results are and how easy the experiment is to achieve and therefore repeat. When considering limitations of an experiment you should consider which procedures could be used and select the one that generates the most valid results for your aim.

Sometimes available laboratory equipment cannot measure a variable, therefore the procedure would need to modified. For example, in a titration if one reactant is significantly more concentrated than the other it may be necessary to dilute one reagent before a titre can be achieved from a 50 cm^3 burette.

False positives

In some experiments the results lead you to incorrect conclusions. For example, when the qualitative test for sulfate is used, it is important to add hydrochloric acid to the reaction sample. Carbon dioxide in the air would dissolve producing carbonate ions. These dissolved ions react to make a white precipitate. This could give a positive result whether sulfate was present or not. This phenomenon is called a **false positive**.

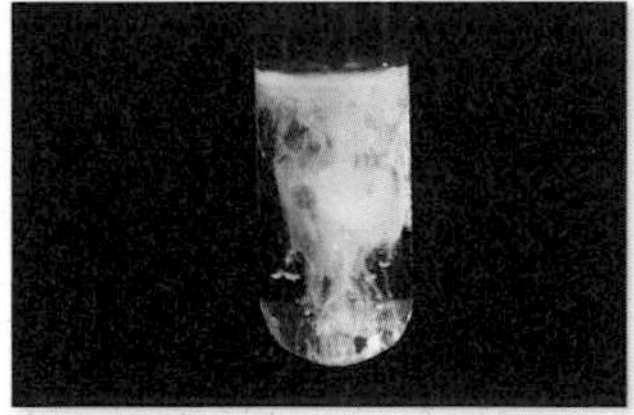

Figure 6 Acidified barium chloride is used to test for sulfate ions. Find out more about this test in topic 3.1.11.

Questions

1. Explain how titration results can be both reliable and accurate.
2. Explain how a false positive can occur with the halide ion test.
3. Explain why the range of a graph is important.

8 Precision and accuracy

By the end of this topic, you should be able to demonstrate and apply your knowledge and understanding of:

- the identification of anomalies in experimental measurements
- precision and accuracy of measurements and data, including margins of error, percentage errors and uncertainties in apparatus
- refining experimental design by suggestion of improvements to the procedures and apparatus

Accuracy and uncertainty

Data that is **accurate** is close to the true value. However, every measured or calculated piece of data will have a level of uncertainty. This is due to two factors:

- Systematic error – the same error in every measurement. This is due to the limits of the equipment; for example, a piece of measuring equipment not being calibrated correctly.
- Random error – an error that may or may not be present and is different every time. This is due to the experience of the scientist; for example, not controlling the draughts in a room.

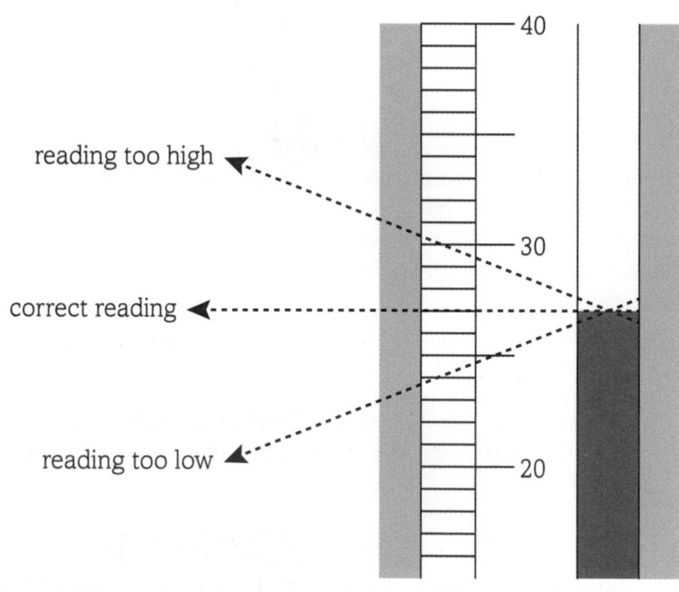

Figure 1 When using a thermometer, always take the reading by looking directly at the scale to avoid a systematic parallax error. (A parallax error can be caused when a scale is read from a different angle, leading to a different result from the accurate value.)

It is important to be able to suggest the causes of error and to consider whether this affects the data to such an extent that the conclusions drawn from it would not be valid. If the data is valid, then it is within the **margin of error** (see opposite).

Anomalies

Anomalies are data points that do not fit the overall trend in the data. These should not be included in the data set that is used to draw conclusions. Anomalies are often highlighted by circling them in a data table, graph or chart.

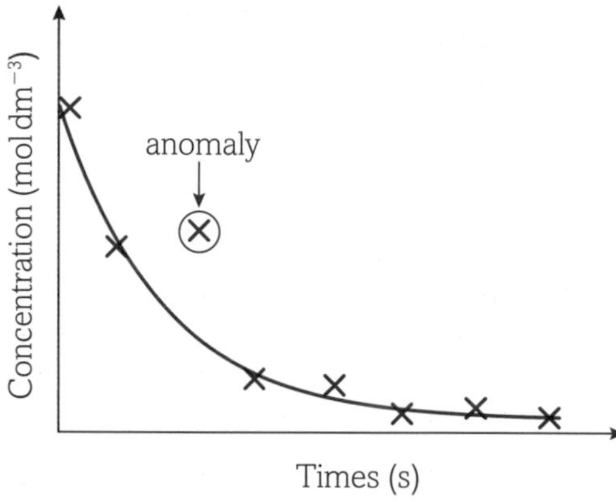

Figure 2 Anomalous data points do not fit the trend and are often circled.

KEY DEFINITIONS

Accurate results are close to the true value.
Margin of error shows the range that a value lies within.
Anomalies are data points that do not fit the overall trend in the data.
Precision is the degree to which repeated values, collected under the same conditions in an experiment, show the same results.
Weighing by difference is a method used to accurately weigh the amount of material transferred. The mass of a container before and after transferring the material is taken and the difference between these values is the mass of material transferred.
Percentage error is a mathematical way of comparing the experimental value with the actual value.

Measurements and data

Uncertainties in apparatus

The term **precision** relates to how close together repeat values are. The smaller the spread, or range, the higher the precision. Precision is often used in conjunction with the term 'resolution', which relates to the smallest quantity that an instrument can measure (see topic 1.1.1). Uncertainties relating to equipment consider the resolution of the equipment. Quantitative data, both observed and mathematically manipulated, should be reported only to the same number of decimal places as the uncertainty.

Once an experiment has been planned it is important to trial it. This should generate preliminary data which will allow you to consider if the experiment is valid, i.e. it generates data which allows you to address the aim.

In your trial you might find that there are limits in terms of the availability of equipment, quantities that have been chosen or the techniques desired. This allows you to determine the most suitable plan and re-draft your method accordingly. For example, if you are short, positioning of apparatus could make it difficult to measure the volume of gas by displacement in a graduated test tube. In this case a gas syringe may be a more suitable piece of equipment.

You may also find that the experiment does not yield any results in the expected time frame and you may need to modify conditions, such as heating the reactants to ensure the experiment occurs swiftly, safely and yields valid data.

WORKED EXAMPLE 1

If a measuring cylinder had an uncertainty of 0.5 cm^3, any volume should be reported to one decimal place only.
Therefore, the volume would be 5.0 cm^3, not 5 cm^3.

Margin of error

Margin of error considers the observational error in an experiment and can be used to define the range that the result has. The larger the margin of error, the less confidence there can be in the data. If there is a very small margin of error, the results will be more accurate and therefore closer to the true value.

There are a number of pieces of equipment that you can choose to measure a value. For example, volume can be measured in:

- a 50 cm^3 measuring cylinder (+/– 0.1 cm^3)
- a 10 cm^3 measuring cylinder (+/– 0.01 cm^3)
- a 50 cm^3 pipette (+/– 0.05 cm^3).

Choosing a more precise piece of equipment will increase the accuracy of the results.

WORKED EXAMPLE 2

A balance can measure +/– 0.1 g; this is the margin of error. Therefore, an accurate mass of 10 g could have a value on the balance from 9.9 g to 10.1 g. The larger the mass measured, the smaller the effect of the margin of error and therefore the more accurate the value becomes.

The error of a mass measurement can be reduced by using the **weighing by difference** method:

- Measure the mass of a clean, dry container.
- Add the chemical and note the mass of the chemical in the container.
- Transfer the chemical to the reaction and re-measure the now empty container.
- Then calculate the mass of the chemical transferred.

Some pieces of equipment, like balances and pH probes, need to be calibrated frequently to ensure they are measuring correctly.

LEARNING TIP

Measurement error values are particular to a piece of equipment, and this information is often marked on the glassware. However, in this chapter we will use typical values for commonly used laboratory equipment.

Percentage error

Percentage error is a mathematical way of comparing the experimental value with the accurate value. The following expressions can be used to calculate the percentage error of a measurement:

- Percentage error of a value
$$= \frac{\text{accepted value} - \text{experimental value}}{\text{accepted value}} \times 100$$
- Percentage error of equipment $= \dfrac{\text{maximum error}}{\text{measured value}} \times 100$

WORKED EXAMPLE 3

A burette has a margin of error of 0.05 cm^3, but when this piece of equipment is used, two volume measurements are made. Therefore, the total error is 0.10 cm^3.
If the burette was used to deliver 5.00 cm^3 of solution:

$$\text{percentage error} = \frac{0.10}{5.00} \times 100 = 2\%$$

If the burette was used to deliver 20.00 cm^3 of solution:

$$\text{percentage error} = \frac{0.10}{20.00} \times 100 = 0.5\%$$

As you can see, the smaller the measured quantity, the higher the percentage error. The total equipment error for an experiment is generated by adding the percentage errors of all the equipment.

Figure 3 Burette readings have an error of 0.10 cm^3. Find out more about titrations in topic 2.1.17.

If reliable results are averaged, then the error is reduced further. In a titration, the mean average of concordant results is used in calculations.

LEARNING TIP

Always present results with the errors. This allows you to comment on their validity.

Questions

1. Calculate the percentage error when measuring 25 cm^3 of water in: (a) a 50 cm^3 measuring cylinder; (b) a 10 cm^3 measuring cylinder; (c) a burette.
2. Explain why larger quantities are more accurate than smaller quantities.

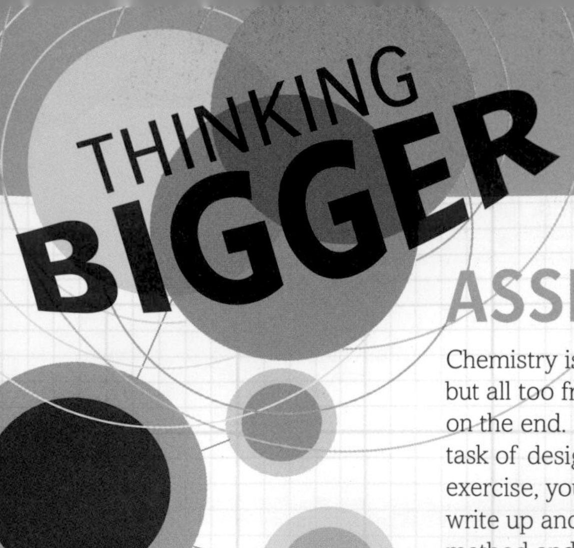

ASSESSING A PRACTICAL WRITE UP

Chemistry is first and foremost a practical subject. This may be a statement of the obvious for some but all too frequently chemistry is presented as a body of theories with the practical material tacked on the end. In the following section a practical problem is presented and the student is given the task of designing their own practical method, carrying it out, and then analysing the results. In this exercise, you will be given the same practical problem and then asked to comment on the student's write up and analysis. At the end of the task you should be able to design your own 'improved' method and be prepared to justify your changes.

The student was presented with the following task.

HOW HARD IS YOUR WATER SUPPLY?

Hardness in water is largely due to dissolved calcium and magnesium salts. The concentration of Ca^{2+} and Mg^{2+} ions in solution can be determined by titrimetric analysis. A sample of water is titrated with a standardised solution of EDTA (ethylenediaminetetraacetic acid). When an indicator called Erichrome black T is added to the sample a reaction between the indicator and the calcium and magnesium ions gives a wine red colour. The sample is then titrated with the standardised solution of EDTA. The EDTA forms a stronger complex with calcium and magnesium ions than the Erichrome black T does. As the EDTA is added it binds to the calcium and magnesium ions replacing the Erichrome black T and once all of them have been complexed with the EDTA the Erichrome black T indicator turns blue. This is the end point of the titration from which the concentration of the calcium and magnesium ions can be calculated. The simplified equation for the reaction is:

$M^{2+}(aq) + EDTA^{2-}(aq) \rightarrow M\ EDTA(aq)$

Indicator red $\rightarrow$ Indicator blue

Design and carry out a procedure to determine the concentration of calcium and magnesium ions in tap water. Your procedure should include:

1. A clear experimental method.
2. A clear appreciation of reliability, precision and percentage error.
3. An evaluation of your procedure in the light of your practical investigation.

The student's write up is shown opposite. Read through the report and then answer the questions that follow.

STUDENT'S WRITE UP

Aim: To find out the amount of hardness in tap water.

Method:

1. Measure out 25 cm^3 of tap water using a measuring cylinder and put it in a conical flask.
2. Add a few drops of the Erichrome black T indicator and record the colour change.
3. Add standardised 0.01 mol dm^{-3} EDTA solution from a burette until a permanent colour change (red to blue) is recorded.
4. Note the volume of the EDTA added.
5. Repeat the experiment a couple of times.
6. Take an average of the titre volumes.
7. Use the average to calculate a value for hardness of water (see calculation and results table).

Results

Titration number	1	2	3	average
Final volume in cm^3	12.9	13	13.4	
Initial volume in cm^3	0.00	0.00	0.00	
Titre	12.9	13	13.4	13.1

Calculation

Moles of EDTA added = $\frac{13.1}{1000} \times 0.01 = 1.31 \times 10^{-4}$ moles

According to the equation same number of moles of Ca^{2+} and Mg^{2+} ions complexed

So concentration = $\frac{1.31 \times 10^{-4}}{0.025} = 5.24 \times 10^{-3}$ mol dm^{-3}

If A_r of Ca = 40 then $5.24 \times 10^{-3} \times 40 = 0.2096$ g or 209.6 mg of calcium per litre of tap water.

Evaluation

My result of 209.6 mg per litre is a good result and suggests that the tap water we tested was hard.

To improve my results I could have repeated the test a few more times and taken an average.

Where else will I encounter these themes?

1.1 | YOU ARE HERE | 2.1 | 2.2

1. Identify as many mistakes as you can that involve issues of reliability and precision in the protocol on the left. How could these be addressed in a revised procedure?
2. A second student used a different practical protocol that involved making an accurate dilution of the tap water using an equal volume of distilled water. Three consecutive titrations of this solution gave values of 28.50, 25.90 and 25.90 cm^3 respectively.
   a. Using this data calculate a suitable average titre value to 2 d.p. and include a percentage error.
   b. Use this information to calculate a value for concentration of Mg^{2+} and Ca^{2+} ions in water in moles per dm^3.
   c. This second student chose not to convert the concentration of mol dm^{-3} into g per dm^3. Suggest why this choice was made.
   d. Comment on the reliability and precision of this second set of data.
3. Critically evaluate the student's own evaluation of their experiment.

Activity

Re-write the practical method including the key issues that you have identified in questions 1, 2 and 3. You can use the second student's results (Q2) to help present your write up but be sure to explain how these results were generated.

DID YOU KNOW?

Whether you happen to have a hard or soft water supply is largely a matter of where you live but there is now some evidence that drinking hard water may protect against heart disease!
A study carried out by Dr Anne Kousa and reported in the Journal of Epidemiology and Community Health (*J Epidemiol Community Health* 2004 58:2 136–139 doi:10.1136/jech.58.2.136) said:

'This study concludes that the incidence of acute myocardial infarction is significantly lower in areas of the country where water is harder.'

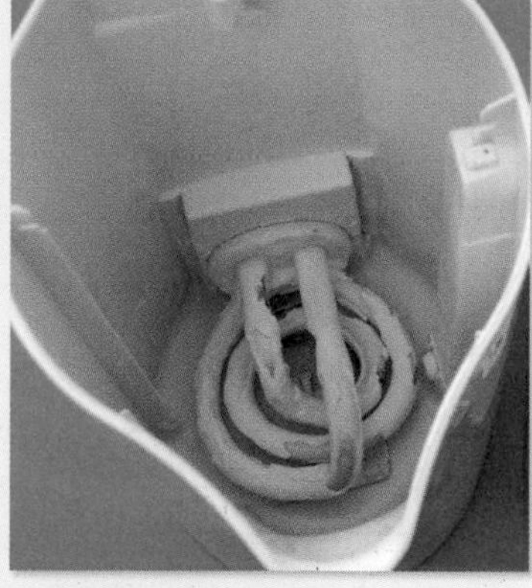

Figure 2 Hard water; good for the heart, not so good for the kettle!

1.1 Practice questions

Section A: Multiple choice questions

In a titrimetric analysis to quantify the amount of iron in iron supplement tablets a student performed four titrations and recorded the following results.

Titration	Rough	1	2	3
Final volume (cm^3)	24	23.55	23.45	25.80
Initial volume (cm^3)	0	0.00	0.00	0.00
Titre (cm^3)	24.0	23.55	23.45	25.80

1. What is the correct average titre value to be used from these results? [1]
   A. 24.20 cm^3
   B. 23.65 cm^3
   C. 23.50 cm^3
   D. 24.27 cm^3

2. In titrimetric analysis, why is the solution to be analysed measured out using a volumetric pipette rather than a measuring cylinder? [1]
   A. A volumetric pipette is more reliable.
   B. A volumetric pipette ensures greater precision.
   C. A volumetric pipette will give a more accurate result.
   D. A volumetric pipette can be reused.

3. In an experiment to measure the effect of temperature on the rate of reaction between hydrochloric acid and sodium thiosulfate solution, what type of variable is temperature? [1]
   A. A dependent continuous variable.
   B. An independent discrete variable.
   C. An independent continuous variable.
   D. A dependent discrete variable.

4. What type of compounds make the best standards for titrimetric analysis? [1]
   A. Compounds with low molar masses that are moderately soluble in water.
   B. Compounds with high molar masses that are very soluble in water.
   C. Compounds with low molar masses that are very soluble in water.
   D. Compounds with low molar masses that undergo hydrolysis in water.

[Total: 4]

Section B: Structured answers

The following question concerns an experiment to measure the enthalpy of combustion for methanol.

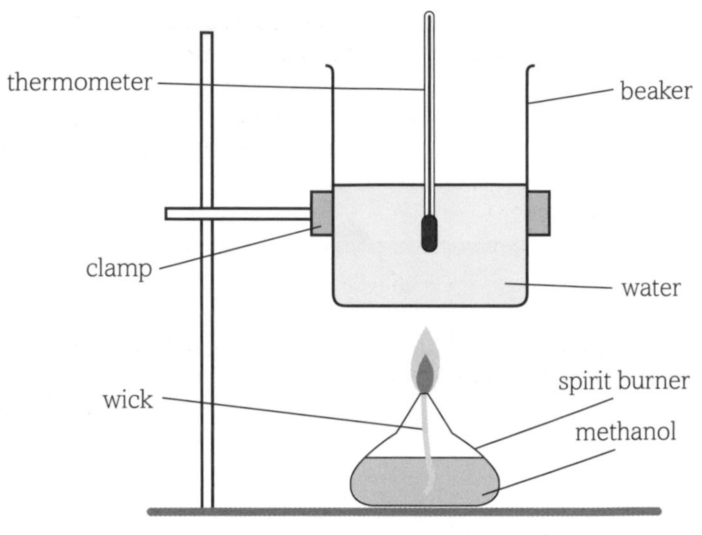

5. (a) Suggest a suitable material from which the calorimeter should be made for this experiment and explain your choice. [2]

A student carried out an experiment to measure the enthalpy of combustion of methanol. The results are shown below.

	Mass of spirit burner (g)	Volume of water (cm^3)	Temperature of water (°C)
Final	26.83	100	39
Initial	26.39	100	21
Difference		0	

(b) Complete the results table above. [1]

(c) (i) Using the data above calculate the amount of heat energy transferred to the water.
(Assume 1 cm^3 of water = 1 g and the specific heat capacity of water = 4.2 $J\,g^{-1}{}^{\circ}C^{-1}$) [1]

(ii) Hence calculate a value for the enthalpy of combustion for methanol. Give a sign and units with your answer. Give your answer to the appropriate number of significant figures. [4]

(d) (i) The value calculated in (c)(ii) is equal to 75.8% of the data book value. Use this information to calculate the data book value. [1]

(ii) Give three reasons for the difference between the experimental value in (c)(ii) and the data book value in (d)(i). [3]

[Total: 12]

Section C: Extended answers

6. Sulfamic acid (formula = NH_2SO_3H) is a white crystalline solid that can be used as a primary standard. It is a monoprotic acid.
   (a) Give procedural details of how you would prepare a standard solution of approximate concentration 0.1 mol dm^{-3}. Note that exact concentration of the solution must be known to 3 d.p. to be suitable as a standard. [4]
   (b) Give procedural details of how you would find out the exact concentration of a solution of KOH that is known to be approximately 1 mol dm^{-3}. Your procedure should include details of any necessary dilutions, all apparatus needed, and information about how reliability and precision are considered. [4]
   (c) Suggest why sulfamic acid is a better choice for a standard than hydrochloric acid. [2]

   [Total: 10]

7. The concentration of a solution of sulfuric acid can be experimentally determined in two different ways. Read through the two diferent methods and then carry out the calculations for each of the two experiments.

   You will then be asked to critically evaluate the two methods. Both methods use the same sample of sulfuric acid.

   Method 1
   1. 100 cm^3 of H_2SO_4 solution was measured out into a 250 cm^3 concical flask using a measuring cylinder.
   2. Excess barium chloride was added to the conical flask.
   3. The mixture was filtered and the precipitate of $BaSO_4$ collected and dried.
   4. The mass of the precipitate was reweighed until a constant mass was recorded.

   Result
   Mass of precipitate = 2.57 g

   (a) Write a balanced equation including state symbols for the formation of barium sulfate from sulfuric acid and barium chloride. [2]
   (b) Calculate the number of moles of $BaSO_4$ formed. [1]
   (c) Calculate the number of moles of H_2SO_4 in the 100 cm^3 of solution and hence calculate the concentration of the sulfuric acid to the appropriate number of significant figures. [2]

   Method 2
   1. 25 cm^3 of H_2SO_4 solution was pipetted into a 250 cm^3 conical flask.
   2. 6 drops of an indicator were added to the conical flask.
   3. A burette was first rinsed with 0.195 mol dm^{-3} NaOH(aq) and then filled to the 0.00 cm^3 mark.
   4. After a rough titration, three subsequent titrations were carried out that were concurrent to within 0.2 cm^3 and gave an average titre of 26.05 cm^3.

   Result
   Average titre = 26.05 cm^3

   (d) Suggest a suitable indicator for this titration and give its colour change at the end point. [1]
   (e) Write a balanced equation for the reaction between NaOH and H_2SO_4 showing state symbols. [2]
   (f) Use the average titre to calculate a value for the concentration of the sulfuric acid giving your answer to an appropriate number of significant figures. [1]

   Evaluation

   (g) Critically evaluate both methods in the light of your results. You should consider issues of reliability, accuracy and precision in your evaluation. [6]

   [Total: 15]

MODULE 2 Foundations in chemistry

CHAPTER 2.1

ATOMS AND REACTIONS

Introduction

Chemists have wondered for centuries about what matter is made from and how substances react. We now have an excellent understanding of atoms and what governs how they behave. Atoms and the particles they are made from, particularly electrons, dictate everything in the world around us, from how plants grow to how we generate the electricity used every day.

A knowledge of atoms and reactions is essential for studying chemistry. People working in the field of chemistry, and indeed many other areas, will use the knowledge covered in this chapter throughout their working lives. Pharmacists needs to know the chemical formulae and behaviour of drugs, a farmer needs to know how to neutralise excess acid in their fields to optimise plant growth, and an industrial chemist needs to know how much reactant to use and how much product to expect when making substances such as ammonia. The clothes you wear, the bricks your home is made from and the food you eat are all made up of atoms. Even the essential process of respiration comes down to atoms and gas volumes. An understanding of atoms and reactions underpins everything!

All the maths you need

To unlock the puzzles of this chapter you need the following maths:

- Units of measurement (*e.g. the mole*)
- Use of standard form and ordinary form (*e.g. representing Avogadro's number*)
- Using an appropriate amount of standard figures (*e.g. the number of moles of a substance*)
- Changing the subject of an equation (*e.g. finding a molecular mass*)
- Substituting numerical values into algebraic equations (*e.g. using* $n = cV$)
- Solving algebraic equations (*e.g. finding the amount of substance in a solution using* $n = cV$)

What have I studied before?

- Atoms, elements, compounds and mixtures
- Protons, neutrons and electrons and where they are found in the atom
- The differences between solids, liquids and gases
- The signs of chemical reactions
- How to represent chemical reactions using word and formulae equations, identifying which substances are reactants and which are products
- Acids and alkalis and what happens when they react together
- Typical reactions of acids with metals, metal oxides and metal carbonates

What will I study later?

- How properties of elements change across periods and down groups (AS)
- How chemical reactions, such as polymerisation, can be made more efficient and sustainable (AS)
- How molecules and compounds can be classified according to the functional groups they contain and the typical reactions these will undergo (AL)
- How rates can be determined for chemical reactions and the factors that influence these rates (AL)
- How the strength of acids is calculated (AL)
- How neutralisation reactions are investigated through titrations and the use of indicators (AL)
- The loss and gain of electrons in electrodes and the use of electrode potentials to consider the feasibility of reactions (AL)

What will I study in this chapter?

- How our understanding of the atom has changed over time
- How the atomic mass of a substance can be determined and how the presence of isotopes affects this
- How to predict the formulae of chemical compounds and ions and how to use these in balanced equations
- How to determine the number of moles in weighed substances, solutions and gases
- Why acids and bases behave as they do and how their behaviour is useful to us
- The use of oxidation numbers to describe what happens to electrons during reactions
- How percentage yields and atom economies are used to consider the efficiency of chemical reactions

2.1 ①

The changing atom

By the end of this topic, you should be able to demonstrate and apply your knowledge and understanding of:

- **atomic structure in terms of the numbers of protons, neutrons and electrons for atoms and ions, given the atomic number, mass number and any ionic charge**

LEARNING TIP

You will not need to recall specific dates of major discoveries but will need to be able to explain how models of the atom have changed over time.

Fifth century BCE (Before the Common Era) – the Greek atom

The Greek philosopher Democritus developed the first idea of the atom. He suggested that you could divide a sample of matter only a certain number of times. Eventually, he believed, you would end up with a particle that could not be split any further. Democritus called this particle 'átomos', which is Greek for 'indivisible'.

Early 1800s – Dalton's atomic theory

In the early 1800s, John Dalton developed his atomic theory. This stated that:

- atoms are tiny particles that make up elements
- atoms cannot be divided
- all atoms of a given element are the same
- atoms of one element are different from those of every other element.

Dalton used his own symbols to represent atoms of different elements. He also developed the first table of atomic masses. Many of Dalton's predictions still hold true and can be applied to chemistry today.

1897–1906 – Joseph John (J.J.) Thomson discovers electrons

Scientists had recently discovered cathode rays, which were emitted from cathode ray tubes. Thomson discovered that cathode rays were a stream of particles with the following properties:

- They had a negative charge.
- They could be deflected by both a magnet and an electric field.
- They had a very, very small mass.

Cathode rays were, in fact, electrons. Thomson concluded that they must have come from within the atoms of the electrodes themselves. The idea that an atom could not be split any further, proposed by the ancient Greeks and by Dalton, had been disproved.

Thomson proposed that atoms are actually made up of negative electrons moving around in a 'sea' of positive charge. This model is commonly called the *plum-pudding atom*. In Thomson's atom, the overall negative charge is the same as the overall positive charge. This means that the atom is neutral with no overall charge.

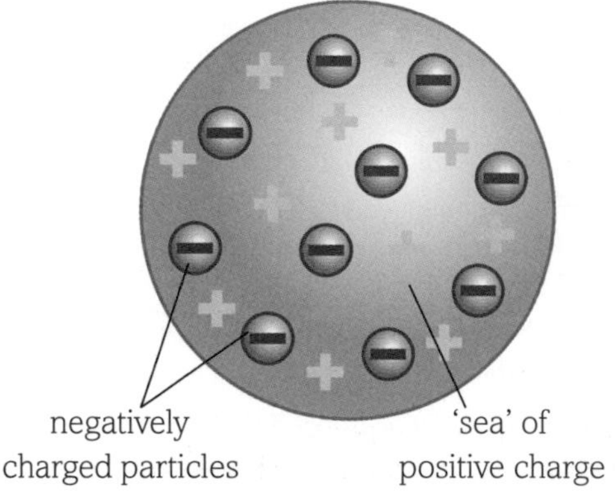

Figure 1 J.J. Thomson's plum-pudding atom.

1909–11 – Ernest Rutherford's gold-leaf experiment

In 1909, Rutherford and two of his students, Hans Geiger and Ernest Marsden, carried out an experiment where they directed α-particles (alpha particles) towards a sheet of very thin gold foil. They measured any deflection (change in direction) of the particles. Rutherford calculated that a plum-pudding atom would hardly deflect α-particles at all.

The results were astonishing:

- Most of the particles, as expected, were not deflected at all.
- However, a small percentage of particles were deflected through large angles.
- A very few particles were actually deflected back towards the source.

In 1911, he proposed the following new model for the atom based on these results:

- The positive charge of an atom and most of its mass are concentrated in a nucleus, at the centre.
- Negative electrons orbit this nucleus, just as the planets orbit the Sun.

- Most of an atom's volume would be the space between the tiny nucleus and the orbiting electrons.
- The overall positive and negative charges must balance.

Rutherford had proposed the nuclear atom and disproved the plum-pudding model.

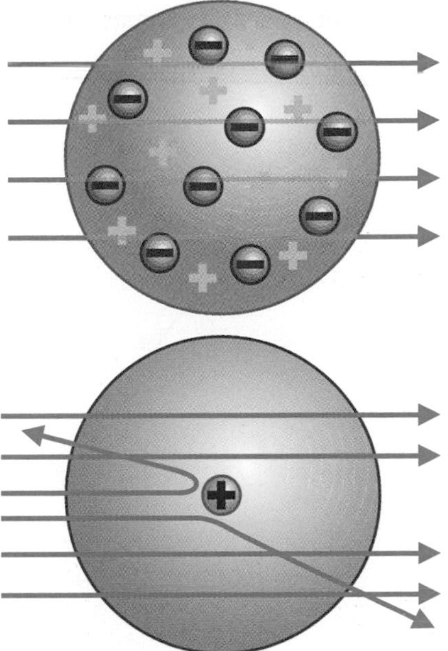

Figure 2 Rutherford's gold-leaf experiment.

Top: Expected results: α-particles passing through the plum-pudding atom undisturbed.

Bottom: Observed results: a few particles were deflected, indicating a small, concentrated positive charge – the nucleus.

1913 – Niels Bohr's planetary model and Henry Moseley's work on atomic numbers

In 1913, the Danish physicist Niels Bohr altered Rutherford's model to allow electrons to follow only certain paths. Otherwise, electrons would spiral into the nucleus. This was the planetary atom, in which electrons orbited a central nuclear 'sun' in 'shells'.

Bohr's model helped to explain some periodic properties, such as:

- spectral lines seen in emission spectra
- the energy of electrons at different distances from the nucleus.

In the same year, Henry Moseley discovered a link between X-ray frequencies and an element's atomic number (i.e. its order in the periodic table). At the time, Moseley couldn't explain this.

1918 – Rutherford discovers the proton

Rutherford's discovery of the proton was able to explain Moseley's finding that an atom's atomic number was linked to X-ray frequencies. We now know that the atomic number tells us the number of protons in an element's atom.

1923–26 – wave and particle behaviour

In 1923, the French physicist Louis de Broglie suggested that particles could have the nature of both a wave and a particle.

In 1926, the Austrian physicist Erwin Schrödinger suggested that an electron had wave-like properties in an atom. He also introduced the idea of atomic orbitals.

1932 – James Chadwick discovers the neutron

In 1932, an English physicist called James Chadwick observed a new type of radiation emitted from some elements. He showed that this new type of radiation was made up of uncharged particles with approximately the same mass as a proton. These uncharged particles became known as neutrons, because they have no charge.

Modern day

It is now thought that protons and neutrons themselves are made up of even smaller particles called quarks. Our understanding of the atom is likely to progress with time as science advances further and further.

Questions

1. Describe the model of the atom that Dalton developed in the early 1800s.
2. Explain how the model of the atom changed following the series of experiments that took place between 1897 and 1911.

2 Atomic structure

By the end of this topic, you should be able to demonstrate and apply your knowledge and understanding of:

* **isotopes as atoms of the same element with different numbers of neutrons and different masses**
* **atomic structure in terms of the numbers of protons, neutrons and electrons for atoms and ions, given the atomic number, mass number and any ionic charge**

Protons, neutrons and electrons

The current model of the atom is as follows:

- Protons and neutrons are found in the nucleus, which is at the centre of the atom.
- Electrons orbit the nucleus in 'shells'.
- The nucleus is tiny compared with the total volume of an atom.
- The nucleus is extremely dense and accounts for almost all of the atom's mass.
- Most of an atom consists of empty space between the tiny nucleus and the electron 'shells'.

The nucleus contains almost all the atom's mass, but it makes up only a tiny portion of the atom's volume.

Table 1 shows the relative mass and charge of a proton, neutron and electron:

- Protons have a positive charge and electrons have a negative charge.
- A neutral atom has the same number of protons as electrons, so an atom is electrically neutral.
- A proton has virtually the same mass as a neutron.
- Protons and neutrons make up almost all of the atom's mass.

Particle	Relative mass	Relative charge
proton	1.0	1+
neutron	1.0	0
electron	$\frac{1}{2000}$	1−

Table 1 Relative masses and charges of atomic particles.

Isotopes

Most elements are made up of a mixture of isotopes. Isotopes of the same element have:

- different masses
- the same number of protons and electrons
- different numbers of neutrons in the nucleus.

In Greek, isotope means 'at the same place'. The word isotope is used for different atoms of the same element that occupy the same position in the periodic table.

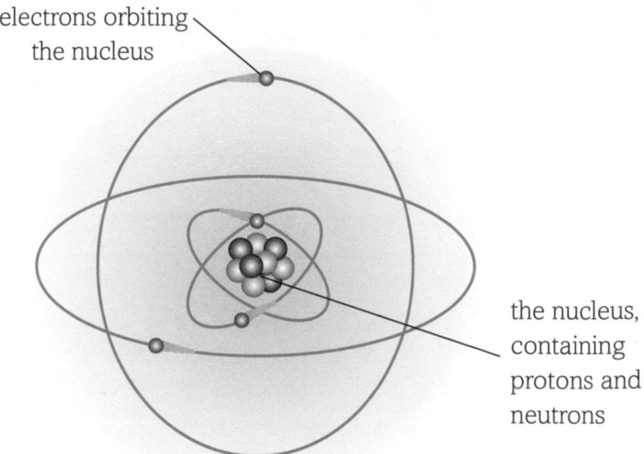

Figure 1 An isotope of beryllium. The beryllium nucleus has four protons (red) and five neutrons (green). Four electrons orbit the nucleus. Scientists now believe that electrons exist as clouds of electrical charge in *atomic orbitals*. We will discuss atomic orbitals in chapter 2.2.

We describe all elements (some of which have isotopes) using two numbers. These are:

- the atomic number, Z (proton number) – the number of protons in the nucleus
- the mass number, A (nucleon number) – the number of particles (protons and neutrons) in the nucleus.

Figure 2 shows how to write the symbol for an element using these numbers.

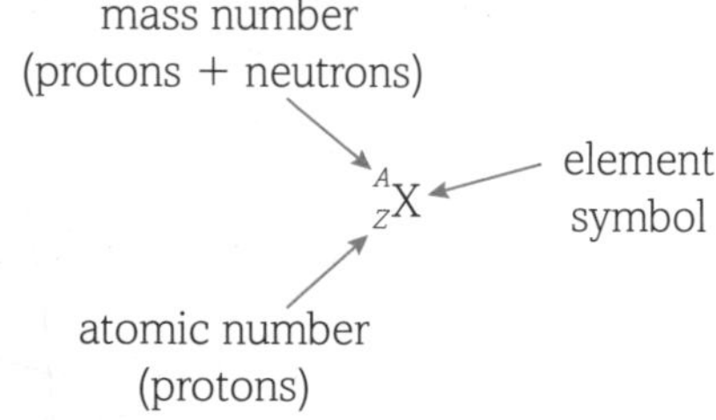

Figure 2 Symbol for an element.

The isotopes of carbon

Carbon exists as a mixture of the isotopes $^{12}_{6}C$, $^{13}_{6}C$ and $^{14}_{6}C$.

- The atomic number of each isotope is six, because each isotope has six protons.
- The overall charge in the atom *must* be zero. So each isotope must also have six electrons, to balance the charge from the six protons.

- The mass numbers are different because each isotope has a different number of neutrons. You can work out the number of neutrons by subtracting the atomic number from the mass number. This is shown in Table 2.

Isotope	$^{12}_{6}C$	$^{13}_{6}C$	$^{14}_{6}C$
Mass number	12	13	14
Atomic number	6	6	6
Number of neutrons	6	7	8

Table 2 Working out the number of neutrons for isotopes of carbon.

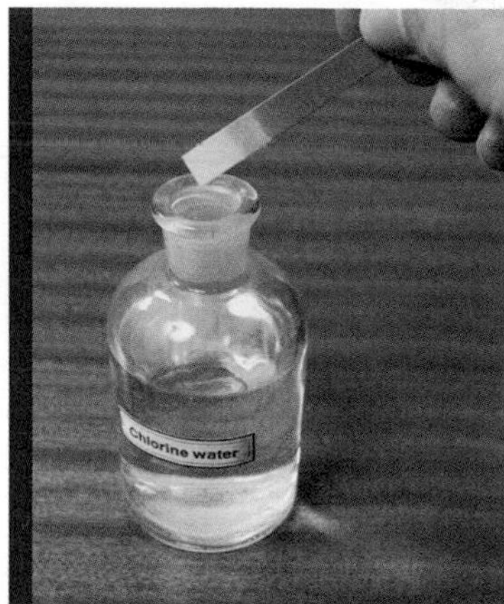

Figure 3 Isotopes of chlorine will all behave and react in the same way, as they all have the same number of electrons.

Reactions of isotopes

Different isotopes of the same element react in the same way. This is because:

- chemical reactions involve electrons, and isotopes have the same number and arrangement of electrons
- neutrons make no difference to chemical reactivity.

Atomic structure of ions

Many atoms react by losing or gaining electrons to form charged particles called ions. Ions are charged because they have unbalanced numbers of protons and electrons, and so the charges no longer cancel each other out.

Table 3 shows the atomic structures of two ions: $^{23}_{11}Na^+$ and $^{35}_{17}Cl$. Note that:

- Na^+ has one electron fewer than its number of protons (it has lost an electron).
- Cl^- has one electron more than its number of protons (it has gained an electron).

Ion	$^{23}_{11}Na^+$	$^{35}_{17}Cl^-$
Atomic number, *Z*	11	17
Mass number, *A*	23	35
Protons	11	17
Neutrons	12	18
Electrons	10	18
Overall charge	11 − 10 = +1	17 − 18 = −1

Table 3 Calculating the numbers of each atomic particle for ions of Na^+ and Cl^-.

LEARNING TIP

When you need to give the atomic structures for ions, only adjust the number of electrons. For example, for a 3+ ion, simply take away three electrons. For a 2− ion, just add two electrons.

Questions

1. How many protons, neutrons and electrons are in the following atoms and ions?
   (a) $^{7}_{3}Li$ (b) $^{24}_{11}Na$ (c) $^{19}_{9}F$ (d) $^{56}_{26}Fe$
   (e) $^{39}_{19}K^+$ (f) $^{19}_{9}F^-$ (g) $^{39}_{20}Ca^{2+}$ (h) $^{17}_{8}O^{2-}$

2. Write down symbols for the following atoms and ions. (Look at how the symbols are written in Question 1 for examples.)
   (a) The atom with 13 protons, 14 neutrons and 13 electrons.
   (b) The ion with 16 protons, 16 neutrons and 18 electrons.

3. Explain why the same products would be formed when both carbon-12 and carbon-13 are combusted in oxygen.

Atomic masses

By the end of this topic, you should be able to demonstrate and apply your knowledge and understanding of:

- explanation of the terms *relative isotopic mass* (mass compared with $\frac{1}{12}$th mass of carbon-12) and *relative atomic mass* (weighted mean mass compared with $\frac{1}{12}$th mass of carbon-12), based on the mass of a ^{12}C atom, the standard for atomic masses
- the terms *relative molecular mass, M_r*, and *relative formula mass* and their calculation from relative atomic masses

Measurement of relative masses

How would you go about weighing something that you cannot see? This is the situation with atoms.

Instead of finding the mass of atoms directly, we compare the masses of different atoms, using the idea of relative mass. The carbon-12 isotope (also written ^{12}C) is the international standard for the measurement of relative mass.

Atomic masses are measured using a unit called the *unified atomic mass unit*, u:

- 1 u is a tiny mass: $1.660\,540\,210 \times 10^{-27}$ kg.
- The mass of an atom of carbon-12 is defined as 12 u.
- So the mass of one-twelfth of an atom of carbon-12 is 1 u.

$$^{12}_{6}C$$

Figure 1 Carbon-12: the standard for measuring atomic masses. An atom of the carbon-12 isotope is made up of 6 protons, 6 neutrons and 6 electrons. The mass of an atom of carbon-12 is 12 atomic mass units, 12 u. All atomic masses are measured compared with the mass of a carbon-12 atom.

Relative isotopic mass

All the atoms in a single isotope are identical. They all have the same mass, measured against carbon-12.

For an isotope, the **relative isotopic mass** is the same as the mass number.

- So, for oxygen-16 (also written as ^{16}O), the relative isotopic mass = 16.0.
- For sodium-23 (also written as ^{23}Na), the relative isotopic mass = 23.0.

Note that we have made two important assumptions:

- We have neglected the tiny contribution that electrons make to the mass of an atom.
- We have taken the masses of both a proton and a neutron as 1.0 u.

Relative atomic mass, A_r

Most elements contain a mixture of isotopes, each in a different amount and with a different mass. We use the term 'weighted mean mass' to account for the contribution made by each isotope to the overall mass of an element.

The contribution made by an isotope to the overall mass depends on:

- the percentage abundance of the isotope
- the relative mass of the isotope.

Chemists then combine the contributions from each isotope to arrive at the **relative atomic mass, A_r**, of an element. Remember, all masses are relative to the mass of carbon-12.

WORKED EXAMPLE

A sample of bromine contains 53.00% bromine-79 and 47.00% bromine-81.

The relative atomic mass of bromine would be calculated as follows:

$$A_r\,(Br) = \frac{(53.00 \times 79.00) + (47.00 \times 81.00)}{100}$$

= 41.87 (*the contribution from ^{79}Br*) +

38.07 (*the contribution from ^{81}Br*)

= 79.94

KEY DEFINITIONS

Relative isotopic mass is the mass of an atom of an isotope compared with one-twelfth of the mass of an atom of carbon-12.

Relative atomic mass, A_r, is the weighted mean mass of an atom of an element compared with one-twelfth of the mass of an atom of carbon-12.

Relative molecular mass, M_r

Many elements and compounds are made up of simple molecules. Common examples are gases found in the air: N_2, O_2 and CO_2. The mass of a molecule is measured as the relative molecular mass by comparison with carbon-12.

You can find the relative molecular mass, M_r, by adding together the relative atomic masses of each atom making up a molecule:

- $M_r(Cl_2) = 35.5 \times 2 = 71.0$
- $M_r(H_2O) = (1.0 \times 2) + 16.0 = 18.0$

Relative formula mass

Compounds with giant structures do not exist as simple molecules. These include ionic compounds, such as NaCl, and covalent compounds, such as SiO_2.

Although the term relative *molecular* mass is often used for these compounds, a better term is relative formula mass. This is because finding the relative molecular mass would mean adding the masses of thousands of atoms together, which would give an extremely large number.

You can find the relative formula mass by adding together the relative atomic masses of each atom making up a formula unit:

- $CaBr_2$: relative formula mass = 40.1 + (79.9 × 2) = 199.9
- Na_3PO_4: relative formula mass = (23.0 × 3) + 31.0 + (16.0 × 4) = 164.0

LEARNING TIPS

- A formula unit is whatever is written in the chemical formula. For example, the formula unit of sodium chloride is one atom of Na and one atom of Cl, because the formula unit is NaCl.
- Make sure you can recall the definitions for relative isotopic mass and relative atomic mass exactly as they are written.
- You do not need to remember the exact definitions of relative formula mass and relative molecular mass. Remember to use the term relative molecular mass when describing simple molecules and relative formula mass when describing giant structures.

Questions

1. Calculate the relative atomic mass, A_r, of the following. Give your answers to four significant figures.
   (a) Boron, containing 19.77% ^{10}B and 80.23% ^{11}B
   (b) Silicon, containing 92.18% ^{28}Si, 4.70% ^{29}Si and 3.12% ^{30}Si
   (c) Chromium, containing 4.31% ^{50}Cr, 83.76% ^{52}Cr, 9.55% ^{53}Cr and 2.38% ^{54}Cr

2. Use A_r values from the periodic table to calculate the relative molecular mass of the following.
   (a) HCl (b) CO_2 (c) H_2S (d) NH_3 (e) H_2SO_4

3. Use A_r values from the periodic table to calculate the relative formula mass of the following.
   (a) Fe_2O_3 (b) Na_2O (c) $Pb(NO_3)_2$ (d) $(NH_4)_2SO_4$ (e) $Ca_3(PO_4)_2$

Determining masses using mass spectrometry

By the end of this topic, you should be able to demonstrate and apply your knowledge and understanding of:

* use of mass spectrometry in:
    (i) the determination of relative isotopic masses and relative abundances of the isotope
    (ii) calculation of the relative atomic mass of an element from the relative abundances of its isotopes

What is mass spectrometry?

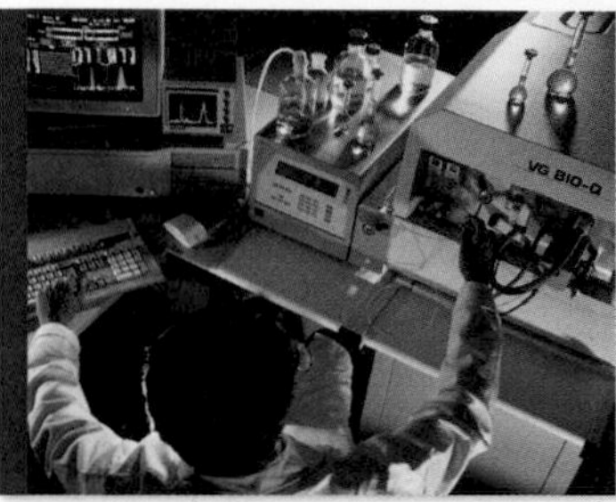

Figure 1 A mass spectrometer is used to determine information about atoms and molecules, such as relative abundances of isotopes.

A mass spectrometer is a piece of apparatus that can be used to find out about molecules. For example, it can be used to:

- identify an unknown compound
- find the relative abundance of each isotope of an element
- determine structural information about molecules.

A mass spectrometer determines the mass of a molecule or isotope by measuring the mass-to-charge ratio of ions. It does this by causing substances to become positive ions. These positive ions are then passed through the apparatus and separated according to their mass and charge. A computer within the mass spectrometer analyses the data on the ions present and produces a mass spectrum. This is similar to a complex bar graph and gives information about the abundance of ions present in the sample.

An example mass spectrum is shown in Figure 2.

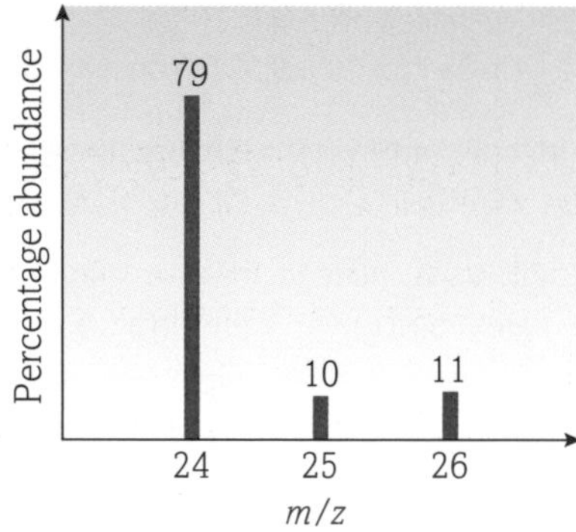

Figure 2 The mass spectrum of magnesium.

Determining relative isotopic mass and abundance

One of the most important uses of mass spectrometry is to determine the isotopes present in an element. You will need to be able to analyse a mass spectrum to find the proportions of each isotope in the element being investigated.

A mass spectrum shows relative or percentage abundance on the y-axis and mass-to-charge ratio on the x-axis.

The mass-to-charge ratio in all mass spectra is shown as m/z

- m is the mass
- z is the charge on the ion, which is usually 1.

Look at the mass spectrum of magnesium shown in Figure 2 as an example.

- There are three peaks in the spectrum, so there are three isotopes of magnesium with isotopic masses of 24, 25 and 26 respectively.
- The heights of the peaks give the relative abundances of the isotopes in the sample, in this case, 79% Mg-24, 10% Mg-25 and 11% Mg-26.

Usually, the mass spectrum will give values for the relative abundances. If there are no values, or no scale on the y-axis, the relative abundances can be worked out by measuring each line and then following the method shown in Worked example 1.

WORKED EXAMPLE 1

Calculating relative abundance from a mass spectrum

Total heights of all peaks on spectra = 2 cm + 8 cm = 10 cm

Height of isotope A peak = 2 cm

$$\% \text{ abundance of isotope A} = \left(\frac{2\text{ cm}}{10\text{ cm}}\right) \times 100 = 20\%$$

Height of isotope B peak = 8 cm

$$\% \text{ abundance of isotope B} = \left(\frac{8\text{ cm}}{10\text{ cm}}\right) \times 100 = 80\%$$

Determining relative atomic mass

Once the individual isotopic masses and the relative abundances of the isotopes have been determined from a mass spectrum, this information can be used to find the relative atomic mass of the sample.

Worked example 2 shows how the data on magnesium from the spectrum shown in Figure 2 would be used.

WORKED EXAMPLE 2

Calculating the relative atomic mass from a mass spectrum of magnesium

Magnesium has three isotopes and their abundance is as follows:

Mg-24 79%

Mg-25 10%

Mg-26 11%

$$\text{Relative atomic mass} = \frac{(24 \times 79) + (25 \times 10) + (26 \times 11)}{100}$$

$$= 24.32$$

LEARNING TIP

Relative atomic mass has no units.

Questions

1. A mass spectrum of lithium showed two peaks. The first peak was at $m/z = 6$ and had an abundance of 7.4%; the second peak was at $m/z = 7$ and had an abundance of 92.6%.

   (a) identify the isotopes of lithium

   (b) calculate the relative atomic mass of the lithium sample to three significant figures.

2. A sample of strontium has four isotopes as shown in Table 1. Sketch a mass spectrum for the strontium sample and calculate its relative atomic mass.

Ion	Percentage abundance
$^{84}Sr^+$	1
$^{86}Sr^+$	9
$^{87}Sr^+$	7
$^{88}Sr^+$	83

Table 1 Strontium's four isotopes.

Ions and the periodic table

By the end of this topic, you should be able to demonstrate and apply your knowledge and understanding of:

* the writing of formulae of ionic compounds from ionic charges, including:
  (i) prediction of ionic charge from the position of an element in the periodic table
  (ii) recall of the names and formulae for the following ions: NO_3^-, CO_3^{2-}, SO_4^{2-}, OH^-, NH_4^+, Zn^{2+} and Ag^+

Predicting ionic charge

You can predict the charge on an element's ion from its position in the periodic table.

The number of electrons in the outer shell is the same as the group number of an element. It is then a simple step to calculate how many electrons need to be lost or gained in order to reach a noble gas electron configuration. From this, you can predict the likely charge on the resulting ion.

- For example, lithium, in group 1, has one electron in its outer shell.
- To form the electron configuration of the nearest noble gas, helium, a lithium atom must lose one outer electron.
- A lithium ion, Li^+, therefore has a charge of 1+.

Elements in the same group of the periodic table have the same number of outer shell electrons and react in similar ways.

Table 1 shows the charges on ions for the elements in periods 2 and 3 of the periodic table. The same principle can be extended to other members of each period further down the periodic table.

LEARNING TIP

A group is a vertical column in the periodic table. Elements in the same group have similar chemical properties and their atoms have the same number of outer shell electrons.

Group	1	2	13	14	15	16	17	18
Number of outer shell electrons	1	2	3	4	5	6	7	8(0)
Period 2 element	Li	Be	B	C	N	O	F	Ne
Period 2 ion	Li^+				N^{3-}	O^{2-}	F^-	
Period 3 element	Na	Mg	Al	Si	P	S	Cl	Ar
Period 3 ion	Na^+	Mg^{2+}	Al^{3+}		P^{3-}	S^{2-}	Cl^-	

Table 1 Ion charges for periods 2 and 3. These show the link between group number and ionic charge.

Remember, metals will generally lose electrons. This means they form positive ions.
Non-metals will generally gain electrons. This means they form negative ions.
Ionic formulae are written with the element symbol, or empirical formula, first, and the size and sign (negative or positive) of the charge in superscript to the right of this, e.g. the silver ion Ag^+ and the zinc ion Zn^{2+}.

Atoms of metals in groups 1–13:

- lose electrons
- form positive ions with the electron configuration of the previous noble gas in the periodic table.

Atoms of non-metals in groups 15–17:

- gain electrons
- form negative ions with the electron configuration of the next noble gas in the periodic table.

Atoms of Be, B, C and Si:

- do not normally form ions
- require too much energy to transfer the outer shell electrons to form ions.

Some elements can form more than one ion, each with a different charge. The oxidation number of the element is written as a Roman numeral, to make it clear which ion has been formed. For example:

- iron(II) for Fe^{2+} and iron(III) for Fe^{3+}
- copper(I) for Cu^{+} and copper(II) for Cu^{2+}.

Molecular ions

Groups of covalently bonded atoms can also lose or gain electrons to form ions. These are called molecular ions. The charge on a molecular ion can be considered as being shared out across the entire molecule. Some common molecular ions are shown in Table 2. You must commit the names and formulae of these ions to memory.

1+	1−	2−
Ammonium NH_4^{+}	Hydroxide OH^{-} Nitrate NO_3^{-}	Carbonate CO_3^{2-} Sulfate SO_4^{2-}

Table 2 Molecular ions, grouped according to their charges.

Predicting ionic formulae

Although an ionic compound is made up of oppositely charged ions, its overall charge must be zero.

In an ionic compound:

total number of + charges from positive ions
= total number of − charges from negative ions.

Working out an ionic formula from the ionic charges

You can write an ionic formula by following some simple rules.

- In an ionic compound, the overall charge is zero.
- Select as many + ions and − ions as required for the charges to balance.

For example, when writing the formula of calcium chloride:

- you can predict that Ca will form a Ca^{2+} ion and Cl will form a Cl^{-} ion
- two Cl^{-} ions will be needed to cancel out the 2+ charge on Ca^{2+}
- the formula for calcium chloride is $CaCl_2$.

For aluminium sulfate, aluminium will form Al^{3+} ions and sulfate molecular ions have the formula SO_4^{2-}.

Therefore, the formula is constructed as shown:

- $2 \times Al^{3+}$ = a positive charge of +6
- $3 \times SO_4^{2-}$ = a negative charge of −6
- Therefore the formula is $Al_2(SO_4)_3$.

Questions

1. Predict the ions formed by the following elements:
   (a) K
   (b) I
   (c) Ca

2. Predict the formulae for the following ionic compounds:
   (a) calcium iodide
   (b) lithium nitride
   (c) aluminium sulfide
   (d) magnesium phosphide.

3. Predict the formulae for the following ionic compounds:
   (a) nickel(II) chloride
   (b) copper(I) oxide
   (c) iron(III) chloride
   (d) chromium(III) oxide
   (e) manganese(VI) oxide
   (f) titanium(IV) chloride.

4. Predict the formulae for the following ionic compounds:
   (a) calcium hydroxide
   (b) iron(III) sulfate
   (c) ammonium sulfate

2.1

6 Amount of substance and the mole

By the end of this topic, you should be able to demonstrate and apply your knowledge and understanding of:

* explanation and use of the terms:
  (i) *amount of substance*
  (ii) *mole* (symbol 'mol'), as the unit for amount of substance
  (iii) the *Avogadro constant*, N_A (the number of particles per mole, 6.02×10^{23} mol^{-1})
  (iv) *molar mass* (mass per mole, units g mol^{-1})
* calculations, using amount of substance in mol, involving mass

Amount of substance

Chemists use a quantity called **amount of substance** for counting atoms. Amount of substance is:

- given the symbol n
- measured using a unit called the **mole** (abbreviated to mol).

Amount of substance is based on a standard count of atoms called the **Avogadro constant**, N_A. The Avogadro constant, N_A, is the number of atoms per mole of the carbon-12 isotope. (See also topic 2.1.8.)

$$N_A = 6.022\,141\,79 \times 10^{23}\ \text{mol}^{-1}$$
$$= 6.02 \times 10^{23}\ \text{mol}^{-1} \text{ to three significant figures}$$

KEY DEFINITIONS

Amount of substance is the quantity that has moles as its unit. Chemists use 'amount of substance' as a way of counting atoms.
A **mole** is the amount of any substance containing as many particles as there are carbon atoms in exactly 12 g of the carbon-12 isotope.
The **Avogadro constant**, N_A, is the number of atoms per mole of the carbon-12 isotope (6.02×10^{23} mol^{-1}).
Molar mass, M, is the mass per mole of a substance. The units of molar mass are g mol^{-1}.

Counting and weighing atoms

Using the definition of one mole, you now know that 12 g of carbon-12 contains 6.02×10^{23} atoms.

Figure 1 shows one mole of atoms of some different elements.

carbon	sulfur	iron	copper	lead
C	S	Fe	Cu	Pb
12.0 g	32.1 g	55.8 g	63.5 g	207.2 g

Figure 1 One mole of atoms of different elements. Each sample contains 6.02×10^{23} atoms or 1 mol of atoms.

The mass of one mole is easy to work out, as it is the same as the relative atomic mass in grams. For example:

- one mole of carbon (C) atoms has a mass of 12.0 g
- one mole of lead (Pb) atoms has a mass of 207.2 g.

You can also work in multiples of 1 mol. For carbon:

- 2.00 mol of carbon (C) atoms has a mass of 24.0 g
- 0.50 mol of carbon (C) atoms has a mass of 6.0 g.

To find out the mass volume of a substance you would need to use a balance. You can either measure the mass of a vessel first (e.g. a weighing boat or a beaker) and then take this away from the final combined mass of the vessel and the substance, or you can 'zero' the display after placing the vessel onto the balance and then add the substance and record its mass directly.

DID YOU KNOW?

Analytical balances

Chemists use analytical balances to weigh out substances. These are balances that can accurately determine the mass of a substance, usually to the nearest 0.000 01 g. Analytical balances have doors on them that allow air flow to be kept away from the substance being weighed, as this could affect the reading.

Figure 2 An analytical mass balance, which allows the mass of substances to be determined to a high degree of accuracy.

Moles of anything!

Amount of substance is not restricted to atoms. You can have an amount of any chemical species. Examples include:

- amount of atoms, for example a mole of C atoms
- amount of molecules, for example a mole of Cl_2 molecules
- amount of ions, for example a mole of OH^- ions.

It is important that you always state the chemical formula to which an amount of substance refers:

- 1 *mole of oxygen* could either refer to oxygen atoms, O, or to oxygen molecules, O_2.
- 1 mole of O or 1 mole of O_2 is clearer.

How big is a mole?

How do you judge the real size of 6.02×10^{23}? It is a huge number, so large that it is almost impossible to imagine its true size. Let's look at some comparisons:

- 6.02×10^{23} (1 mol) of drink cans would cover the surface of the Earth to a depth of over 300 km.
- Even if you were able to count atoms at the rate of 10 million per second, it would still take you about 2 billion years to count the atoms in one mole of atoms.

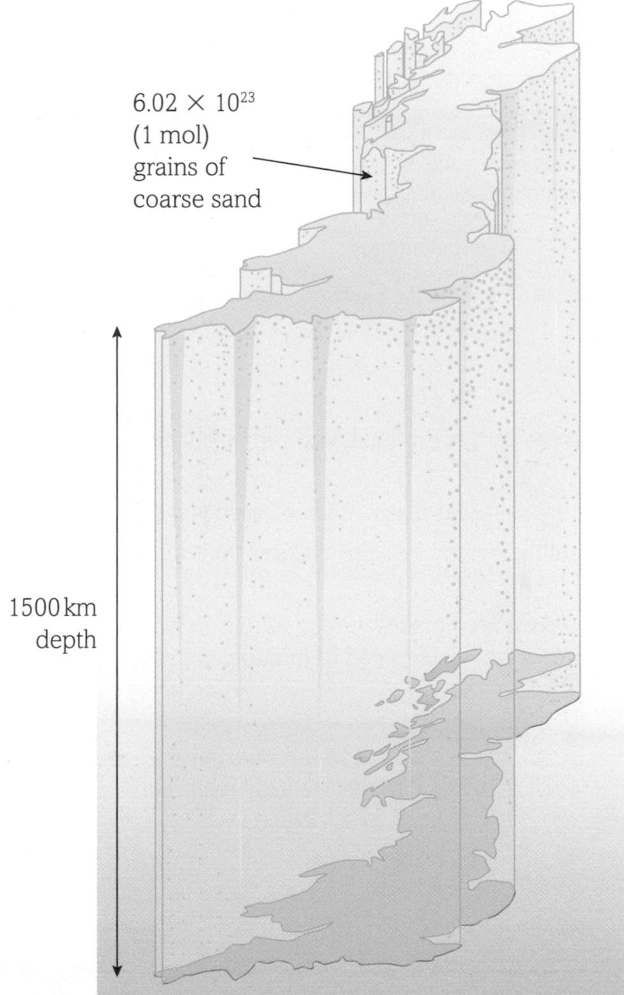

Figure 3 One mole of coarse sand grains would cover the surface of Great Britain to a depth of 1500 km.

Molar mass

Molar mass, *M*, is an extremely useful term as it can be applied to any chemical substance, whether an element, molecule or ion.

You can find the molar mass by adding together all the relative atomic masses for each atom that makes up a formula unit.

- For carbon: $M(C) = 12.0 = 12.0\ g\ mol^{-1}$
- For carbon dioxide: $M(CO_2) = 12.0 + (16.0 \times 2) = 44.0\ g\ mol^{-1}$

LEARNING TIP

The unit $g\ mol^{-1}$ is the same as g/mol and is read as 'grams per mole'.

The amount of substance, *n*, mass, *m*, and molar mass, *M*, are all linked by the formula:

$$\text{No. of moles (mol)} = \frac{\text{mass (g)}}{\text{molar mass (g mol}^{-1})} \quad \text{or } n = \frac{m}{M}$$

For example, the number of moles in 48.0 g of C can be calculated as follows:

$$n(C) = \frac{48.0\ g}{12.0\ g\ mol^{-1}} = 4.00\ mol$$

And the number of moles in 11.0 g of CO_2:

$$n(CO_2) = \frac{11.0\ g}{44.0\ g\ mol^{-1}} = 0.250\ mol$$

LEARNING TIP

Throughout your chemistry studies it is essential that you know this equation:

$$\text{number of moles} = \frac{\text{mass}}{\text{molar mass}}$$

You should also be able to rearrange it to find out the mass or molar mass of a substance.

Questions

1. Calculate the mass, in g, of the following:
   (a) 3.00 mol SiO_2 (b) 0.500 mol Fe_2O_3 (c) 0.250 mol Na_2CO_3
2. Calculate the amount, in mol, of the following:
   (a) 8.00 g BeF_2 (b) 23.7 g $KMnO_4$ (c) 10.269 g $Al_2(SO_4)_3$
3. Calculate the molar mass, in $g\ mol^{-1}$, of the following:
   (a) 0.05 mol of a substance with a mass of 2 g
   (b) 0.20 mol of a substance with a mass of 13 g
4. Which elements had been measured in Question 3?

Types of formulae

By the end of this topic, you should be able to demonstrate and apply your knowledge and understanding of:

* use of the terms:
  (i) *empirical formula* (the simplest whole number ratio of atoms of each element present in a compound)
  (ii) *molecular formula* (the number and type of atoms of each element in a molecule)
* calculations of empirical and molecular formulae, from composition by mass or percentage compositions by mass and relative molecular mass

What is an empirical formula?

An empirical formula is the simplest way of showing a chemical formula. It shows the ratio between elements, rather than actual numbers of atoms of each element (although sometimes the empirical formula will match this).

Figure 1 shows part of the structure of sodium chloride. If you wanted to show the formula for sodium chloride using the actual number of ions bonded together, it would run into millions of millions – remember how big the mole is! The formula would also depend on the size of the salt crystal. Structures like this are called giant structures.

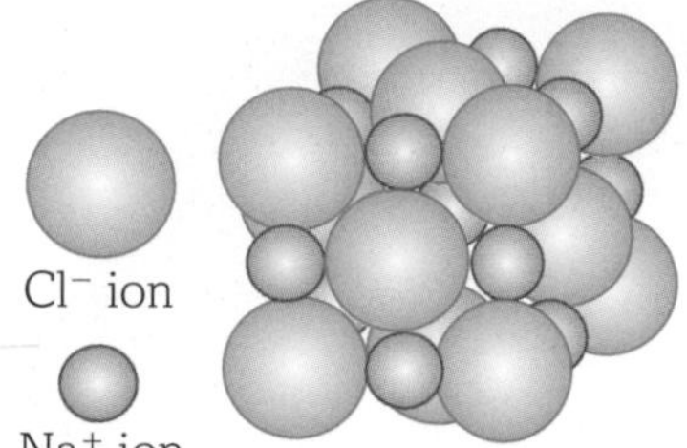

Figure 1 Structure of sodium chloride. The size of the structure depends on the number of ions making up the crystal. We use the empirical formula, NaCl, because we know that there is always one Na^+ ion for every Cl^- ion in a crystal of sodium chloride.

For this reason, empirical formulae are always used for compounds with giant structures, including:

- ionic compounds, such as NaCl
- giant covalent compounds, such as SiO_2.

INVESTIGATION

You can find the empirical formula of magnesium oxide by heating magnesium and oxygen and a crucible. You will need to find out the mass of magnesium before the reaction and the mass of magnesium oxide after the reaction. You can then use the formula number of moles = $\frac{\text{mass}}{\text{molar mass}}$ to determine how many moles of each substance have reacted together.

Once you know how many moles have reacted, you can determine the empirical formula for magnesium oxide.

Calculating empirical formulae

If you know the mass of each element in a sample of a compound, you can calculate its empirical formula, using the following steps:

1. Divide the amount of each element present by its molar mass – this will give you the molar ratio.
2. Divide the answer for each element by the smallest number – this ensures your ratio is in the format 1 : x.
3. If necessary, multiply the answer by a suitable value to make sure the ratio is in whole numbers only. For example, to have a whole number ratio you would need to multiply the ratio 1 : 1.5 by 2 to make it 2 : 3.

LEARNING TIP

Did you notice that Step 1 in the investigation involves using the equation

number of moles = $\frac{\text{mass}}{\text{molar mass}}$?

Make sure you remember this equation!

WORKED EXAMPLE 1

Analysis showed that 0.6075 g magnesium combines with 3.995 g of bromine to form a compound.
[A_r: Mg, 24.3; Br, 79.9]

Find the molar ratio of atoms: Mg : Br.

$$\frac{0.6075}{24.3} : \frac{3.995}{79.9}$$

$$0.025 : 0.050$$

Divide by smallest number (0.025):

$$1 : 2$$

Empirical formula is $MgBr_2$

WORKED EXAMPLE 2

Analysis of a compound showed the following percentage composition by mass:

Na: 74.19%; O: 25.81%. [A_r: Na, 23.0; O, 16.0]

100.0 g of the compound contains 74.19 g of Na and 25.81 g of O.

Find the molar ratio of atoms: Na : O

$$\frac{74.19}{23.0} : \frac{25.81}{16.0}$$

$$3.226 : 1.613$$

Divide by smallest number (1.613):

$$2 : 1$$

Empirical formula is Na_2O

What is a molecular formula?

Molecular formulae are used for compounds that exist as simple molecules. A molecular formula tells you the number of each type of atom that make up a molecule.

Figure 2 shows a propane molecule, which contains three carbon atoms and eight hydrogen atoms.

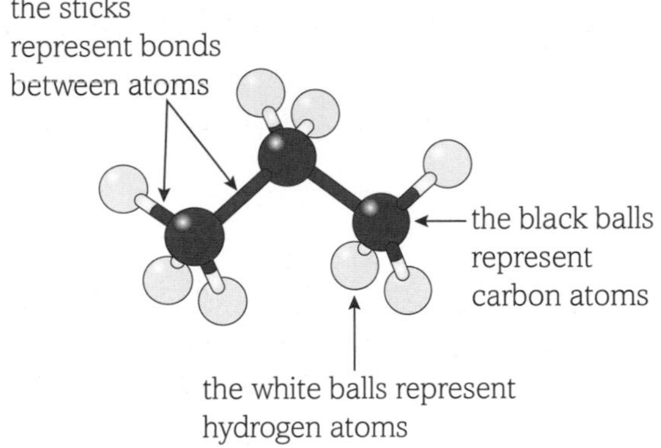

Figure 2 Molecule of propane.

The molecular formula of propane is C_3H_8.

The simplest ratio of atoms is 3 C to 8 H, so the empirical formula of propane is also C_3H_8.

Figure 3 shows a butane molecule, which contains four carbon atoms and 10 hydrogen atoms. The simplest ratio of atoms is 2 C to 5 H. So, the molecular formula is C_4H_{10} but the empirical formula is C_2H_5.

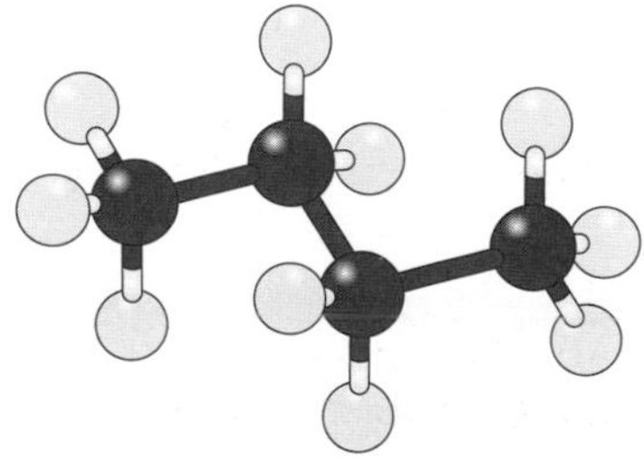

Figure 3 A molecule of butane, C_4H_{10}.

For many molecules, the molecular formula is adequate. In organic chemistry, however, the molecular formula does not tell you the order in which the atoms are bonded to each other. Other types of formulae can be used to describe the structure of organic molecules. You will learn about these in topic 4.1.2.

Calculating molecular formulae

We usually find empirical formulae and relative molecular masses of compounds through experiments. We can use the experimental results to find out the molecular mass of a compound.

The molecular mass of a compound must always be the same as the mass of the empirical formula, or a whole-number multiple of it.

WORKED EXAMPLE 3

A compound has an empirical formula of CH_2 and a relative molecular mass, M_r, of 56.0. What is its molecular formula?

empirical formula mass of $CH_2 = 12.0 + (1.0 \times 2) = 14.0$

number of CH_2 units in a molecule: $= \frac{56.0}{14.0} = 4$

molecular formula: $(4 \times CH_2) = C_4H_8$

Questions

1 Determine the empirical formulae of the following:

(a) The compound formed when 6.54 g of zinc reacts with 1.60 g of oxygen.

(b) The compound formed when 5.40 g of aluminium reacts with 4.80 g of oxygen.

(c) The compound containing: Ag, 69.19%; S, 10.29%; and O, 20.52%.

2 Determine the molecular formulae of the following:

(a) The compound of nitrogen and oxygen containing 0.49 g of nitrogen combined with 1.12 g of oxygen, M_r 92.0.

(b) The compound consisting of carbon, hydrogen and oxygen atoms containing 1.80 g of carbon combined with 0.30 g of hydrogen and 1.20 g of oxygen, M_r 88.0.

(c) The compound of carbon, hydrogen and oxygen with the composition by mass: C, 40.0%; H, 6.7%; O, 53.3%; M_r 180.0.

3 Determine the molecular formula of an organic compound with a molecular mass of 112 g mol^{-1} and an empirical formula of CH_2.

2.1 8

Moles and gas volumes

By the end of this topic, you should be able to demonstrate and apply your knowledge and understanding of:

* explanation and use of the term *molar gas volume* (gas volume per mole, units $dm^3\,mol^{-1}$)
* calculations, using amount of substance in mol, involving gas volume
* the techniques and procedures required during experiments requiring the measurement of gas volumes
* the ideal gas equation: $pV = nRT$

Avogadro's hypothesis

In 1811, the Italian Amedeo Avogadro put forward a hypothesis. It stated that under the same conditions of temperature and pressure, a mole of any gas would fill the same volume of space.

This idea, now known as Avogadro's law, is important because it means that we can compare the number of molecules in different gases by comparing their volumes. It does not matter which type of gas is being studied. By measuring the volume of any gas, we are indirectly counting the number of molecules.

At room temperature and pressure (RTP):

- one mole of gas molecules occupies approximately $24.0\,dm^3$ ($24\,000\,cm^3$)
- the volume per mole of gas molecules is $24.0\,dm^3\,mol^{-1}$.

Molar gas volume is the volume per mole of gas molecules.

KEY DEFINITION

Molar gas volume is the volume per mole of a gas. The units of molar volume are $dm^3\,mol^{-1}$. At room temperature and pressure, the molar volume is approximately $24.0\,dm^3\,mol^{-1}$.

At RTP, 1 mol of H_2(g), O_2(g) and CO_2(g) each have a molar volume of $24.0\,dm^3\,mol^{-1}$, as indeed 1 mol of any gas would. To give an idea of scale, $24.0\,dm^3$ is equivalent to the amount of air contained in four footballs.

Air consists mainly of N_2 and O_2 gases, which have similar relative molecular masses. Look at the diagram in Figure 1.

- The same volume of H_2(g) would be much lighter than air.
- The same volume of CO_2(g) would be much heavier than air.

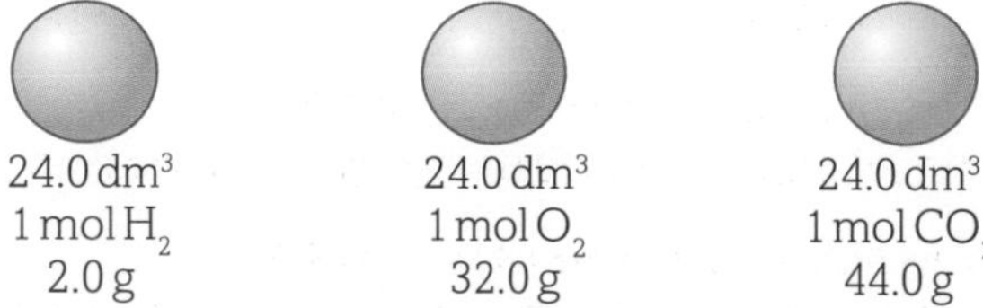

Figure 1 One mole of different gases. Note that all the gases have the same volume but different masses, so H_2(g), O_2(g) and CO_2(g) have different densities.

The differences in the individual particles have no effect on the overall volume because the particles are so spread out that any differences become unimportant.

DID YOU KNOW?

Chemists can measure gas volumes using a gas syringe. The plunger moves out as gas is produced, and the volume of gas can be measured by reading the scale on the side.

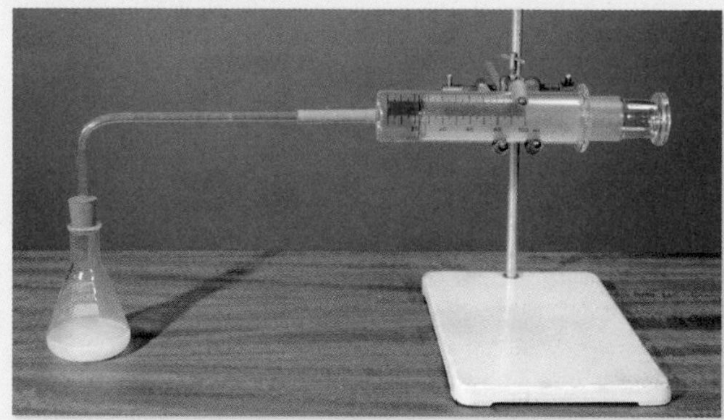

You can make a simpler version of this by using a gas delivery tube leading into an upturned measuring cylinder in a water bath.

Calculating amounts using gas volumes

For a gas at room temperature and pressure, you can use the equations below to convert between:

- the amount of gas molecules, n, in mol; and
- the gas volume, V, in dm^3 or cm^3.

You can work out the amount, in mol, by dividing the volume by the molar volume.

At room temperature and pressure:

if volume V is in dm^3: $n = \frac{V}{24.0}$ mol

if volume V is in cm^3: $n = \frac{V}{24\,000}$ mol

LEARNING TIP

To work out amount, in moles, in any volume of a gaseous substance, learn these equations:

$$n = \frac{V\,(\text{in } dm^3)}{24.0}$$

$$n = \frac{V\,(\text{in } cm^3)}{24\,000}$$

Also remember that a volume in cm^3 can be converted to dm^3 by dividing it by 1000.

WORKED EXAMPLE 1

What amount, in mol, of gas molecules is in 72 cm^3 of any gas at RTP?

$$n = \frac{V\,(\text{in cm}^3)}{24\,000} = \frac{72}{24\,000} = 0.0030\text{ mol}$$

WORKED EXAMPLE 2

What is the volume, in cm^3, of 2.130×10^{-3} mol of a gas at RTP?

$$n = \frac{V\,(\text{in cm}^3)}{24\,000}$$

So, $V = n \times 24\,000 = 2.130 \times 10^{-3} \times 24\,000 = 51.12\text{ cm}^3$

The ideal gas equation

Gases are assumed to behave in an ideal way:

- they are in continuous motion and do not experience any intermolecular forces
- they exert pressure when they collide with each other and the walls of containers
- all collisions between gas molecules and between gas molecules and container walls are elastic – they do not cause kinetic energy to be lost
- the kinetic energy of gases increases with increasing temperature
- gas molecules are so small compared to the size of any container they are found in that any differences in sizes of different gas molecules can be ignored (they can be imagined as identical perfect spheres).

The relationships between the volume, pressure, temperature and number of moles of an ideal gas can be described by the ideal gas equation:

$pV = nRT$

where p is pressure

V is volume

n is the number of moles

R is the gas constant, with a value of 8.314 J mol^{-1} K^{-1}

T is temperature

LEARNING TIP

The ideal gas equation uses SI units, so you must remember to convert units where necessary.

The SI unit for pressure is Pascal, Pa. 1 atm = 101325 Pa

The Si unit for volume is cubic metres, m^3. 1 m^3 = 1000 dm^3

The SI unit for temperature is Kelvin, K. 0 °C = 273 K

You do not need to remember the value of R.

Converting units in the ideal gas equation is often an area where errors occur. Make sure you have converted all the units into SI units *before* you substitute them into the equation, then you can be confident that any missing value will have the correct units. But be careful, you may need to convert them back again before you give your answers!

The ideal gas equation can be used to find out how the pressure and volume of gases will change as temperatures and numbers of moles vary.

WORKED EXAMPLE 3

Find the volume of 2 g O_2 when it is heated to 50 °C at 1 atm of pressure. Give your answer in dm^3, to 2 significant figures.

Step 1

Calculate the number of moles of O_2

$$n = \frac{m}{M_r} = \frac{2\text{ g}}{32\text{ g mol}^{-1}} = 0.0625\text{ mol}$$

Step 2

Rearrange the ideal gas equation to find volume

$pV = nRT$

so, $V\,(\text{m}^3) = \frac{nRT}{p}$

Step 3

Input values into the equation, remembering to convert values into SI units where necessary.

$$V = \frac{0.0625 \times 8.314 \times 323}{101\,325}$$

$V = 1.66 \times 10^{-3}\text{ m}^3$

$V = 1.70\text{ dm}^3$

DID YOU KNOW?

From hypothesis to law

Most scientific advances start off with a hypothesis – a tentative explanation for an observation, possibly only an educated guess. A hypothesis can be tested by further investigation. Once the hypothesis has been tested repeatedly and is found to hold true by several different scientists, it may become accepted as a law.

Avogadro's hypothesis was at first neglected by scientists, partly because it contradicted Dalton's view of matter only consisting of atoms. It was not until after Avogadro's death in 1856 that his brilliant deduction was accepted by the scientific community. Avogadro's *hypothesis* had become a *law*.

In recognition of his contribution to scientific knowledge, the number of fundamental particles per mole of substance is called the Avogadro constant. Avogadro had no idea of the actual number of molecules in equal volumes of gases, and he did nothing to measure it. But his original hypothesis did lead to the eventual determination of this number as 6.02×10^{23} mol^{-1}.

Questions

1 (a) Calculate the amount, in mol, of gas molecules that are in the following gas volumes at RTP:
(i) 36 dm^3 (ii) 1080 dm^3 (iii) 4.0 dm^3
(b) Calculate the volume of the following at RTP:
(i) 6 mol SO_2(g) (ii) 0.25 mol O_2(g) (iii) 20.7 g NO_2(g)

2 Calculate the mass of the following at RTP:
(a) 0.6 dm^3 N_2(g) (b) 1920 cm^3 C_3H_8(g) (c) 84 cm^3 N_2O(g)

3 Calculate the volume of the following at RTP:
(a) 1.282 g SO_2(g) (b) 1.485 g HCN(g) (c) 1.26 g C_3H_6(g)

4 At 0 °C and 1 atm, 10 g of CO_2 has a volume of 5.15 dm^3. Calculate the temperature, in °C, at which the volume would reach 7.00 dm^3. Give your answer to 3 significant figures.

2.1 

Moles and solutions

By the end of this topic, you should be able to demonstrate and apply your knowledge and understanding of:

* explanation and use of the terms:
  (i) *amount of substance*
  (ii) *mole* (symbol 'mol'), as the unit for amount of substance
  (iii) *molar mass* (mass per mole, units g mol^{-1})
* calculations, using amount of substance in mol, involving solution volume and concentration
* the techniques and procedures required during experiments requiring the measurement of mass and volumes of solutions
* the techniques and procedures used when preparing a standard solution of required concentration

Concentration

In a solution, the solute dissolves in the solvent. The concentration of a solution tells you how much solute is dissolved in a given amount of solvent. Concentrations are measured in moles per cubic decimetre, mol dm^{-3}.

In a solution with a concentration of 2 mol dm^{-3}, there are 2 moles of solute dissolved in every 1 dm^3 of solution. This means that in this solution:

- 1 dm^3 contains 2 mol of dissolved solute
- 2 dm^3 contains 4 mol of dissolved solute
- 0.25 dm^3 contains 0.5 mol of dissolved solute.

If you know the concentration in mol dm^{-3}, you can find the amount, in mol, in any volume of a solution. If the volume of the solution is measured in dm^3:

$$n = c \times V(\text{in dm}^3)$$

where:

- n = amount of substance, in mol
- c = concentration of solute, in mol dm^{-3}
- V = volume of solution, in dm^3.

In smaller volumes of the solution, it is more convenient to measure volumes in cm^3. The expression then becomes:

$$n = c \times \frac{V(\text{in cm}^3)}{1000}$$

LEARNING TIP

To work out amount, in moles, in a solution, learn these equations:

$$n = c \times V(\text{in dm}^3)$$

$$n = c \times \frac{V(\text{in cm}^3)}{1000}$$

Don't forget to look at the volumes you are working with to decide whether you need to convert to or from cm^3.

WORKED EXAMPLE 1

What is the amount, in mol, of NaOH dissolved in 25.0 cm^3 of an aqueous solution of concentration 0.0125 mol dm^{-3}?

$$n(\text{NaOH}) = c \times \frac{V(\text{in cm}^3)}{1000}$$

$$= 0.0125 \times \frac{25.0}{1000}$$

$$= 3.125 \times 10^{-4} \text{ mol}$$

Standard solutions

A standard solution has a known concentration. Standard solutions are important in certain types of investigation, for example a titration to find out the concentration of another substance.

If you need to make up a standard solution, you will need to weigh out a substance. To do this, you will need to:

- know the concentration and volume of the solution you need to make
- work out the amount, in mol, of solute needed
- convert this amount of solute into a mass, in g, so that you know how much to weigh out.

KEY DEFINITION

The **concentration** of a solution is the amount of solute, in mol, dissolved per 1 dm^3 (1000 cm^3) of solution.
A **standard solution** is a solution of known concentration. Standard solutions are normally used in titrations to determine unknown information about another substance.

INVESTIGATION

A volumetric flask is used to accurately make up a standard solution. Two classes of flask are available, depending on the precision required. Class A flasks measure volumes with a higher degree of precision than class B flasks. The class and precision values are marked on the volumetric flask, along with an etched line that shows where the solution must be made up to.

Figure 1 A volumetric flask. **Figure 2** The correct position of the meniscus.

To make up a standard solution, follow these steps.

1. Using the weigh by difference method described in topic 2.1.6, weigh out the solute.
2. Completely dissolve the solute in solvent in a beaker. Transfer the solution to the flask and rinse the beaker repeatedly, using more solvent, adding the rinsings to the flask.
3. Add solvent to the flask, but do not fill it all the way up to the graduation line.
4. Carefully add solvent drop by drop up to the line on the flask, until the bottom of the meniscus sits exactly on the graduation mark on the flask (as shown in Figure 2). If the solution goes over the meniscus line, you must throw it away and start again.

Figure 3 Mixing the solution.

5. Finally, you must mix your solution thoroughly, by inverting the flask several times.

WORKED EXAMPLE 2

Find the mass of potassium hydroxide required to prepare 250 cm^3 of a 0.200 $mol\ dm^{-3}$ solution.

1. Find the amount of KOH, in mol, required in the solution:

$$n(\text{KOH}) = c \times \frac{V(\text{in cm}^3)}{1000}$$

$$= 0.200 \times \frac{250}{1000} = 0.0500 \text{ mol}$$

2. Convert moles to grams:
molar mass, M (KOH) = 39.1 + 16.0 + 1.0 = 56.1 $g\ mol^{-1}$

$$\text{amount } n = \frac{\text{mass } m}{\text{molar mass } M}$$

Hence, $m = n \times M = 0.0500 \times 56.1 = 2.805$ g
So, the mass of KOH required is 2.805 g.

Mass concentrations

The mass concentration of a solute is the mass dissolved in 1 dm^3 of solution.

Mass concentrations are measured in units of $g\ dm^{-3}$.

- The solution in Worked example 2 contains 2.805 g of dissolved KOH in 250 cm^3 of solution.
- This would be 11.220 g of dissolved KOH in 1000 cm^3 (1 dm^3) of solution.
- We can refer to this concentration of KOH as 11.220 $g\ dm^{-3}$.

Concentrated and dilute solutions

The terms *concentrated* and *dilute* are descriptive terms for the amount, in mol, of dissolved solute in a solution.

- Concentrated – a large amount of solute per dm^3.
- Dilute – a small amount of solute per dm^3.

Normal bench solutions of acids usually have concentrations of 1 $mol\ dm^{-3}$ or 2 $mol\ dm^{-3}$. These are dilute solutions.

Concentrated acids usually have a concentration greater than 10 $mol\ dm^{-3}$.

LEARNING TIP

Remember, the concentration of a solution is *not* the same as its strength. Concentrated solutions have large amounts of solute per dm^3. Dilute solutions have small amounts of solute per dm^3.
You will learn about strengths when you cover acids and their reactions in topic 2.1.14.

Molar solutions

You will often see 'M' used on solutions in chemistry.

- 'M' means *Molar* and refers to a solution with a concentration in moles per cubic decimetre, $mol\ dm^{-3}$.
- So, 2 $mol\ dm^{-3}$ and 2 M mean the same thing: 2 mol of solute in 1 dm^3 of solution.
- You should quote concentrations using the units $mol\ dm^{-3}$.

Questions

1. Calculate the amount, in moles, of solute dissolved in the following solutions:
(a) 4 dm^3 of a 2 $mol\ dm^{-3}$ solution
(b) 25.0 cm^3 of a 0.150 $mol\ dm^{-3}$ solution.

2. Calculate the concentration, in $mol\ dm^{-3}$, of the following solutions:
(a) 6 moles dissolved in 2 dm^3 of solution
(b) 8.75×10^{-3} moles dissolved in 50.0 cm^3 of solution.

3. Calculate the mass concentration, in $g\ dm^{-3}$, of the following solutions:
(a) 0.042 moles of HNO_3 dissolved in 250 cm^3 of solution
(b) 3.56×10^{-3} moles of H_2SO_4 dissolved in 25 cm^3 of solution.

4. Calculate the mass of $CuSO_4$ required to prepare 250 cm^3 of a 0.10 $mol\ dm^{-3}$ solution. Give your answer to 2 decimal places.
[A_r: Cu = 63.5, S = 32.1, O = 16]

10 Chemical equations

By the end of this topic, you should be able to demonstrate and apply your knowledge and understanding of:

* construction of balanced chemical equations (including ionic equations), including state symbols, for reactions studied and for unfamiliar reactions given appropriate information

Reactants and products

In a chemical reaction, the number and type of atom is conserved. A chemical reaction does not create or destroy atoms, or change one type of atom into another. A chemical reaction only rearranges the atoms that were originally present into different combinations.

A chemical reaction starts with the reactants, which change chemically to form the products.

reactants → products

You will use equations throughout your Chemistry course. It is essential that you can construct and interpret these equations correctly.

When writing an equation, the first thing you need to do is identify the reactants and products.

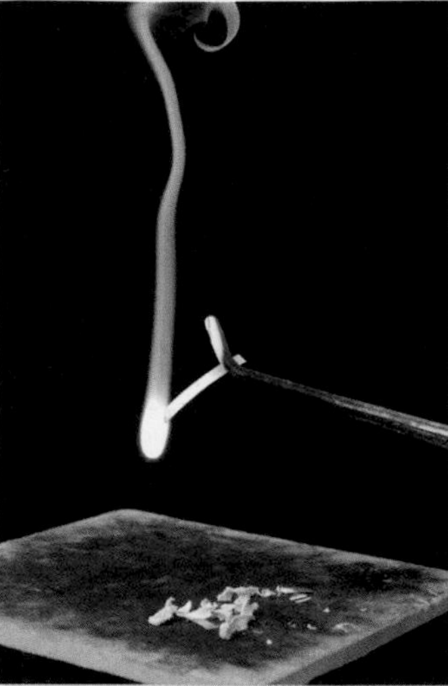

Figure 1 Magnesium burning in air: magnesium and oxygen are the reactants; magnesium oxide is the product.

Species in equations

A chemical equation is a symbolic representation of the chemical reaction taking place. In a chemical equation, you can show various chemical *species* in different ways:

- A species is a type of particle that takes part in a reaction.
- A species could be an atom, ion, molecule, empirical formula or electron.

Molecules

In equations, a simple molecule is shown by its molecular formula, for example: N_2, O_2, H_2, H_2O, CH_4.

Giant structures

A giant structure is formed when many atoms or ions bond together in a repeating fashion. The number of atoms or ions depends on the size of the crystalline structure. For this reason, a compound with a giant structure is represented by its empirical formula. This gives the simplest whole-number ratio of atoms or ions in the structure. Giant structures are discussed in detail in topics 2.2.4, 2.2.5 and 2.2.8.

Examples of giant structures include:

- ionic compounds, such as NaCl, $CaCl_2$, $Mg(OH)_2$
- giant covalent structures, such as SiO_2.

Some elements have giant structures, for example:

- all metals
- some non-metals, such as carbon, silicon and boron.

In equations:

- a compound with a giant structure is shown by its empirical formula
- an element with a giant structure is shown by its symbol – effectively the empirical formula of the element. For example, iron is just shown as Fe.

State symbols in equations

State symbols are added to equations to provide more information about the species in reactions. The common four state symbols are:

- (s) solid
- (l) liquid
- (g) gaseous
- (aq) aqueous.

Balancing a chemical equation is a convenient way of accounting for all the atoms. You balance an equation in order to have the same number of atoms for each element on either side of the equation. Remember, atoms are not created or destroyed during a chemical reaction, so what is present in the reactants must be present in the products.

In the following equations you will be balancing atoms only. Later, we will discuss equations in which electrons must also be balanced.

WORKED EXAMPLE

Balancing an equation

To balance an equation, first write down the equation using formulae and state symbols, for example:

nitrogen + hydrogen → ammonia

$N_2(g) + H_2(g) \rightarrow NH_3(g)$

Then, check on the balancing:

- Nitrogen is not balanced (there are more nitrogen atoms on the left side of the equation).
- Hydrogen is not balanced (there are more hydrogen atoms on the right side of the equation).

$N_2(g) + H_2(g) \rightarrow NH_3(g)$

N: 2 1 ✗

H: 2 3 ✗

Now balance the equation.

You balance an equation to give the *same number* of atoms of each element on each side of the equation.

- Nitrogen can be balanced by placing a '2' in front of NH_3.
- Hydrogen is still unbalanced:

$N_2(g) + H_2(g) \rightarrow 2NH_3(g)$

N: 2 2 ✓

H: 2 6 ✗

Hydrogen can be balanced by placing a '3' in front of H_2.

The equation is now balanced: there are the same numbers and types of atoms on both sides of the equation.

$N_2(g) + 3H_2(g) \rightarrow 2NH_3(g)$

N: 2 2 ✓

H: 6 6 ✓

LEARNING TIP

When balancing equations, remember these rules:

- In a formula, a subscript number applies only to the symbol immediately before it, so SO_2 means 1 S and 2 O atoms.
- When balancing an equation, you must not change any formulae.
- You can only add a balancing number in front of a formula.
- Everything in the formula is multiplied by the balancing number. For example:

$2\ SO_2$ means:	$3\ K_2SO_4$ means:
S: $2 \times 1 = 2$	K: $3 \times 2 = 6$
O: $2 \times 2 = 4$	S: $3 \times 1 = 3$
	O: $3 \times 4 = 12$

- With brackets, the subscript number applies to everything in the bracket. For example:

  $Cu(NO_3)_2$ means:
  Cu: 1
  N: $(1 \times 2) = 2$
  O: $(3 \times 2) = 6$

Questions

1. Balance each of the following equations.
   (a) $Li(s) + O_2(g) \rightarrow Li_2O(s)$
   (b) $Al(s) + Cl_2(g) \rightarrow Al_2Cl_6(s)$
   (c) $Al(s) + H_2SO_4(aq) \rightarrow Al_2(SO_4)_3(aq) + H_2(g)$
   (d) $C_3H_8(g) + O_2(g) \rightarrow CO_2(g) + H_2O(l)$
   (e) $Zn(s) + HNO_3(aq) \rightarrow Zn(NO_3)_2(aq) + H_2O(l) + NO_2(g)$
   (f) $Cu(s) + HNO_3(aq) \rightarrow Cu(NO_3)_2(aq) + H_2O(l) + NO(g)$

2. Write balanced equations, with state symbols, for the following reactions.
   (a) Calcium reacting with oxygen to form calcium oxide, CaO.
   (b) Magnesium reacting with an aqueous solution of silver nitrate, $AgNO_3$, to form silver and a solution of magnesium nitrate, $Mg(NO_3)_2$.
   (c) Lead(II) nitrate, $Pb(NO_3)_2$, decomposing when heated to form a solid, lead(II) oxide, PbO, and a mixture of two gases, nitrogen dioxide and oxygen.
   (d) Copper reacting with concentrated sulfuric acid, $H_2SO_4(l)$, to form water, a blue solution and a gas.

2.1

11 Moles and reactions

By the end of this topic, you should be able to demonstrate and apply your knowledge and understanding of:

* use of stoichiometric relationships in calculations

Stoichiometry and reacting quantities

Stoichiometry studies the amounts of substances that are involved in a chemical reaction. You can use a balanced chemical equation to find the stoichiometry of a reaction. This tells you the number of moles of each species that will react together, and how many moles of each product are formed. It also tells you:

- the reacting quantities that are needed to prepare a required quantity of a product
- the quantities of products formed by reacting together known quantities of reactants.

First, you work out the stoichiometry (molar reacting quantities) from a balanced equation. Then, you can scale these molar quantities up or down to work with any quantity.

Worked examples using stoichiometry

The worked examples in this topic show you how to use a balanced equation using moles. The three examples are:

- reacting masses
- reacting masses and gas volumes
- reacting masses, gas volumes and solution volumes.

Together, these provide worked examples for finding molar quantities from masses, gas volumes and solution volumes.

LEARNING TIP

There are three ways to work out an amount in moles.

From mass:

$$n = \frac{m}{M}$$

From gas volumes:

$$n = \frac{V\,(\text{in dm}^3)}{24.0}$$

$$n = \frac{V\,(\text{in cm}^3)}{24\,000}$$

From solutions:

$$n = c \times V\,(\text{in dm}^3)$$

$$n = c \times \frac{V\,(\text{in cm}^3)}{1000}$$

You will need to use these formulae throughout your Chemistry studies. Make sure that you commit them to memory and practise using them!

Reacting masses

WORKED EXAMPLE 1

Sodium and chlorine react to form sodium chloride:

$2Na(s) + Cl_2(g) \rightarrow 2NaCl(s)$

In this example, we will calculate the masses of Na and Cl_2 that would form 2.925 g of NaCl.

[A_r: Na, 23.0; Cl, 35.5]

(a) Calculate the amount, in mol, in 2.925 g of NaCl.

$$n = \frac{m}{M} = \frac{2.925}{(23.0 + 35.5)} = \frac{2.925}{58.5} = 0.0500 \text{ mol}$$

(b) Calculate the amounts, in mol, of Na and Cl_2 that would form 2.925 g NaCl.

equation: $2Na(s) + Cl_2(g) \rightarrow 2NaCl(s)$

amounts in equation: 2 mol : 1 mol : 2 mol

amounts required: 0.0500 mol : 0.0250 mol : 0.0500 mol

(c) Calculate the masses of Na and Cl_2 that would form 2.925 g NaCl.

Na: $m = n \times M = 0.0500 \times 23.0 = 1.150$ g

Cl_2: $m = n \times M = 0.0250 \times 71.0 = 1.775$ g

LEARNING TIP

For (a) above, we need to use: amount, $n = \frac{\text{mass, } m}{\text{molar mass, } M}$ because only masses are involved.

For (b) above, we use the equation to work out the stoichiometry between reacting quantities.

This means we look for the ratio between the species in the equation.

For (c) above, we rearrange: $n = \frac{m}{M}$ because we now know n and M.
So, $m = n \times M$

Reacting masses and gas volumes

WORKED EXAMPLE 2

A student heated 2.55 g of sodium nitrate, $NaNO_3$, which fully decomposes as in the equation below.

[A_r: Na, 23.0; N, 14.0; O, 16.0]

$2NaNO_3(s) \rightarrow 2NaNO_2(s) + O_2(g)$

(a) Calculate the amount, in mol, of $NaNO_3$ that decomposed.

$$n = \frac{m}{M} = \frac{2.55}{(23.0 + 14.0 + (16.0 \times 3))} = \frac{2.55}{85.0} = 0.0300 \text{ mol}$$

(b) Calculate the amount, in mol, of O_2 that formed.

$2NaNO_3(s) \rightarrow 2NaNO_2(s) +$	$O_2(g)$
2 mol	1 mol
0.0300 mol	0.0150 mol

(c) Calculate the volume of O_2 that formed.

$V = n \times 24.0 = 0.0150 \times 24.0 = 0.360 \text{ dm}^3$

LEARNING TIP

For (a), we use amount, $n = \frac{\text{mass, } m}{\text{molar mass, } M}$

For (b), we use the equation to work out the reacting quantities. Remember that if a species has no number in front of it, this can be read as '1 mole'.

For (c), we rearrange:

amount, $n = \frac{V \text{ (in dm}^3)}{24.0}$ because we need to find out the volume of O_2 gas produced.

So, $V = n \times 24.0$

Reacting masses, gas and solution volumes

WORKED EXAMPLE 3

A chemist reacted 0.23 g of sodium with water to form 250 cm^3 of aqueous sodium hydroxide. Hydrogen gas was also produced. The equation is shown below.

[A_r: Na, 23.0; H, 1.0; O, 16.0]

$2Na(s) + 2H_2O(l) \rightarrow 2NaOH(aq) + H_2(g)$

(a) Calculate the amount, in mol, of Na that reacted.

$n = \frac{m}{M} = 0.010$ mol

(b) Calculate the volume of H_2 formed at room temperature and pressure.

$2Na(s) + 2H_2O(l) \rightarrow 2NaOH(aq) + H_2(g)$

2 mol　　　　1 mol

0.010 mol　　　　0.0050 mol

Convert this amount of H_2 to the volume formed.

$V = n \times 24\,000 = 0.0050 \times 24\,000 = 120 \text{ cm}^3$

(c) Calculate the concentration, in mol dm^{-3}, of NaOH(aq) formed.

$2Na(s) + 2H_2O(l) \rightarrow 2NaOH(aq) + H_2(g)$

2 mol　　　　2 mol

0.010 mol　　　　0.010 mol

Finally, work out the concentration, in mol dm^{-3}, of NaOH(aq).

$c(\text{NaOH}) = \frac{n \times 1000}{V \text{ (in cm}^3} = \frac{0.010 \times 1000}{250} = 0.040 \text{ mol dm}^{-3}$

LEARNING TIP

For (a), use: amount, $n = \frac{\text{mass, } m}{\text{molar mass, } M,}$ because only the mass and A_r of the species are known.

For (b), we first use the balanced equation to work out the amount of H_2 that forms. Notice that we find the value of H_2 by multiplying 1×0.005, because $\frac{0.010 \text{ mol Na}}{2 \text{ mol Na}} = 0.005$.

We then find the volume of this amount of H_2.

We rearrange: $n = \frac{V \text{ (in cm}^3)}{24\,000}$ because we want to find the volume of gas.

So, $V = n \times 24\,000$

We use the equation with '24 000 cm^3' rather than 24.0 dm^3 as the question gave a value in cm^3.

For (c), we again use the balanced equation, but this time we need to work out the amount of NaOH that forms.

This amount of NaOH is dissolved in 250 cm^3 of solution.

We rearrange: $n = c \times \frac{V \text{ (in cm}^3)}{1000}$

So, $c = \frac{n \times 1000}{V \text{ (in cm}^3)}$

Questions

1 (a) Balance the equation below.

$NaHCO_3(s) \rightarrow Na_2CO_3(s) + CO_2(g) + H_2O(l)$

(b) Calculate the volume of CO_2 (at RTP) formed by the decomposition at RTP of 5.04 g of $NaHCO_3$.

2 (a) Balance the equation below.

$Pb(NO_3)_2(s) \rightarrow PbO(s) + NO_2(g) + O_2(g)$

(b) Calculate the total volume of gas (at RTP) formed by the complete decomposition of 2.12 g of $Pb(NO_3)_2$.

3 (a) Balance the equation below.

$MgCO_3(s) + HNO_3(aq) \rightarrow Mg(NO_3)_2(aq) + H_2O(l) + CO_2(g)$

(b) 2.529 g of $MgCO_3$ reacts with an excess of HNO_3.

(i) Calculate the volume of CO_2 (at RTP) formed.

(ii) The final volume of the solution is 50.0 cm^3. Calculate the concentration, in mol dm^{-3}, of $Mg(NO_3)_2$ formed.

Percentage yields

By the end of this topic, you should be able to demonstrate and apply your knowledge and understanding of:

* calculations to determine the percentage yield of a reaction or related quantities

Percentage yield

When writing a fully balanced chemical equation, it is assumed that all of the reactants will be converted into products. If this were the case, the yield would be 100%.

However, in practical work, yields of 100% are rarely obtained for various reasons:

- the reaction may be at equilibrium and may not go to completion
- side reactions may occur, leading to by-products
- the reactants may not be pure
- some of the reactants or products may be left behind in the apparatus used in the experiment
- separation and purification may result in the loss of some of the product.

The **percentage yield** can be calculated to measure the success of a laboratory preparation.

KEY DEFINITION

Percentage (%) **yield** $= \frac{\text{actual amount, in mol, of product}}{\text{theoretical amount, in mol, of product}} \times 100$

WORKED EXAMPLE 1

A student prepared some ethanoic acid, CH_3COOH, by oxidising ethanol, C_2H_5OH.
The equation for the reaction is shown below:

$CH_3CH_2OH + 2[O] \rightarrow CH_3COOH + H_2O$ (where [O] is the oxidising agent)

The student used a mixture of excess sulfuric acid and potassium dichromate (VI) as the oxidising agent.
She reacted 9.20 g of ethanol.
The student obtained 4.35 g of ethanoic acid.
What is the percentage yield of ethanoic acid?

Answer

1. Calculate the amount, in mol, of ethanol that was used:

   amount of ethanol, $n = \frac{\text{mass, } m}{\text{molar mass, } M} = \frac{9.20}{46.0} = 0.200$ mol

2. Using the equation, calculate the amount, in mol, of ethanoic acid product expected.
   - Looking at the equation, you can see that one mole of ethanol reacts to produce one mole of ethanoic acid.
   - So, 0.200 mol of ethanol should react to give 0.200 mol of ethanoic acid.
   - This is the theoretical amount of ethanoic acid product, in mol.

3. Find the actual amount, in mol, of ethanoic acid product made in the experiment.

   amount of ethanoic acid made, $n = \frac{m}{M} = \frac{4.35}{60.0} = 0.0725$ mol

4. Using your answers from steps 1 and 3, calculate the percentage yield.

   $\%\text{ yield} = \frac{\text{actual amount, in mol, of product}}{\text{theoretical amount, in mol, of product}} \times 100$

   $= \frac{0.0725}{0.200} \times 100 = 36.25\%$

LEARNING TIP

If a chemical is used in excess, you do not need to worry about how many mols of it there are. There is enough for the other chemical to react with completely (with some left over – hence 'excess'). Therefore, the number of moles of any expected product will be dictated solely by the chemical that is not in excess. The chemical that is not in excess will be the limiting reagent.

In some questions you have to calculate the amount, in mol, of each reactant to find the limiting reagent. The theoretical amount, in mol, of product is calculated based on the amount, in mol, of the limiting reagent. Worked example 2 uses this principle.

WORKED EXAMPLE 2

A student prepared propyl methanoate, $HCOOCH_2CH_2CH_3$, from propan-1-ol, $CH_3CH_2CH_2OH$, and methanoic acid, HCOOH. The equation for the reaction is:

$$CH_3CH_2CH_2OH + HCOOH \rightarrow HCOOCH_2CH_2CH_3 + H_2O$$

The student reacted 3.00 g of propan-1-ol with 2.50 g of methanoic acid in the presence of a sulfuric acid catalyst. He was disappointed to obtain only 1.75 g of propyl methanoate.
What is the percentage yield of propyl methanoate?

Answer

1. Calculate the amount, in mol, of each reagent.
   - In this example, two chemicals are reacting together, but we do not know which is the limiting reagent and which is in excess.
   - We need to find out the amount, in mol, of each reagent.
   - We will then know which reagent is the limiting reagent, to be used in Step 2.

   amount of propan-1-ol, $n = \dfrac{\text{mass, } m}{\text{molar mass, } M} = \dfrac{3.00}{60.0} = 0.0500 \text{ mol}$

   amount of methanoic acid, $n = \dfrac{m}{M} = \dfrac{2.50}{46.0} = 0.0543 \text{ mol}$

   - Propan-1-ol was the limiting reagent and was used up first in the reaction, because there are fewer moles of it and the reagents react in a 1 : 1 ratio.
   - Once we know the limiting reagent, we use this for step 2.
2. Using the equation, calculate the amount, in mol, of propyl methanoate product expected.
   - The equation tells us that one mole of propan-1-ol reacts to form one mole of propyl methanoate.
   - Therefore, 0.0500 mol of propan-1-ol should react to form 0.0500 mol of propyl methanoate.
   - The theoretical amount, in mol, of propyl methanoate product = 0.0500 mol.
3. Find the actual amount, in mol, of propyl methanoate product made in the experiment.

   amount of propyl methanoate made, $n = \dfrac{m}{M} = \dfrac{1.75}{88.0} = 0.0199 \text{ mol}$
4. Using your answers from steps 1 and 3, calculate the percentage yield.

   $\% \text{ yield} = \dfrac{\text{actual amount, in mol, of product}}{\text{theoretical amount, in mol, of product}} \times 100 = \dfrac{0.0199}{0.0500} \times 100 = 39.8\%$

LEARNING TIP

Notice the importance of working in moles.

- the mass of propan-1-ol used was 3.00 g, greater than the mass of methanoic acid used, 2.50 g. So it appears the propan-1-ol is in excess.
- However, there are more moles of methanoic acid. Although less mass of methanoic acid was used, its molar mass is also less. You will only discover this by converting grams into moles.

Questions

1. 3.925 g of 2-chloropropane reacted with an excess of aqueous sodium hydroxide. 2.955 g of propan-2-ol was formed, as shown in the equation below.

   $$CH_3CH(Cl)CH_3 + NaOH \rightarrow CH_3CH(OH)CH_3 + NaCl$$

   Calculate the percentage yield of propan-2-ol.

2. Ethanol reacts with ethanoic acid to produce an ester and water, as shown in the equation below.

   $$C_2H_5OH + CH_3COOH \rightarrow CH_3COOC_2H_5 + H_2O$$

   5.5 g of ethyl ethanoate is produced from 4.0 g of ethanol and 4.5 g of ethanoic acid.
   Calculate the percentage yield of ethyl ethanoate.

13 Atom economy

By the end of this topic, you should be able to demonstrate and apply your knowledge and understanding of:

* calculations to determine the atom economy of a reaction

Atom economy

In the previous topic (2.1.12) we discussed how percentage yield is used to assess the efficiency of a chemical reaction. However, calculating the percentage yield only tells part of the story, since a reaction may produce by-products as well as the desired product.

What do we do with the by-products?

By-products are often considered as waste and usually have to be disposed of. This is costly, poses potential environmental problems and wastes valuable resources. By-products may be sold on or used elsewhere in the chemical plant. This practice is likely to increase in the future, as we become increasingly concerned about preserving the Earth's resources and minimising waste.

Atom economy considers not only the desired product, but also all the by-products of a chemical reaction. It describes the efficiency of a reaction in terms of all the atoms involved. A reaction with a high atom economy uses atoms with minimal waste.

KEY DEFINITION

Atom economy $= \frac{\text{molecular mass of the desired product}}{\text{sum of molecular masses of all products}} \times 100$

How atom economy can benefit society

We are now much more aware of our environment. By using processes with a higher atom economy, chemical companies can reduce the amount of waste produced. This is good news, as it has been suggested that about 5–10% of the total expenditure of a chemical company goes on waste treatment. Reactions with high atom economies make processes much more sustainable – that is, they can be maintained at a productive level without completely depleting resources.

Calculating atom economy

WORKED EXAMPLE 1

The reaction between propene and bromine is shown in Figure 1. The desired product is 1,2-dibromopropane. Calculate the atom economy.

$CH_2{=}CHCH_3 + Br_2 \longrightarrow CH_2BrCHBrCH_3$

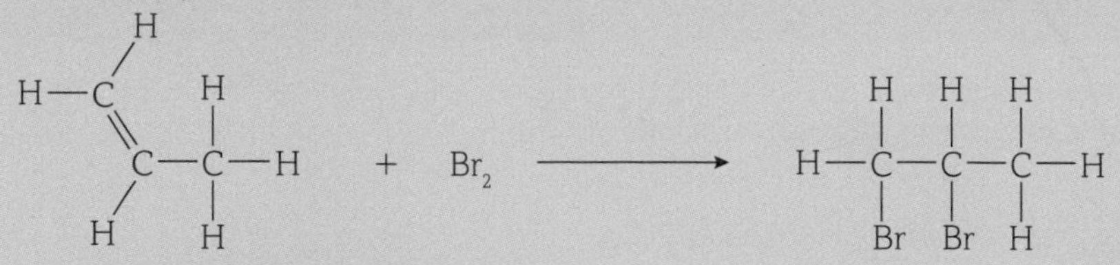

Figure 1 The reaction between propene and bromine.

To calculate the atom economy you will need to:

- calculate the molecular mass of the desired product
- divide this by the sum of the molecular masses of all products.

In this reaction, the desired product is the only product! So the atom economy is 100%.

Molecular mass of desired product: $C_3H_6Br_2 = (3 \times 12.0) + (5 \times 1.0) + (2 \times 79.9) = 200.8$

All products: $C_3H_6Br_2$

Molecular masses of all products = 200.8

$$\text{atom economy} = \frac{\text{molecular mass of the desired products}}{\text{sum of molecular masses of all products}} \times 100 = \frac{200.8}{200.8} \times 100 = 100\%$$

WORKED EXAMPLE 2

A student decides to prepare a sample of butan-1-ol by reacting 1-bromobutane with aqueous sodium hydroxide. The equation for the reaction is given in Figure 2. Calculate the atom economy of the reaction.

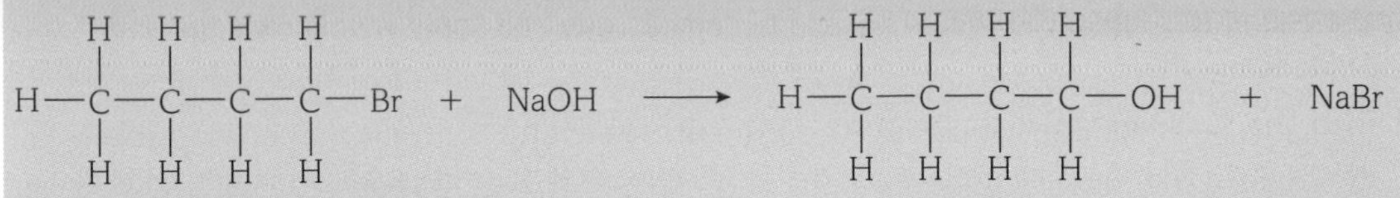

$$\underset{\text{1-bromobutane}}{CH_3CH_2CH_2CH_2Br} + NaOH \longrightarrow \underset{\text{butan-1-ol}}{CH_3CH_2CH_2CH_2OH} + NaBr$$

Figure 2 Preparation of butan-1-ol from 1-bromobutane.

To calculate the atom economy, you will need to calculate the molecular mass of the desired product and divide this by the sum of the molecular masses of all products.

Molecular mass of desired product: $C_4H_9OH = (4 \times 12.0) + (10 \times 1.0) + (1 \times 16.0) = 74.0$

All products: $C_4H_9OH + NaBr$

Molecular masses of all products = 74.0 + (23.0 + 79.9) = 176.9

$$\text{atom economy} = \frac{\text{molecular mass of the desired product}}{\text{sum of molecular masses of all products}} \times 100 = \frac{74.0}{176.9} \times 100 = 41.8\%$$

The reaction has an atom economy of 41.8%.

- This means the majority of the starting materials is turned into waste.
- Even if the reaction proceeded with a 100% yield, more than half the mass of atoms used would be wasted.

LEARNING TIP

Remember that percentage yield and atom economy are different.

Percentage yield tells you the efficiency with which reactants are converted into products.

Atom economy tells you the proportion of desired products compared with all the products formed.

Atom economy and type of reaction

From the worked examples above, it is clear that the type of reaction used for a chemical process is a major factor in achieving a higher atom economy.

- Addition reactions have an atom economy of 100%.
- Reactions involving substitution or elimination have an atom economy less than 100%.

This is not to say that we should never carry out substitution or elimination reactions. However, to improve their atom economy, we need to find a use for all the products of a reaction.

In Worked example 2, we need to find a use for the NaBr product, rather than simply throwing it away. If the undesired products are toxic, then we have an even bigger problem!

LEARNING TIP

Addition reactions involve two or more reactants joining together. The atom economy for addition reactions is always 100%.

Substitution reactions involve one atom or molecule 'swapping places' with another. The atom economy for substitution reactions is always less than 100%.

Elimination reactions involve the removal of a small molecule, usually water, from another molecule. The atom economy of elimination reactions is always less than 100%.

Questions

1. 2-iodopropane C_3H_7I can be hydrolysed using aqueous sodium hydroxide to form the product propan-2-ol. Write a balanced equation for this reaction and calculate the atom economy. (Tip: refer to Worked example 2 to help you write an equation for this reaction.)

2. The following reaction was carried out to produce a sample of but-2-ene.

   $CH_3CH_2CH(OH)CH_3 \rightarrow CH_3CH{=}CHCH_3 + H_2O$

   Calculate the atom economy of the reaction.

2.1

14 Acids and bases

By the end of this topic, you should be able to demonstrate and apply your knowledge and understanding of:

* the formulae of the common acids (HCl, H_2SO_4, HNO_3 and CH_3COOH) and the common alkalis ($NaOH$, KOH and NH_3) and explanation that acids release H^+ ions in aqueous solution and alkalis release OH^- ions in aqueous solution
* qualitative explanation of strong and weak acids in terms of relative dissociations
* the benefits for sustainability of developing chemical processes with a high atom economy

Acids

When studying Chemistry, you will come across acids in both practical and theoretical work. The word 'acid' comes from the Latin 'acidus' meaning sour. In water, acids give a solution with a pH of less than 7.0.

In the laboratory, the common acids that you will come across are:

- sulfuric acid, H_2SO_4
- hydrochloric acid, HCl
- nitric acid, HNO_3.

LEARNING TIP

You must learn the names and formulae of these acids.

Figure 1 Everyday acids. These all contain hydrogen ions, H^+, in aqueous solutions.

You may be familiar with sulfuric acid and hydrochloric acid without realising it. Sulfuric acid is the battery acid used in cars. Hydrochloric acid is the acid present in your stomach that helps you digest food.

You will also come across weaker acids that occur naturally; for example, ethanoic (acetic) acid, CH_3COOH, in vinegar.

If you look at the formulae of the acids we have listed above, you will notice that they all contain the element hydrogen.

DID YOU KNOW?

All acids contain hydrogen. When added to water, acids release this hydrogen as H^+ ions. H^+ ions are also known as protons. This is because an atom of H contains one proton and one electron. After losing an electron to become the positive H^+ ion, hydrogen is literally just one proton.

Acids in aqueous solution

When an acid is added to water, the acid releases H^+ ions (also known as protons) into solution:

- hydrochloric acid, HCl: $HCl(g) \rightarrow H^+(aq) + Cl^-(aq)$
- sulfuric acid, H_2SO_4: $H_2SO_4(l) \rightarrow H^+(aq) + HSO_4^-(aq)$

The H^+ ion (proton) is the active ingredient in acids:

- An $H^+(aq)$ ion is responsible for all acid reactions.
- One definition of an acid is a proton donor.

Acids and dissociations

Strong acids are very good at giving up H^+ ions. We say that they fully, or almost fully, dissociate.

Weak acids are not very good at giving these H^+ ions away. Once H^+ ions are released from weak acids, they are quickly taken back again. Weak acids are excellent at accepting these H^+ ions back, whereas strong acids are not. We say weak acids only partially dissociate.

Bases

You will also meet bases frequently when studying Chemistry. Common bases are metal oxides and hydroxides:

- Metal oxides: MgO, CuO
- Metal hydroxides: $NaOH$, $Mg(OH)_2$.

Ammonia is also a base, as are organic compounds called amines:

- Ammonia: NH_3
- Amines: CH_3NH_2.

You may be familiar with some other bases used in everyday life:

- Magnesium hydroxide, $Mg(OH)_2$, is also known as 'milk of magnesia' and is used for treating acid indigestion.
- Calcium hydroxide, $Ca(OH)_2$, is also known as lime and is used for reducing the acidity of acid soils.

A base is the opposite of an acid:

- One definition of a base is a proton, H^+, acceptor.
- Bases neutralise acids.

Alkalis

The word 'alkali' comes from the Arabic 'al kali' meaning 'the ashes'. The ash of burnt plant materials is alkaline.

In Chemistry, an alkali is any substance that gives a solution with a pH greater than 7.0 when dissolved in water. Alkalis release $OH^-(aq)$ ions when they are dissolved in water (an aqueous solution).

In the laboratory, you will come across common alkalis such as:

- sodium hydroxide, NaOH
- potassium hydroxide, KOH
- ammonia, NH_3.

LEARNING TIP

You must learn the names and formulae of these alkalis.

Figure 2 Everyday alkalis. These items are all types of base that will dissolve in water and release OH^- ions, giving a pH above 7.0.

You may have come across sodium hydroxide without realising it, as it is used in oven cleaners and paint strippers.

Alkalis are very corrosive, even more so than many acids:

- The alkali sodium hydroxide is corrosive even when as dilute as $0.5\ \text{mol dm}^{-3}$.
- Hydrochloric acid needs to be stronger than $6.5\ \text{mol dm}^{-3}$ before it is classified as being corrosive.

Alkalis in aqueous solution

An alkali is a special type of base that is able to dissolve in water to form aqueous hydroxide ions, $OH^-(aq)$, for example:

$$NaOH(s) + aq \rightarrow Na^+(aq) + OH^-(aq)$$

In solution, the hydroxide ions from alkalis neutralise the protons from acids, forming water:

$$H^+(aq) + OH^-(aq) \rightarrow H_2O(l)$$

Ammonia as a weak base

Ammonia (NH_3) is a gas that dissolves in water to form a weak alkaline solution.

Dissolved NH_3 reacts with water as follows:

$$NH_3(aq) + H_2O(l) \rightleftharpoons NH_4^+(aq) + OH^-(aq)$$

Ammonia is a weak base because only a small proportion of the dissolved NH_3 reacts with water. This is shown by the equilibrium sign: $\rightleftharpoons$

You will study equilibrium reactions later in topic 3.2.10.

LEARNING TIP

Learn the balanced equation that shows how ammonia is able to react with water to form a weak base.

Amphoteric substances

Some substances can behave as acids and bases. These are known as *amphoteric* substances. An example of this is an amino acid molecule, such as glycine, which contains:

- a *carboxyl* acid group, COOH, which is able to donate a proton
- an *amino* basic group, NH_2, which could accept a proton.

$$\textbf{base}\quad H_2N-\overset{\displaystyle H}{\underset{\displaystyle H}{\overset{|}{\underset{|}{C}}}}-COOH\quad \textbf{acid}$$

Figure 3 The amino acid glycine. An amino acid has both acid and base parts.

Questions

1 Write down the formulae for the following:
(a) sulfuric acid
(b) nitric acid
(c) ethanoic acid
(d) potassium hydroxide
(e) calcium hydroxide
(f) ammonia.

2 Define the terms:
(a) acid
(b) base
(c) alkali.

3 Explain why ammonia is able to react with acids. Include a balanced equation in your answer.

15 Salts

By the end of this topic, you should be able to demonstrate and apply your knowledge and understanding of:

* neutralisation as the reaction of:
  (i) H^+ and OH^- to form H_2O
  (ii) acids with bases, including carbonates, metal oxides and alkalis (water-soluble bases), to form salts, including full equations

Salts

A salt is an ionic compound with the following features:

- The positive ion, or cation, in a salt is usually a metal ion or an ammonium ion, NH_4^+.
- The negative ion, or anion, in a salt is derived from an acid.
- The formula of a salt is the same as that of the parent acid, except that an H^+ ion has been replaced by the positive ion.

Examples

- Sulfuric acid, $H_2SO_4 \rightarrow$ sulfate salts, for example potassium sulfate, K_2SO_4
- Hydrochloric acid, HCl $\rightarrow$ chloride salts, for example sodium chloride, NaCl
- Nitric acid, $HNO_3 \rightarrow$ nitrate salts, for example calcium nitrate, $Ca(NO_3)_2$

Acid salts

Sulfuric acid has two H^+ ions that can be replaced by a different positive ion. It is an example of a diprotic acid. If one H^+ ion is replaced, an acid salt is formed, for example sodium hydrogensulfate, $NaHSO_4$:

- $H_2SO_4 \rightarrow NaHSO_4$ Notice that one of the H^+ ions has been replaced by an Na^+ ion.

The acid salt can itself behave as an acid, because the other H^+ ion can be replaced to form a conventional salt, for example sodium sulfate, Na_2SO_4:

- $NaHSO_4 \rightarrow Na_2SO_4$ Notice that the second H^+ has now been replaced by a second Na^+ ion.

Formation of salts

Salts can be produced by neutralising acids with bases (proton acceptors), such as:

- carbonates (these contain the CO_3^{2-} ion)
- metal oxides
- alkalis (a soluble base that forms OH^- ions).

Examples of neutralisation reactions are shown below. Hydrochloric acid has been used to illustrate these reactions, but any acid will react in a similar way.

In each example, a second equation (called an ionic equation) shows the important role of the H^+ ion in the reaction. You will notice that H^+ ions react with OH^- ions to form H_2O. This gives the ionic equation for neutralisation: $H^+(aq) + OH^-(aq) \rightarrow H_2O(l)$.

DID YOU KNOW?

Salts from carbonates

Acids react with carbonates to form a *salt*, *carbon dioxide* and *water*. You will see bubbles of CO_2.

$$2HCl(aq) + CaCO_3(s) \longrightarrow CaCl_2(aq) + H_2O(l) + CO_2(g)$$

$$2H^+(aq) + CaCO_3(s) \longrightarrow Ca^{2+}(aq) + H_2O(l) + CO_2(g)$$

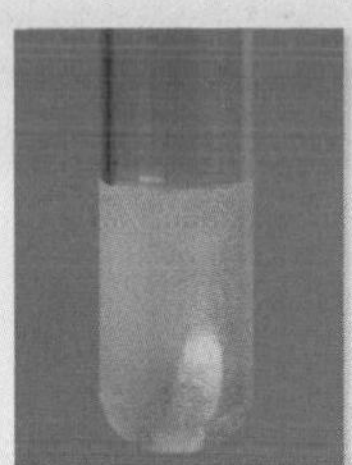

Figure 1 Calcium carbonate reacting with hydrochloric acid. The equation for this reaction is: $CaCO_3(s) + 2HCl(aq) \longrightarrow CaCl_2(aq) + H_2O(l) + CO_2(g)$. This reaction occurs when acidic rain water erodes limestone buildings and statues, as limestone is mainly made of calcium carbonate.

Salts from metal oxides

Acids react with metal oxides to form a *salt* and *water*. In the example below, you would see solid CaO decrease in size or disappear.

$$2HCl(aq) + CaO(s) \longrightarrow CaCl_2(aq) + H_2O(l)$$

$$2H^+(aq) + CaO(s) \longrightarrow Ca^{2+}(aq) + H_2O(l)$$

Salts from alkalis

Acids react with alkalis to form a *salt* and *water*.

$$HCl(aq) + NaOH(aq) \longrightarrow NaCl(aq) + H_2O(l)$$

$$H^+(aq) + OH^-(aq) \longrightarrow H_2O(l)$$

Salts from metals

Acids react with metals to form a *salt* and *hydrogen* gas.

$$2Li(s) + 2HCl(aq) \longrightarrow 2LiCl(s) + H_2(g)$$

The reaction between metals and acids is an example of a 'redox' reaction. You will learn about redox reactions in topic 2.1.19.

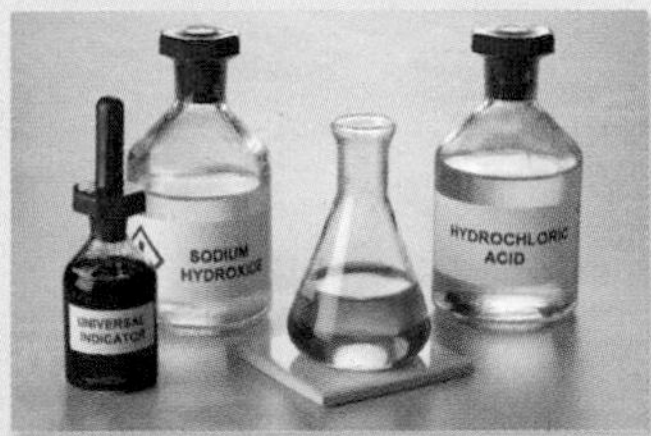

Figure 2 Acids react with alkalis to produce a salt and water. Indicators can be used to see when the neutralisation has happened.

LEARNING TIP

Learn the general equations below for how salts are formed by reactions of acids:

acid + carbonate ⟶ salt + carbon dioxide + water

acid + metal oxide ⟶ salt + water

acid + alkali ⟶ salt + water

acid + metal ⟶ salt + hydrogen

Practise writing examples of each, using both HCl and H_2SO_4 as the acid.

DID YOU KNOW?

You will notice that the products and equations are the same when acids react with metal oxides and with alkalis. This is because alkalis and metal oxides are both types of bases.

Ammonium salts

When acids are neutralised by aqueous ammonia, ammonium salts are formed, containing the ammonium ion, NH_4^+. For example, ammonium nitrate is formed when ammonia and nitric acid react:

$$NH_3(aq) + HNO_3(aq) \longrightarrow NH_4NO_3(aq)$$

When ammonium nitrate is in solution, it is found as two ions: $NH_4^+(aq)$ and $NO_3^-(aq)$.

Questions

1. Write balanced equations for the following acid reactions:
   (a) hydrochloric acid and potassium hydroxide
   (b) nitric acid and calcium hydroxide
   (c) sulfuric acid and sodium hydroxide
   (d) nitric acid and magnesium carbonate, $MgCO_3$
   (e) phosphoric acid, H_3PO_4, and sodium carbonate, Na_2CO_3
   (f) sulfuric acid and copper oxide, CuO.

2. Write balanced equations for the formation of ammonium salts from ammonia and:
   (a) hydrochloric acid
   (b) sulfuric acid
   (c) phosphoric acid.

16 Formulae for crystals and salts

By the end of this topic, you should be able to demonstrate and apply your knowledge and understanding of:

* the terms *anhydrous, hydrated* and *water of crystallisation* and calculation of the formula of a hydrated salt from given percentage composition, mass composition or based on experimental results

Hydrated crystals

Some compounds contain water within their structures; others don't. Chemists use the following terms to describe these two forms:

- **Hydrated** – the crystalline form containing water.
- **Anhydrous** – the form containing no water.

For example, the blue copper sulfate crystals you may be familiar with are the hydrated form. They have the formula $CuSO_4{\cdot}5H_2O$.

Anhydrous copper sulfate is a white powder, with the formula $CuSO_4$.

Figure 1 Copper sulfate can exist in a hydrated form and an anhydrous form.

Why are some compounds hydrated?

If a compound crystallises within water, the water can become part of the resulting crystalline structure. Water that becomes involved in this way is known as **water of crystallisation**.

KEY DEFINITIONS

Hydrated refers to a crystalline compound containing water molecules.
Anhydrous refers to a substance that contains no water molecules.
Water of crystallisation refers to water molecules that form an essential part of the crystalline structure of a compound.

Dot formulae

Hydrated crystals can contain different amounts of water. The amount of water contained is shown by a dot formula. This dot formula gives the ratio between the number of compound molecules and the number of water molecules within the crystalline structure. They are written as follows:

- The empirical formula of the compound is separated from the water of crystallisation by a dot.
- The relative number of water molecules of crystallisation is shown after the dot.
- So, for hydrated copper sulfate (full name, copper sulfate pentahydrate), which has five H_2O molecules for each formula unit of $CuSO_4$, its dot formula is written as $CuSO_4{\cdot}5H_2O$.

Other examples are:

- $CoCl_2{\cdot}6H_2O$, cobalt chloride hexahydrate
- $Na_2SO_4{\cdot}10H_2O$, sodium sulfate decahydrate.

LEARNING TIP

Water of crystallisation is one of the rare instances in chemistry in which a number without a subscript is used within a formula, for example the '5' in $CuSO_4{\cdot}5H_2O$.

This practice is very unusual and is only used in special cases, such as with hydrated crystals.

A number without a subscript is normally reserved for balancing a chemical equation, and it is placed before a formula. For example:

$$2Na(s) + Cl_2(g) \rightarrow 2NaCl(s)$$

Determining the formula of a hydrated salt

The dot formula of hydrated salts can be determined using:

- the empirical formula, gained from percentage or mass compositions
- experimental results.

Using empirical formulae

In topic 2.1.7 you learnt how to determine an empirical formula from data on percentage or mass composition. Once you know how many atoms of each element are present, you can work out which ones are part of water molecules and which ones are part of the main compound.

WORKED EXAMPLE 1

The empirical formula for hydrated magnesium chloride is $MgCl_2H_{10}O_5$. Determine the dot formula.

Step 1: Use the number of hydrogen atoms to work out how many water molecules are present.
There are 10 H atoms, so there must be 5 water molecules.
These 5 water molecules would also require 5 O atoms.
The end of the dot formula must be $\cdot 5H_2O$

Step 2: Use the remaining atoms to determine the formula of the main salt, and how many molecules are present.
Magnesium chloride has the formula $MgCl_2$. There are only 1 × Mg and 2 × Cl remaining, so there must only be 1 $MgCl_2$ in the dot formula.

Step 3: Construct the dot formula.
The dot formula must be $MgCl_2{\cdot}5H_2O$

INVESTIGATION

Using experimental results

When hydrated salts are heated, the water of crystallisation is driven off by evaporation.

The amount of water lost corresponds to how much water of crystallisation was in the salt sample.

Figure 3 shows suitable apparatus for determining the amount of water of crystallisation.

Experimental results are used to determine:

- the mass of the hydrated salt, containing water of crystallisation
- the mass of the anhydrous salt, without the water
- and therefore the mass of water that was in the hydrated salt.

Figure 2 Heating hydrated copper sulfate. Although this experimental method is easy to carry out, it will only work if the anhydrous salt is not decomposed by heat. Unfortunately, many salts are decomposed by heat. This is especially true for nitrates.

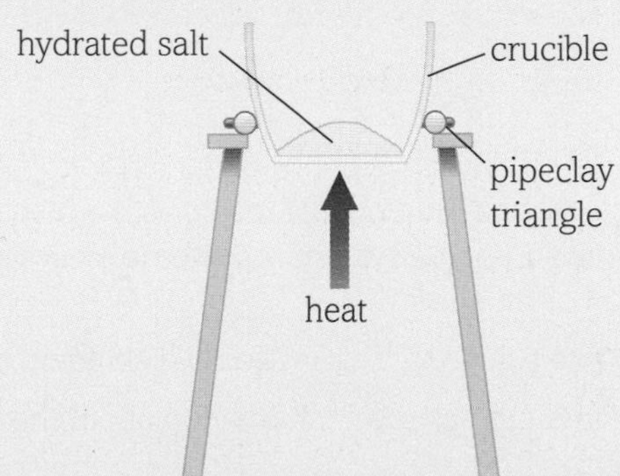

Figure 3 Apparatus for the determination of water of crystallisation.

LEARNING TIP

When you use an empirical formula to construct a dot formula for a hydrated salt, make sure you keep track of the oxygen molecules present. The oxygen atoms may be shared out between the water of crystallisation and the salt. For example:

- Hydrated sodium carbonate has the empirical formula $Na_2CH_{20}O_{13}$ and the dot formula $Na_2CO_3{\cdot}10H_2O$
- Hydrated calcium nitrate has the empirical formula $CaN_2H_8O_{10}$ and the dot formula $Ca(NO_3)_2{\cdot}4H_2O$

WORKED EXAMPLE 2

From an experiment to determine the formula of hydrated magnesium sulfate:

mass of the hydrated salt, $MgSO_4{\cdot}xH_2O$ = 4.312 g

mass of the anhydrous salt, $MgSO_4$ = 2.107 g

So, mass of H_2O in $MgSO_4{\cdot}xH_2O$ = 2.205 g

- First, calculate the amount, in mol, of anhydrous $MgSO_4$:

  [A_r: Mg, 24.3; S, 32.1; O, 16.0]

  molar mass, $M(MgSO_4) = 24.3 + 32.1 + (16.0 \times 4) = 120.4\ \text{g mol}^{-1}$

  $n(MgSO_4) = \frac{2.107}{120.4} = 0.0175\ \text{mol}$

- Then calculate the amount, in mol, of water ($M = 18.0\ \text{g mol}^{-1}$):

  $n(H_2O) = \frac{2.205}{18.0} = 0.1225\ \text{mol}$

- Finally, determine the formula of the hydrated salt:

  Molar ratio $MgSO_4 : H_2O = 0.0175 : 0.1225 = 1 : 7$ (divide by the smaller number)

  So, formula of the hydrated salt = $MgSO_4{\cdot}7H_2O$; i.e. $x = 7$.

LEARNING TIP

The x in the formula of the salt stands for the unknown number of moles of water of crystallisation. Yet again we are using the formula $n = \frac{m}{M}$

Questions

1. You are supplied with three empirical formulae. Write down the dot formula of each salt to show the water of crystallisation.

   (a) $BaCl_2H_4O_2$ (b) $ZnSH_{14}O_{11}$ (c) $FeN_3H_{12}O_{15}$

2. From the experimental results shown below, work out the formula of the hydrated salt.

   [A_r: Ca, 40.1; Cl, 35.5; H, 1.0; O, 16.0]

   Mass of $CaCl_2{\cdot}xH_2O$ = 6.573 g

   Mass of $CaCl_2$ = 3.333 g

17 Titrations

By the end of this topic, you should be able to demonstrate and apply your knowledge and understanding of:

* calculations, using amount of substance in mol, involving solution volume and concentration
* the techniques and procedures required during experiments requiring the measurement of volumes of solutions
* the techniques and procedures used when carrying out acid–base titrations
* structured and non-structured titration calculations, based on experimental results of familiar and non-familiar acids and bases

Acid–base titrations

During volumetric analysis, you measure the volume of one solution that reacts with a measured volume of a different solution. An acid–base titration is a special type of volumetric analysis, in which you react a solution of an acid with a solution of an alkali.

- You must know the concentration of one of the two solutions. This is usually a standard solution.
- In the analysis, you use this standard solution to find out unknown information about the substance dissolved in the second solution.

The unknown information may be:

- the concentration of the solution
- a molar mass
- a formula
- the number of molecules of water of crystallisation.

INVESTIGATION

How to carry out a titration

1. Using a *pipette*, add a measured volume of one solution to a conical flask. Add a suitable indicator.
2. Place the other solution in a *burette*.
3. Add the solution in the burette to the solution in the conical flask until the reaction has *just* been completed. This is called the *end point* of the titration. Measure the volume of the solution added from the burette.
4. You now know the volume of one solution that *exactly* reacts with the volume of the other solution.

Indicators

We identify the end point using an *indicator*. The indicator must be a different colour in the acidic solution than in the basic solution.

Table 1 lists the colours of some common acid–base indicators. It also shows the colour at the end point. Notice that this end point colour is in between the colours in the acidic and basic solutions.

Indicator	Colour in acid	Colour in base	End point colour
methyl orange	red	yellow	orange
bromothymol blue	yellow	blue	green
phenolphthalein	colourless	pink	pale pink*

Table 1 Colours of some common acid–base indicators.

*This assumes that the aqueous base has been added from the burette to the aqueous acid. If acid is added to base, the titration is complete when the solution goes colourless.

Figure 1 This acid–base titration is using methyl orange as an indicator. Methyl orange is coloured red in acidic solutions and yellow in basic solutions. The end point is the colour in between – orange. The solution in the conical flask above has reached the end point.

Calculating unknowns from titration results

Analysis of titration results follows a set pattern, as shown in the worked examples:

- The first two steps are always the same.
- The third step may be different, depending on the unknown that you need to work out.

WORKED EXAMPLE 1

Calculating an unknown concentration

In a titration, 25.0 cm^3 of 0.150 mol dm^{-3} sodium hydroxide, NaOH(aq), reacted exactly with 23.40 cm^3 of sulfuric acid, H_2SO_4(aq).

$2NaOH(aq) + H_2SO_4(aq) \rightarrow Na_2SO_4(aq) + 2H_2O(l)$

(a) Calculate the amount, in mol, of NaOH that reacted.

$$n(\text{NaOH}) = c \times \frac{V}{1000} = 0.150 \times \frac{25.0}{1000} = 3.75 \times 10^{-3}\text{ mol}$$

(b) Calculate the amount, in mol, of H_2SO_4 that was used.

equation	2NaOH(aq)	+	H_2SO_4(aq) →
moles from equation	2 mol	+	1 mol →
actual moles	3.75×10^{-3} mol		1.875×10^{-3} mol

(c) Calculate the concentration, in mol dm^{-3}, of the sulfuric acid.

$$c(\text{H}_2\text{SO}_4) = n \times \frac{1000}{V}$$

$$= 1.875 \times 10^{-3} \times \frac{1000}{23.40}$$

$$= 0.080\text{ mol dm}^{-3}$$

LEARNING TIP

For (a), we use: amount, $n = c \times \frac{V(\text{in cm}^3)}{1000}$

We need to divide by 1000 because values are in cm^3.

For (b), we use the balanced equation to work out the reacting quantities of the acid and alkali.

2 mol NaOH reacts with 1 mol H_2SO_4

For (c), we rearrange: $n = c \times \frac{V(\text{in cm}^3)}{1000}$

Hence, $c = n \times \frac{1000}{V}$

WORKED EXAMPLE 2

Calculating an unknown molar mass

A student dissolved 2.794 g of an acid, HX, in water and made the solution up to 250 cm^3. The student titrated 25.0 cm^3 of this solution against 0.0614 mol dm^{-3} sodium carbonate, Na_2CO_3(aq). 23.45 cm^3 of Na_2CO_3(aq) were needed to reach the end point.

The equation for this reaction is:

$Na_2CO_3(aq) + 2HX(aq) \rightarrow 2NaX(aq) + CO_2(g) + H_2O(l)$

(a) Calculate the amount, in mol, of Na_2CO_3 that reacted.

$$n(\text{Na}_2\text{CO}_3) = c \times \frac{V}{1000} = 0.0614 \times \frac{23.45}{1000} = 1.44 \times 10^{-3}\text{ mol}$$

(b) Calculate the amount, in mol, of HX that was used in the titration.

equation	Na_2CO_3(aq)	+	2HX(aq) →
moles from equation	1 mol	+	2 mol →
actual moles	1.44×10^{-3} mol		2.88×10^{-3} mol

(c) Calculate the amount, in mol, of HX that was used to make up the 250 cm^3 solution.

25.0 cm^3 HX(aq) contains 2.88×10^{-3} mol

So, the 250 cm^3 solution contains $10 \times 2.88 \times 10^{-3} = 2.88 \times 10^{-2}$ mol

(d) Calculate the concentration, in g mol^{-1}, of the acid HX.

$$n = \frac{m}{M} \quad \text{so,} \quad M = \frac{m}{n} = \frac{2.794}{2.88 \times 10^{-2}} = 97.0\text{ g mol}^{-1}$$

LEARNING TIP

For (a) and (b), the steps are the same as in the first worked example. In this worked example, however, there are two further steps.

For (c), we scale up by a factor of 10 to find the amount, in mol, of HX in the 250 cm^3 solution that was made up.

For (d), we rearrange:

$$\text{amount, } n = \frac{\text{mass, } m}{\text{molar mass, } M}$$

Hence, $M = \frac{m}{n}$

In this example, the mass of HX is 2.794 g.

Questions

1. Use the method in Worked example 1 to calculate the unknown concentration below.

   In a titration, 25.0 cm^3 of 0.125 mol dm^{-3} aqueous sodium hydroxide reacted exactly with 22.75 cm^3 of hydrochloric acid.

   $HCl(aq) + NaOH(aq) \rightarrow NaCl(aq) + H_2O(l)$

   Find the concentration of the hydrochloric acid.

2. Use the method in Worked example 2 to calculate the molar mass of the acid H_2X.

   A student dissolved 1.571 g of an acid, H_2X, in water and made the solution up to 250 cm^3. She titrated 25.0 cm^3 of this solution against 0.125 mol dm^{-3} sodium hydroxide, NaOH(aq). 21.30 cm^3 of NaOH(aq) were needed to reach the end point.

   The equation for this reaction is:

   $2NaOH(aq) + H_2X(aq) \rightarrow Na_2X(aq) + 2H_2O(l)$

18 Oxidation numbers

By the end of this topic, you should be able to demonstrate and apply your knowledge and understanding of:

* rules for assigning and calculating oxidation number for atoms in elements, compounds and ions
* writing formulae using oxidation numbers
* use of a Roman numeral to indicate the magnitude of the oxidation number when an element may have compounds/ions with different oxidation numbers

Oxidation numbers

Oxidation numbers are used extensively in Chemistry. You will also come across the term *oxidation state*, which essentially means the same thing.

Oxidation numbers cannot be measured directly by experiment. Chemists use oxidation numbers to keep track of how electrons are being used in bonding.

In a chemical formula, each atom can be assigned an oxidation number. An atom uses electrons to bond with atoms of other elements. The oxidation number is the number of electrons that an atom uses to bond with atoms of another element.

Oxidation numbers are worked out by following a set of rules. These rules are shown in Table 1. You must learn these rules and how to apply them correctly.

Species	Oxidation number	Examples
uncombined element	0	C; Na; O_2; P_4
combined oxygen	−2	H_2O; CaO
combined oxygen in peroxides	−1	H_2O_2
combined hydrogen	+1	NH_3, H_2S
combined hydrogen in metal hydrides	−1	LiH
simple ion	charge on ion	Na^+, +1; Mg^{2+}, +2; Cl^-, −1
combined fluorine	−1	NaF, CaF_2, AlF_3

Table 1 Oxidation number rules.

As always, there are exceptions to the rules. For example, when bonded to fluorine, oxygen has an oxidation number of +2.

Oxidation numbers in formulae

LEARNING TIP

It is essential to apply oxidation numbers to each atom within a formula. A good habit is to write down the oxidation number separately for each atom above the formula and the totals underneath.
For example, for carbon dioxide:

+4 −2 *individual oxidation numbers*

CO_2

+4 −4 total = 0 *sum total of oxidation numbers must match the overall charge*

Compounds

For example, in sulfur dioxide, SO_2, the overall charge on the molecule is zero.

- The sum of the oxidation numbers must equal the overall charge of zero.
- There are two oxygen atoms, each with an oxidation number of −2.
- This gives a total contribution of −4.
- The oxidation number of sulfur must be +4 to balance and give the overall oxidation number of zero.

Atom	SO_2
O	−2
O	−2
S	+4
overall	0

Molecular ions

For example, in a carbonate ion, CO_3^{2-}, the overall charge is −2.

- The sum of the oxidation numbers must equal the overall charge of −2.
- There are three oxygen atoms, each with an oxidation number of −2.
- This gives a total contribution of −6.
- The oxidation number of carbon must be +4 to give the overall charge of −2.

Atom	CO_3^{2-}
O	−2
O	−2
O	−2
C	+4
overall	−2

Oxidation numbers in chemical names

Some elements can have more than one stable oxidation number. When this is the case, the oxidation number of the element is included in the name of the compound or ion as a Roman numeral. Without the oxidation number assigned in this way, you wouldn't know which oxidation state the element was in.

Compounds of transition elements

Transition elements form ions with different oxidation numbers. Without the oxidation number, the chemical name could apply to more than one compound of a transition element. Some common examples are shown below:

- $FeCl_2$ iron(II) chloride Fe: oxidation number +2
- $FeCl_3$ iron(III) chloride Fe: oxidation number +3
- Cu_2O copper(I) oxide Cu: oxidation number +1
- CuO copper(II) oxide Cu: oxidation number +2

Oxyanions

Oxyanions are negative ions that contain an element along with oxygen.

- Examples: SO_4^{2-}, NO_3^-, CO_3^{2-}
- The names of oxyanions usually end in *-ate*, to indicate oxygen; for example carbonate.

As with transition element ions, an element may form oxyanions in which the element has different oxidation numbers. Without the oxidation number, the name would be ambiguous. Some common examples are:

- NO_2^- nitrate(III) N: oxidation number +3
- NO_3^- nitrate(V) N: oxidation number +5
- SO_3^{2-} sulfate(IV) S: oxidation number +4
- SO_4^{2-} sulfate(VI) S: oxidation number +6

Questions

1 Deduce the oxidation state of each element in the following:
(a) K (b) Br_2 (c) NH_4^+ (d) CaF_2
(e) Al_2O_3 (f) NO_3^- (g) PO_4^{3-}

2 Deduce the oxidation state of nitrogen in the following:
(a) N_2O (b) N_2O_5 (c) N_2H_4 (d) NO_3^- (e) HNO_2

3 Deduce the oxidation state of each element in bold.
(a) Na**Cl**O_3 (b) Na_2**S**O_3 (c) K**Mn**O_4 (d) K_2**Mn**O_4 (e) K_2**Cr_2**O_7

4 Use oxidation number rules to work out the formula for the following ions.
(a) chlorate(III) with an overall charge of −1.
(b) chlorate(VII) with an overall charge of −1.
(c) phosphate(III) with an overall charge of −3.
(d) chromate(VI) with an overall charge of −2.

19 Redox reactions

By the end of this topic, you should be able to demonstrate and apply your knowledge and understanding of:

* oxidation and reduction in terms of:
  (i) electron transfer (ii) changes in oxidation number
* redox reactions of metals with acids to form salts, including full equations
* interpretation of redox equations for reactions of metals with acids to form salts, and unfamiliar redox reactions, to make predictions in terms of oxidation numbers and electron loss/gain

Oxidation and reduction

We originally used the terms oxidation and reduction to describe reactions of substances with oxygen. The definitions were simple:

- Oxidation is the gain of oxygen.
- Reduction is the loss of oxygen.

Oxidation and reduction are now used to describe any reactions in which electrons are transferred. The definitions are now more general:

- Oxidation is the loss of electrons.
- Reduction is the gain of electrons.

The substance that is *reduced* takes electrons from the substance that is *oxidised*.

Reduction *must always* be accompanied by oxidation. A reaction in which both *red*uction and *ox*idation take place is called a redox reaction.

LEARNING TIP

Use this phrase as a way to help you remember the difference between oxidation and reduction:
OILRIG:
Oxidation **I**s **L**oss
Reduction **I**s **G**ain
If one species gains electrons (reduction), another species loses the same number of electrons (oxidation).

Electron transfer in redox reactions

Magnesium reacts with chlorine to form magnesium chloride.

$Mg + Cl_2 \rightarrow MgCl_2$

This is a redox reaction, but this is not always obvious from looking at the overall equation.

At an electronic level, electrons have been transferred from the magnesium atoms to the chlorine atoms. The half equations below show this clearly.

$Mg \rightarrow Mg^{2+} + 2e^-$ oxidation (loss of electrons)

$Cl_2 + 2e^- \rightarrow 2Cl^-$ reduction (gain of electrons)

Metals tend to be oxidised, losing electrons to form positive ions.

Non-metals tend to be reduced, gaining electrons to form negative ions.

Oxidation numbers in redox reactions

Oxidation and reduction can also be described in terms of oxidation number:

- Reduction is a decrease in oxidation number.
- Oxidation is an increase in oxidation number.

You can identify the oxidation and reduction processes using oxidation numbers. Assign an oxidation number to each atom in a redox reaction. Then follow any changes to these oxidation numbers.

The oxidation and reduction processes for the reaction of magnesium and chlorine are shown in Table 1.

$Mg + Cl_2 \longrightarrow MgCl_2$	
$0 \longrightarrow +2$	oxidation (oxidation number increases)
$0 \longrightarrow -1$ $0 \longrightarrow -1$	reduction (oxidation number decreases)

Table 1 Redox reaction for magnesium and chlorine.

LEARNING TIP

Always assign the oxidation number to *each atom*.
For example, in $MgCl_2$ you may correctly spot that Mg has an oxidation state of +2 and that overall Cl is −2. However, remember that *each* Cl has an oxidation state of −1, and it is this that you must show clearly.

Redox reactions of metals with acids

Reactive metals undergo redox reactions with many acids:

- The metal is oxidised, forming positive metal ions.
- The hydrogen in the acid is reduced, forming the element hydrogen as a gas (H_2).

The oxidation and reduction processes for the reactions of magnesium metal with either dilute hydrochloric acid or dilute sulfuric acid are shown in Tables 2 and 3, respectively.

A salt of the acid is formed together with hydrogen.

LEARNING TIP

When considering oxidation numbers in the reaction between metals and acids, remember that the metal is an element and the oxidation number of its atoms is ZERO.
You can work out the oxidation number of the metal in the salt from its ionic charge.

$Mg(s)$ +	$2HCl(aq)$	⟶	$MgCl_2(aq)$ +	$H_2(g)$	
0		⟶	+2		oxidation
	+1	⟶		0	
	+1	⟶		0	reduction

Table 2 Reaction of magnesium metal with dilute hydrochloric acid.

$Mg(s)$ +	$H_2SO_4(aq)$	⟶	$MgSO_4(aq)$ +	$H_2(g)$	
0		⟶	+2		oxidation
	+1	⟶		0	
	+1	⟶		0	reduction

Table 3 Reaction of magnesium metal with dilute sulfuric acid.

Figure 1 Magnesium ribbon (Mg) reacting with hydrochloric acid (HCl).

The equation for this reaction is: $Mg(s) + 2HCl(aq) \longrightarrow MgCl_2(aq) + H_2(g)$.
Magnesium is oxidised and hydrogen is reduced.

Acids react with reactive metals as shown below:

- metal + acid ⟶ salt + hydrogen

The equation can also be written to show the role of the hydrogen ion, H^+ (also called a proton). This equation is the same for most acids you will encounter:

- $Mg(s) + 2H^+(aq) \longrightarrow Mg^{2+}(aq) + H_2(g)$

LEARNING TIP

If one species decreases its oxidation number (reduction), another species increases its oxidation number (oxidation). The total increase in oxidation number equals the total decrease in oxidation number.

Using oxidation numbers with equations

You can assign oxidation numbers to each atom in any equation. You can then identify whether a redox reaction has taken place. You can also deduce what has been oxidised and what has been reduced. Table 4 gives an example.

$MnO_2(s)$ +	$4HCl(aq)$	⟶	$MnCl_2(aq)$ +	$2H_2O(l)$ +	$Cl_2(g)$	
+4 −2	+1 −1		+2 −1	+1 −2	0	oxidation numbers for each atom involved
−2	+1 −1		−1	+1 −2	0	
	+1 −1			+1		
	+1 −1			+1		
+4		⟶	+2			reduction (Mn)
	−1	⟶			0	
	−1	⟶			0	
						oxidation (Cl)

Table 4 Assigning oxidation numbers.

Questions

1 (a) Write equations for the reaction of:
(i) iron and hydrochloric acid, forming the salt iron(II) chloride and hydrogen gas
(ii) aluminium with sulfuric acid, forming the salt $Al_2(SO_4)_3$ and hydrogen gas.
(b) In each reaction, identify what has been oxidised and what has been reduced.

2 In the redox reactions below, use oxidation numbers to show what has been oxidised and what has been reduced.
(a) $Cl_2 + 2KBr \longrightarrow Br_2 + 2KCl$
(b) $2SO_2 + O_2 \longrightarrow 2SO_3$
(c) $2HBr + H_2SO_4 \longrightarrow SO_2 + Br_2 + 2H_2O$

THINKING BIGGER

ELEMENTAL FINGERPRINTS

How can we possibly know what stars in distant galaxies are made from? The following extract, taken from *From Stars to Stalagmites: How Everything Connects* by Paul Braterman (2012), will help you understand.

ELEMENTAL 'FINGERPRINTS'

In 1825, the French Philosopher Auguste Comte said that there are some things we will never know, among them the chemical composition of the stars. This was an unfortunate example. We do not need actual material from the stars in order to analyse them. What we do need, of course, is information, and this they send us plentifully, in the form of light. To reveal this information, we must separate the light into its different wavelengths or colours.

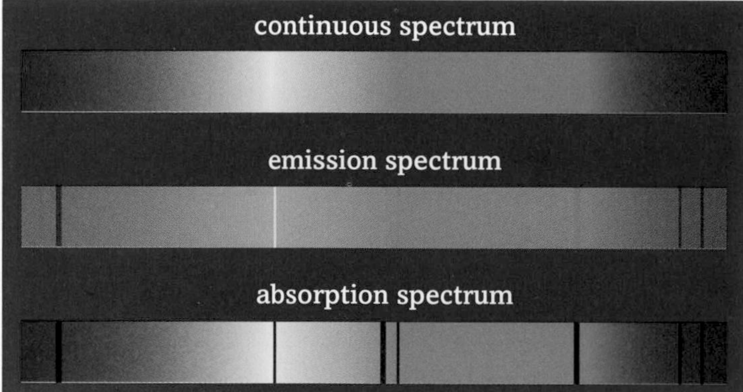

Figure 1

Many years earlier Isaac Newton had passed sunlight through a glass prism, and found that this gave him all the colours of the rainbow. In 1802 William Wollaston in England… repeated the experiment, using an up-to-date high quality glass prism, and discovered that the continuous spectrum was interrupted by narrow dark lines. (Rather unfairly these are now known as Frauenhofer lines, after the German physicist Joseph von Frauenhofer, who confirmed and extended Wollaston's observations.) These lines later proved to match exactly the light given out by different elements when heated or in electric discharges. A now familiar example is provided by the element sodium, whose yellow emission is used in street lamps. The lines can be used as a kind of fingerprint for each element. We can take this fingerprint using electrical discharges here on Earth and compare it with the lines found in the spectrum of the Sun. In this way we can get a good chemical analysis of the surface layer of the Sun or of any other star whose light we can collect and astronomers now extend the same process to the most distant galaxies.

- Extract from *From Stars to Stalagmites: How Everything Connects* by Paul Braterman. Pub: World Scientific Publishing (2012) printed in Singapore.

DID YOU KNOW?

Many images of galaxies taken by the Hubble Space Telescope use the line emission spectra of elements such as hydrogen, sulfur and oxygen to build up a colour image of the galaxy. The different elements are assigned red, green and blue frequencies to build up a full colour image.

Where else will I encounter these themes?

1.1 | 2.1 | YOU ARE HERE | 2.2

Let us start by considering the nature of the writing in the article.

1. Do you think this article is aimed at scientists, the general public, or people who are not scientists but have an interest in science? Look back through the extract and find examples to support your answer.
2. Explain how the book's tagline 'How Everything Connects' is supported by the extract.

As well as considering the complexity of the scientific ideas, think about the level of detail in which they are discussed. Notice how many different branches of natural science are mentioned in the extract. Are all of the scientists mentioned chemists or did they work in different scientific disciplines?

Now we will look at the chemistry in, or connected to, this article. Don't worry if you are not ready to give answers to these questions yet. You may like to return to the questions once you have covered other topics later in this book. Use the timeline at the bottom of the page to help you put this work in context with what you have already learned and what is ahead in your course.

3. Describe the structure of the sodium atom in terms of protons, neutrons and electrons.
4. Give the electronic structure of a sodium atom using s, p, d notation.
5. What is giving rise to (i) the absorption and (ii) the emission spectrum of sodium?
6. Can you suggest why sodium vapour is used in preference to potassium vapour in street lamps?
7. Would you expect heavier isotopes of sodium to give a different 'fingerprint'? Explain your answer.
8. The element helium was first discovered in the outer layer of the Sun. Suggest why it was discovered there before it was discovered here on Earth.

Think back to work you have done on identifying alkali metals using flame tests.

Activity

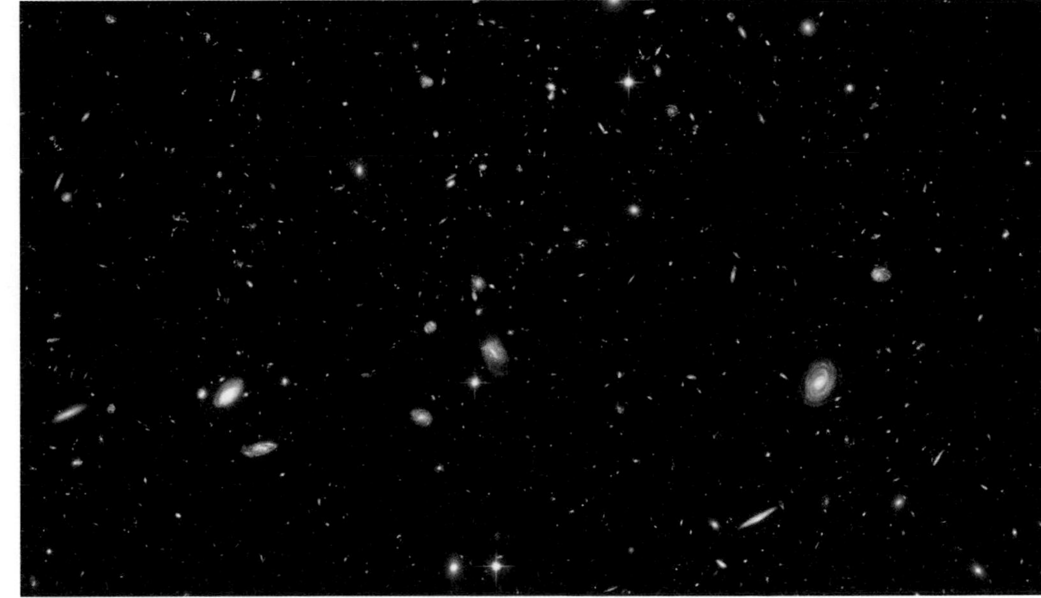

Figure 2 A picture taken by the Hubble Space Telescope showing over 10 000 galaxies!

Today, the elemental fingerprints in the light collected from distant galaxies have provided evidence for an expanding universe. The elemental fingerprints are subject to a phenomenon called 'red shift'.

Write a 300–500 word essay explaining how red shift has provided evidence for an expanding universe. Try to structure the essay so that the concepts can be understood by an audience of 16-year-old GCSE students.

2.1 Practice questions

1. Which of the following statements about $^{25}_{12}Mg$ is true? [1]
   A. It has the same number of neutrons and protons.
   B. It can be oxidised to form an Mg^{2+} ion.
   C. It is the most common isotope of magnesium.
   D. It has a higher ionisation energy than $^{24}_{12}Mg$.

2. Which of the following is the best explanation for the fact that the relative atomic mass of carbon on the periodic table is given as 12.011 to 3 d.p.? [1]
   A. All carbon atoms have a mass slightly greater than 12.
   B. There are at least 3 isotopes of carbon.
   C. The relative atomic mass takes all carbon isotopes into account.
   D. The amount of ^{13}C in the atmosphere fluctuates.

3. What is the total number of ions (to 2 d.p.) in 25 cm^3 of a 0.05 mol dm^{-3} solution of Na_2SO_4?
   Avogadro's number = 6.02×10^{23}. [1]
   A. 2.26×10^{21}
   B. 7.53×10^{20}
   C. 7.53×10^{23}
   D. 1.51×10^{21}

4. What volume of 0.05 mol dm^{-3} KOH(aq) is required to exactly neutralise 24.50 cm^3 of 0.11 mol dm^{-3} H_2SO_4, according to the equation: $2KOH(aq) + H_2SO_4 \rightarrow K_2SO_4(aq) + 2H_2O(l)$? [1]
   A. 12.25 cm^3
   B. 26.95 cm^3
   C. 13.48 cm^3
   D. 107.8 cm^3

[Total: 4]

5. With increasing numbers of computers being sold the demand for silicon has never been greater. Luckily silicon is the second most abundant element in the Earth's crust after oxygen and is extracted industrially by a reaction with carbon:

   $$2C + SiO_2 \rightarrow 2CO + Si$$

   (a) Calculate the minimum mass of carbon required reduce 50 tonnes of silicon dioxide. (1 tonne = 1000 kg) [2]
   (b) What is the atom economy for this reaction? [1]
   (c) What volume of carbon monoxide, measured at STP, would be produced in this reaction? (1 mole of gas occupies 24 dm^3 at STP) [2]

[Total: 5]

6. Hydrogen chloride gas can be prepared by the following reaction:

   $$NaCl + H_2SO_4 \rightarrow NaHSO_4 + HCl$$

   What masses of sodium chloride and concentrated sulfuric acid are needed to make 5.0 dm^3 of hydrogen chloride measured at STP? [4]

[Total: 4]

7. (a) 0.54 g of aluminium combine with 4.8 g of bromine. Calculate the emprical formula of the compound. [2]
   (b) Write a balanced equation for the reaction. [2]
   (c) If the molecular mass of the compound is 534 g mol^{-1} calculate the molecular formula for the molecule. [1]

[Total: 5]

8. A 3.50 g sample of calcium reacted with excess water and all of the gas produced was collected.
   (a) Write a balanced equation, including state symbols for the reaction of calcium with water. [3]
   (b) If the total volume of gas collected at STP was 1.95 dm^3, calculate the percentage purity of the calcium sample. [3]
   (c) Suggest what the impurity might be giving a reason for your answer. [2]

[Total: 8]

9. Europium, atomic number 63, is used in some television screens to highlight colours. A chemist analysed a sample of europium using mass spectrometry. The results are shown in the **Table 1.1** below.

Isotope	Relative isotopic mass	Abundance (%)
^{151}Eu	151.0	47.77
^{153}Eu	153.0	52.23

Table 1.1

   (a) Define the term *relative isotopic mass*. [2]
   (b) Using **Table 1.1**, calculate the relative atomic mass of the europium sample. Give your answer to **two** decimal places. [2]
   (c) Isotopes of europium have differences and similarities.
      (i) In terms of protons, neutrons and electrons, how is an atom of ^{151}Eu **different** from an atom of ^{153}Eu? [1]
      (ii) In terms of protons, neutrons and electrons, how is an atom of ^{151}Eu **similar** to an atom of ^{153}Eu? [1]

(d) Modern plasma television screens emit light when mixtures of noble gases, such as neon and xenon, are ionised.
The first ionisation energies of neon and xenon are shown in the table below.

element	1st ionisation energy/kJ mol^{-1}
neon	+2081
xenon	+1170

Explain why xenon has a lower first ionisation energy than neon. [3]

[Total: 9]

[Q1, F321 Jan 2010]

10. A student carries out experiments using acids, bases and salts.

(a) Calcium nitrate, $Ca(NO_3)_2$, is an example of a salt.
The student prepares a solution of calcium nitrate by reacting dilute nitric acid, HNO_3, with the base calcium hydroxide, $Ca(OH)_2$.

(i) Why is calcium nitrate an example of a salt? [1]

(ii) Write the equation for the reaction between dilute nitric acid and calcium hydroxide. Include state symbols. [2]

(iii) Explain how the hydroxide ion in aqueous calcium hydroxide acts as a base when it neutralises dilute nitric acid. [1]

(b) A student carries out a titration to find the concentration of some sulfuric acid.
The student finds that 25.00 cm^3 of 0.0880 mol dm^{-3} aqueous sodium hydroxide, NaOH, is neutralised by 17.60 cm^3 of dilute sulfuric acid, H_2SO_4.

$$H_2SO_4(aq) + 2NaOH(aq) \rightarrow Na_2SO_4(aq) + 2H_2O(l)$$

(i) Calculate the amount, in moles, of NaOH used. [1]

(ii) Determine the amount, in moles, of H_2SO_4 used. [1]

(iii) Calculate the concentration, in mol dm^{-3}, of the sulfuric acid. [1]

(c) After carrying out the titration in (b), the student left the resulting solution to crystallise. White crystals were formed, with a formula of $Na_2SO_4 \bullet xH_2O$ and a molar mass of 322.1 g mol^{-1}.

(i) What term is given to the '$\bullet \boldsymbol{x}H_2O$' part of the formula? [1]

(ii) Using the molar mass of the crystals, calculate the value of x. [2]

[Total: 10]

[Q2, F321 Jan 2010]

11. A 20 g sample of a white powder contains unknown proportions of magnesium oxide and magnesium carbonate only.

(a) Write balanced equations, including state symbols, for the reaction of hydrochloric acid with:

(i) MgO

(ii) $MgCO_3$ [2]

(b) Design an experimental procedure that would allow you to determine the proportions of magnesium oxide and magnesium carbonate in the sample. You should include:

(i) a detailed method including all of the apparatus used

(ii) an indication of how you would ensure reliability and precision

(iii) an indication of how you would use your results to determine the proportions of each compound in the mixture. [5]

[Total: 7]

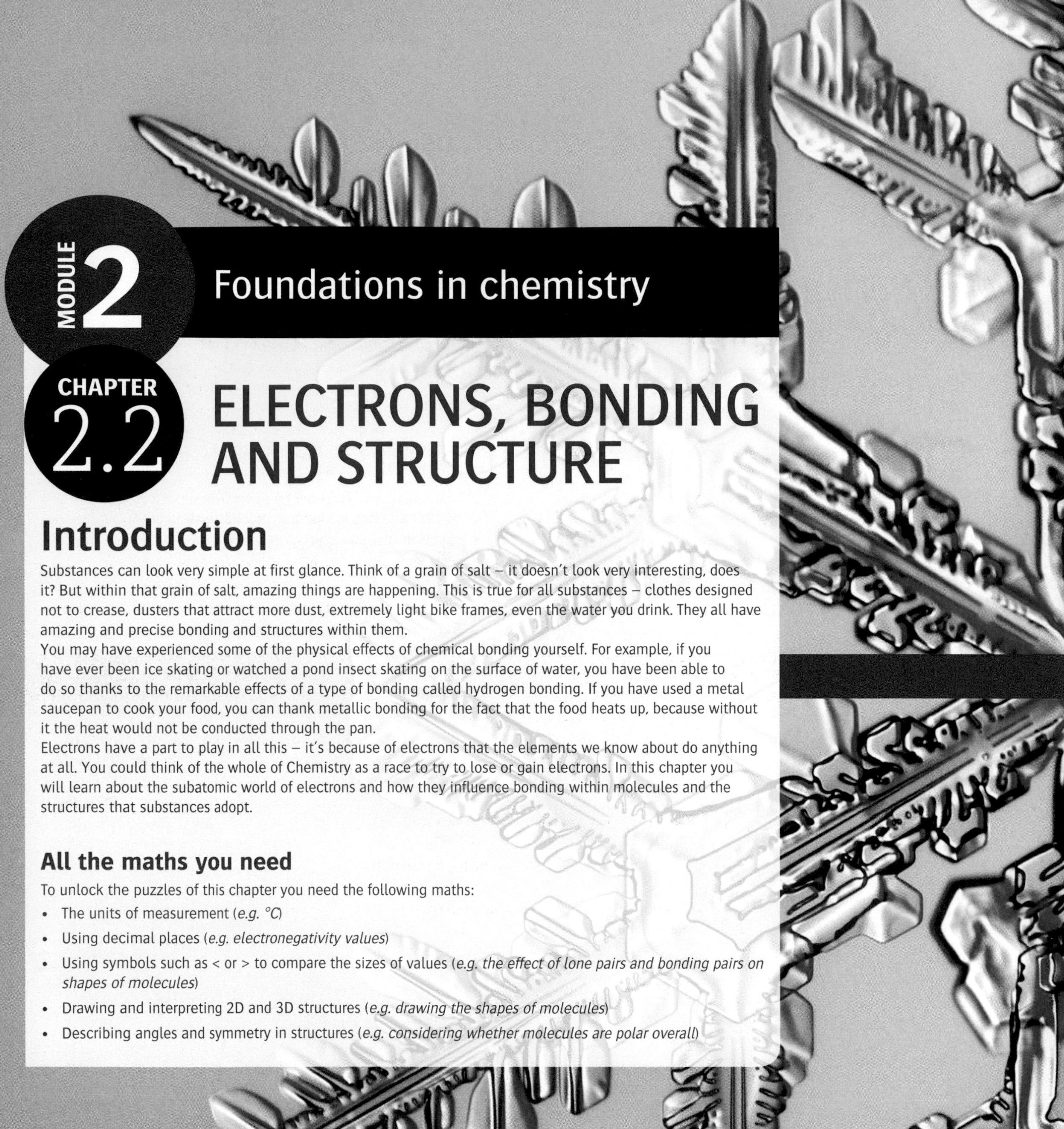

MODULE 2

Foundations in chemistry

CHAPTER 2.2

ELECTRONS, BONDING AND STRUCTURE

Introduction

Substances can look very simple at first glance. Think of a grain of salt – it doesn't look very interesting, does it? But within that grain of salt, amazing things are happening. This is true for all substances – clothes designed not to crease, dusters that attract more dust, extremely light bike frames, even the water you drink. They all have amazing and precise bonding and structures within them.

You may have experienced some of the physical effects of chemical bonding yourself. For example, if you have ever been ice skating or watched a pond insect skating on the surface of water, you have been able to do so thanks to the remarkable effects of a type of bonding called hydrogen bonding. If you have used a metal saucepan to cook your food, you can thank metallic bonding for the fact that the food heats up, because without it the heat would not be conducted through the pan.

Electrons have a part to play in all this – it's because of electrons that the elements we know about do anything at all. You could think of the whole of Chemistry as a race to try to lose or gain electrons. In this chapter you will learn about the subatomic world of electrons and how they influence bonding within molecules and the structures that substances adopt.

All the maths you need

To unlock the puzzles of this chapter you need the following maths:

- The units of measurement (*e.g. °C*)
- Using decimal places (*e.g. electronegativity values*)
- Using symbols such as < or > to compare the sizes of values (*e.g. the effect of lone pairs and bonding pairs on shapes of molecules*)
- Drawing and interpreting 2D and 3D structures (*e.g. drawing the shapes of molecules*)
- Describing angles and symmetry in structures (*e.g. considering whether molecules are polar overall*)

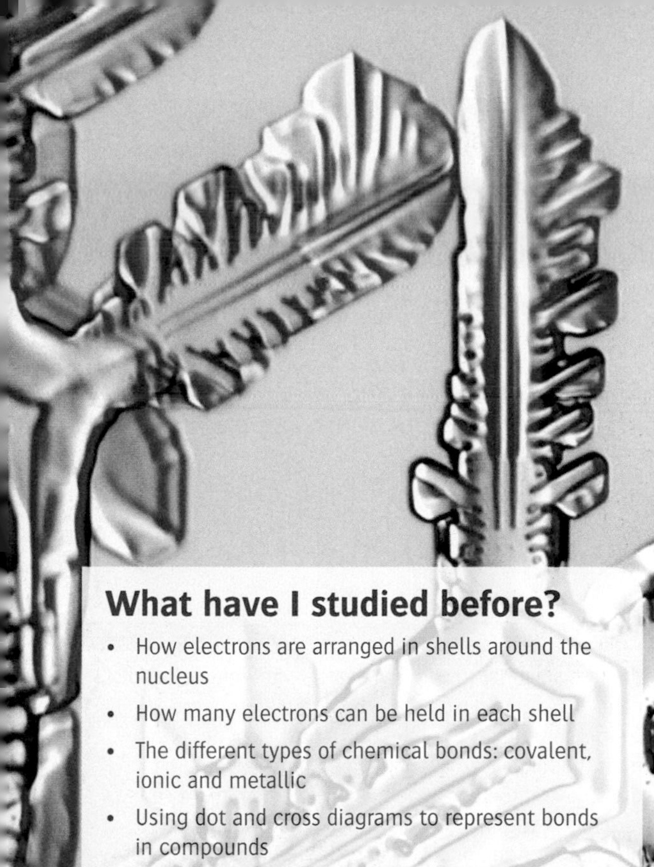

What have I studied before?

- How electrons are arranged in shells around the nucleus
- How many electrons can be held in each shell
- The different types of chemical bonds: covalent, ionic and metallic
- Using dot and cross diagrams to represent bonds in compounds
- Substances that have allotropes, such as carbon, which exists as diamond, graphite and buckminsterfullerenes
- The alloys of metals

What will I study later?

- How an element's electronic structure is linked to its position in the periodic table (AS)
- How an element's reactivity is dependent on its electronic configuration (AS)
- Why bonding within elements changes across the periodic table (AS)
- How electronic structure influences how easily ionic substances dissolve (AL)
- The variable oxidation states and catalytic behaviour of transition metals (AL)
- The ways in which bonds are broken and re-formed, and the different mechanisms by which substances can react (AL)

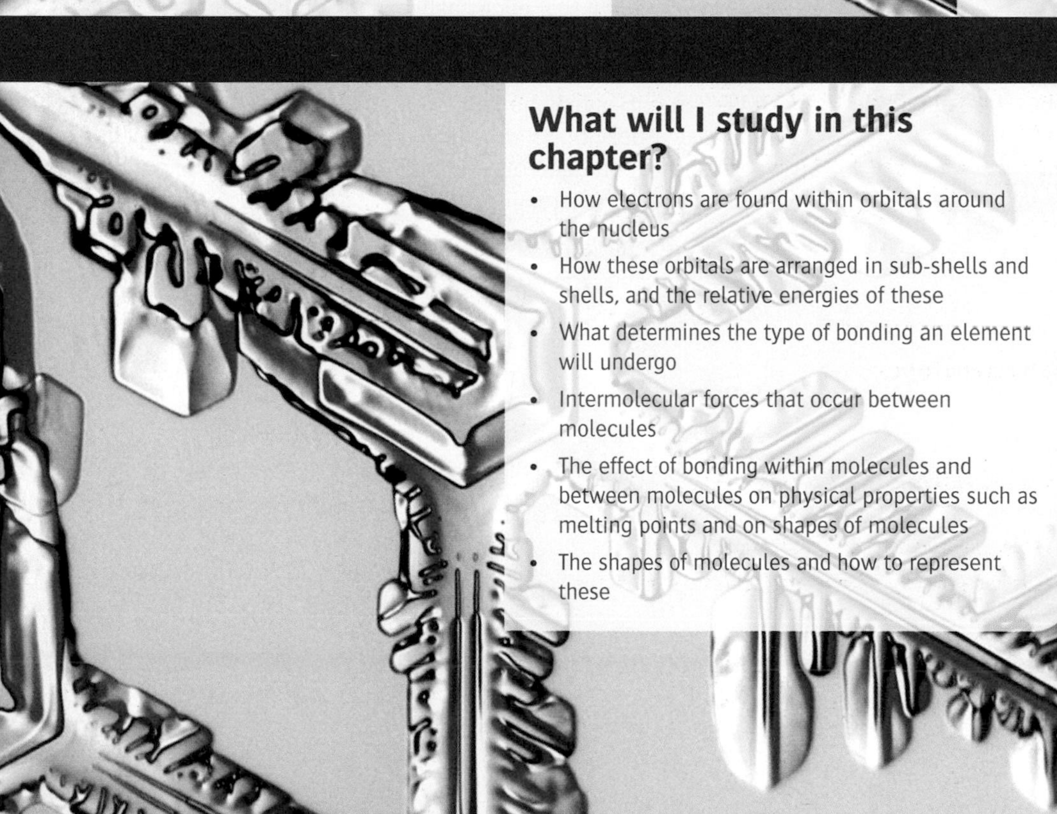

What will I study in this chapter?

- How electrons are found within orbitals around the nucleus
- How these orbitals are arranged in sub-shells and shells, and the relative energies of these
- What determines the type of bonding an element will undergo
- Intermolecular forces that occur between molecules
- The effect of bonding within molecules and between molecules on physical properties such as melting points and on shapes of molecules
- The shapes of molecules and how to represent these

2.2 

Shells and orbitals

By the end of this topic, you should be able to demonstrate and apply your knowledge and understanding of:

* the number of electrons that can fill the first four shells
* atomic orbitals, including:
  (i) as a region around the nucleus that can hold up to two electrons, with opposite spins
  (ii) the shapes of s- and p-orbitals
  (iii) the number of orbitals making up s-, p- and d-sub-shells, and the number of electrons that can fill s-, p- and d-sub-shells

Energy levels or shells

Electrons orbit the nucleus. They are not free to go wherever they like, however. They exist at certain energy levels, or shells.

Quantum numbers are used to describe the electrons in atoms:

- The principal quantum number, n, indicates the shell that the electrons occupy.
- Different shells have different principal quantum numbers.
- The larger the value of n, the further the shell is from the nucleus and the higher the energy level.

Throughout this book, we will use the word *shell*, but its meaning is equivalent to the energy level of the principal quantum number.

The first four shells hold different numbers of electrons, as shown in Table 1. Each shell holds up to $2n^2$ electrons. Look at Table 1; can you see the pattern?

n	Shell	Number of electrons
1	1st shell	2
2	2nd shell	8
3	3rd shell	18
4	4th shell	32

Table 1 Numbers of electrons in the first four shells.

LEARNING TIP

A shell is a group of atomic orbitals with the same principal quantum number, n. Also known as a main energy level.
Principal quantum number, n, is a number representing the relative overall energy of each orbital, which increases with distance from the nucleus. The sets of orbitals with the same n value are referred to as electron shells or energy levels.

Atomic orbitals

There are rules governing where electrons are allowed to be within each shell. Scientists once thought that electrons orbited around the nucleus, much like very fast planets orbiting the Sun. But electrons are not solid particles and this planetary electron idea has now been replaced.

DID YOU KNOW?

Niels Bohr was the first scientist to suggest that the different behaviour of different elements was related to the way their electrons were arranged in the atom. He thought that electrons must exist at certain fixed distances from the nucleus. His work earned him a Nobel Prize. Bohr also helped disprove the plum-pudding model, which proposed that electrons were found within a sea of positive charge (see topic 2.1.1). Calculating the ionisation energy values for removing successive electrons from atoms supported Bohr's model, as the values clearly showed that electrons were in distinct regions at set distances from the nucleus. We now call these shells.

It is now thought that each shell is made up of a number of atomic orbitals. Each atomic orbital can hold a maximum of two electrons with opposite spins. An orbital is a region of space where electrons may be found. Think of the idea in these terms: you may usually be found anywhere within your house, which has a set shape; someone outside may know you are inside but they wouldn't know where.

There are four different types of orbital – s, p, d and f. Each has a different shape. You do not need to learn about f-orbitals for your course.

s-orbitals

An s-orbital has a spherical shape, as illustrated in Figure 1. From $n = 1$ upwards, each shell contains one s-orbital.

- This gives a total of $1 \times 2 = 2$ s electrons in each shell.

Figure 1 An s-orbital is spherical in shape. Each shell contains one s-orbital.

p-orbitals

A p-orbital has a three-dimensional dumb-bell shape, as shown in Figure 2. From $n = 2$ upwards (that is, from the second shell), each shell contains three p-orbitals, p_x, p_y and p_z, at right angles to one another. The $_x$, $_y$ and $_z$ are used to describe the plane each orbital lies on. p_x orbitals can be visualised as lying from side to side, p_y orbitals from top to bottom and p_z orbitals from front to back.

As with all orbitals, each p-orbital can hold up to two electrons. This gives a possible total of $3 \times 2 = 6$ p electrons.

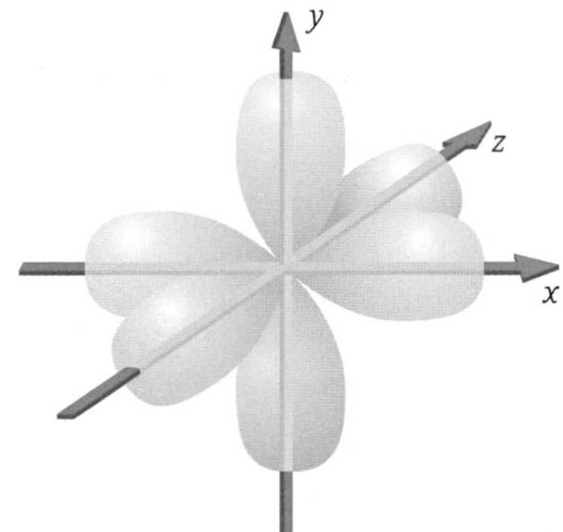

Figure 2 Three p-orbitals. Each p-orbital has a shape like a dumb-bell, with its centre at the nucleus.

d-orbitals and f-orbitals

The structures of d- and f-orbitals are more complex. From $n = 3$ upwards, each shell contains five d-orbitals. This gives a possible total of $5 \times 2 = 10$ d electrons.

From $n = 4$ upwards, each shell contains seven f-orbitals. This gives a possible total of $7 \times 2 = 14$ f electrons.

LEARNING TIP

Remember that each shell has a *maximum* number of electrons it can hold. The *actual* number of electrons found at each quantum level will depend on the number of electrons the atom contains.

Representing electrons in orbitals

Orbitals have different types and different shapes, so chemists often use box diagrams to represent electrons in orbitals. Each box represents an individual orbital. Therefore, each box can hold two electrons. Chemists call this representation '*electrons in boxes*'. An example of this is shown in Figure 3.

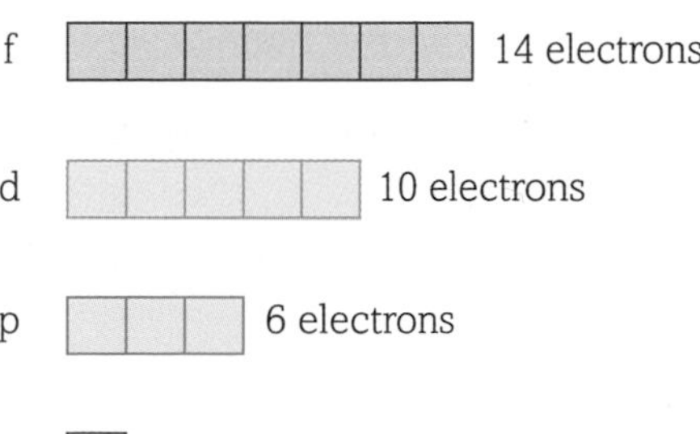

Figure 3 Electrons in boxes. It is convenient to represent an orbital as a box. This diagram shows boxes for the orbitals in each shell. Each orbital can hold a maximum of two electrons.

Each electron has a negative charge, so we would expect the two electrons in an orbital to repel one another. This does not happen, however, because an electron has a property called spin. The two electrons in an orbital must have opposite spins. We can represent the opposite spins of an electron using arrows, either 'up' or 'down' (see Figure 4).

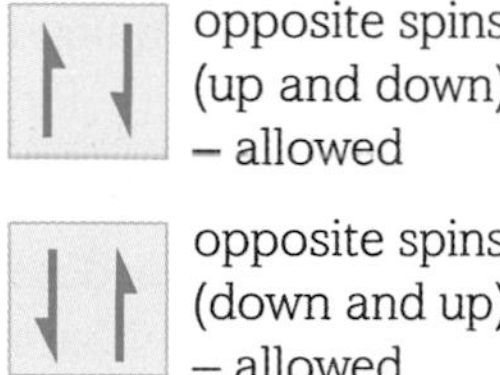

Figure 4 Representing electrons with opposite spin.

But what is an electron? The answer to this question is not easy. In some ways an electron behaves like a particle but in others it behaves like a wave. We can't even be certain where exactly an electron is within an atomic orbital. So it is very difficult, if not impossible, to visualise exactly what an electron is! However, one thing is certain – the idea that an electron is like a tiny planet whizzing around the nucleus is not true.

Luckily, chemists are not too concerned with the exact properties of an electron at this level of detail. They can still use information about atomic orbitals and electrons to describe the behaviour of atoms, how elements react and the structure of the Periodic Table.

Questions

1. How many electrons can be held by the first four shells ($n = 1$ to $n = 4$)?
2. (a) What is meant by the term orbital?
   (b) How many electrons can an orbital hold?
   (c) State the number of s-, p-, and d-orbitals within a shell.
3. Explain why two negatively charged electrons are able to exist in the same orbital without repelling one another.

2 Sub-shells and energy levels

By the end of this topic, you should be able to demonstrate and apply your knowledge and understanding of:

* filling of orbitals:
  (i) for the first three shells and the 4s and 4p orbitals, in order of increasing energy
  (ii) for orbitals with the same energy, occupation singly before pairing
* deduction of the electron configurations of:
  (i) atoms, given the atomic number, up to $Z = 36$
  (ii) ions, given the atomic number and ionic charge, limited to s- and p-blocks up to $Z = 36$

Sub-shells

An electron shell is made up of atomic orbitals with the same principal quantum number, *n*. Within each shell, orbitals of the same type are grouped together as a sub-shell. Each sub-shell is made up of one type of atomic orbital only. So, there are s, p, d and f sub-shells.

Figure 1 shows the orbitals available in the first three shells:

- Each orbital is shown as a box that can hold a maximum of two electrons.
- Each shell gains a new type of sub-shell.

***n* = 1 shell: 2 electrons**

Sub-shell	1s
Orbital	□
Electrons	2

***n* = 2 shell: 8 electrons**

Sub-shell	2s	2p
Orbital	□	□□□
Electrons	2	2 2 2

***n* = 3 shell: 18 electrons**

Sub-shell	3s	3p	3d
Orbital	□	□□□	□□□□□
Electrons	2	2 2 2	2 2 2 2 2

Figure 1 An 'electrons in boxes' diagram showing sub-shells within each shell.

DID YOU KNOW?

Refining the model of the atom

Many chemists and physicists carried out investigations into the detailed structure of the atom and the behaviour of the electrons within it, after the atomic model had been proposed. Notable examples include Erwin Schrödinger, Niels Bohr, Eugene Wigner, Goeppert Mayer and Hans Jensen, many of whom won Nobel Prizes for their work.

Much about the electronic structure of elements was found from spectra: characteristic lines of absorption or emission that elements make when they interact with light. Niels Bohr developed an atomic model that could explain these spectral lines. The lines observed were directly related to the position of electrons within shells and sub-shells.

Spectral lines were once classified using the names of ***s**harp*, ***p**rincipal*, ***d**iffuse* and ***f**undamental*. Notice the first letters of these names – s, p, d and f – from the orbitals that we now know collectively form sub-shells.

Electron energy levels

The sub-shells within a shell have different energy levels. The order of these energy levels is shown in Figure 2. Within a shell, the sub-shell energies increase in the order s, p, d and f.

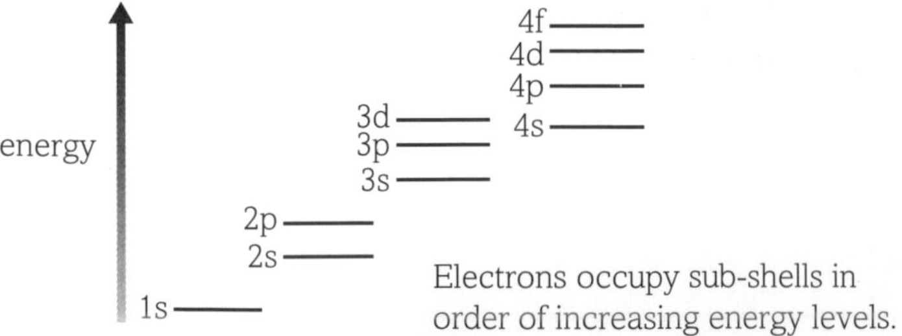

Figure 2 Shells, sub-shells and energy levels.

Filling shells and sub-shells

The electron configuration is the arrangement of electrons in an atom or ion. You can work out the electron configuration of an atom by following a set of rules. These rules are sometimes called the Aufbau principle, from the German *Aufbauprinzip* – building-up principle.

The electrons in the shells of an atom are arranged as follows:

- Electrons are added, one at a time, to 'build up' the atom.
- The lowest available energy level is filled first. You can consider this level as being the closest to the nucleus.
- Each energy level must be full before the next, higher energy level starts to fill.

Sub-shells are made up of several orbitals, each with the same energy level:

- When a sub-shell is built up with electrons, each orbital is filled singly before pairing starts.
- The 4s orbital is at a slightly lower energy level than the 3d orbital. This means 4s will fill before 3d.

LEARNING TIP

The electron in the highest energy position would be the first electron to be removed during ionisation. Use the saying 'Last in, first out' to help you remember this.

Filling the orbitals

Figure 3 shows how the electron configuration is built up for the elements B, C, N and O. Notice that:

- orbitals fill from the lowest energy level upwards
- the 2p-orbitals are filled singly before pairing starts at oxygen
- paired electrons have opposite spins.

Electron configuration

We use a special shorthand to show electron configurations (see figure 3 and Table 1).

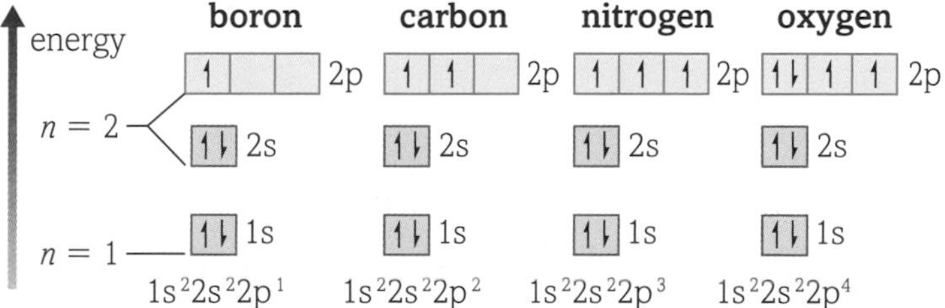

Figure 3 How the electron configuration builds up for the elements B, C, N and O.

Electron configuration of atoms

Table 1 shows the electron configurations for atoms of B, C, N and O. Notice that each occupied sub-shell is written in the form nx^y, where:

- *n* is the shell number
- *x* is the type of orbital
- *y* is the number of electrons in the orbitals making up the sub-shell.

Element	Orbitals occupied	Electron configuration
B	$1s^22s^22p_x^1$	$1s^22s^22p^1$
C	$1s^22s^22p_x^12p_y^1$	$1s^22s^22p^2$
N	$1s^22s^22p_x^12p_y^12p_z^1$	$1s^22s^22p^3$
O	$1s^22s^22p_x^22p_y^12p_z^1$	$1s^22s^22p^4$

Table 1 Electron configurations.

LEARNING TIP

You need to be able to deduce the electron configurations for the first 36 atoms. i.e. up to krypton, which has the configuration $1s^22s^22p^63s^23p^64s^23d^{10}4p^6$.

Electron configuration of ions

Ions are formed when atoms lose or gain electrons. If atoms are ionised to become positive ions, then the electrons found in the highest energy levels are lost first. The electron configuration will therefore show fewer electrons at the highest energy levels. When atoms gain electrons and become negative ions, the extra electrons will continue to fill the sub-shells in the same way as described above.

For example, lithium has the electronic configuration $1s^22s^1$. When it loses an electron to form Li^+, its electronic configuration becomes $1s^2$.

Fluorine has the electronic configuration $1s^22s^22p^5$. When it gains an electron to become F^-, its electron configuration becomes $1s^22s^22p^6$.

Questions

1. Draw diagrams similar to those in Figure 3 to show electrons in orbitals for F and Ne.
2. Write the electron configuration in terms of sub-shells for the elements H to Ne.
3. Iron is able to form several ions. Give the electron configuration for iron (atomic number 26) when it forms a 2+ ion.

An introduction to chemical bonding

By the end of this topic, you should be able to demonstrate and apply your knowledge and understanding of:

* ionic bonding as electrostatic attraction between positive and negative ions
* covalent bond as the strong electrostatic attraction between a shared pair of electrons and the nuclei of the bonded atoms

Why do elements react and bond together?

Chemical reactions are often accompanied by electron transfers. Elements involved in chemical reactions will often become more stable by combining with other elements or transferring electrons. The most stable and unreactive elements are the noble gases. These elements are stable because they contain a full outer shell of electrons. Other elements react in ways that allow them to end up with the electron configuration of a noble gas.

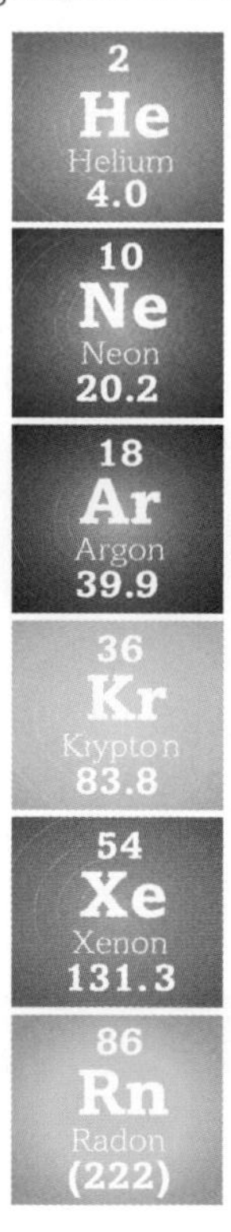

Figure 1 The noble gases, found in group 18 (group 0) of the periodic table. All the elements in this group are stable because they contain a full outer shell of electrons.

The eight electrons in the outer shell of a noble gas (except helium) are made up of two in the s-orbital and two each in the three p-orbitals. During reactions, other elements will bond with a tendency to acquire this noble gas configuration. Although other orbitals can be used for bonding, the s- and p-orbitals are perhaps the most important, especially for the formation of compounds involving the first 18 elements.

LEARNING TIP

Remember, elements have a tendency to bond so that they acquire a noble gas configuration. A noble gas configuration is energetically stable.

Types of chemical bonding

Chemical bonds are classified into three main types: ionic, covalent and metallic. A compound is formed when atoms of different elements are chemically bonded together. In a compound, the atoms of the different elements are always in the same proportions. For example, water, H_2O, always has two hydrogen atoms to one oxygen atom.

Ionic bonding

In general, ionic bonding occurs in compounds consisting of a *metal* and a *non-metal*. If we imagine a bond forming between atoms, electrons are *transferred* from the metal atom to the non-metal atom to form oppositely charged ions that attract each other. Examples are NaF, MgO, Fe_2O_3.

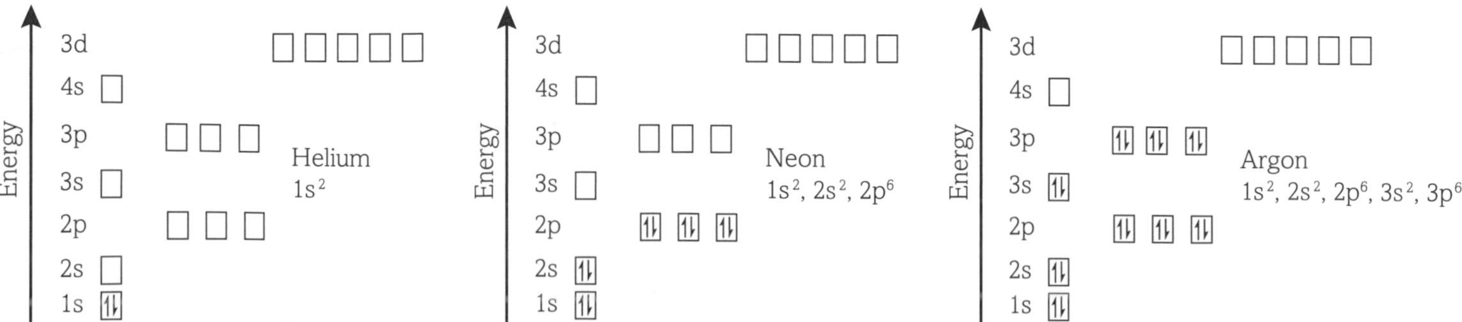

Figure 2 Electrons in boxes diagrams of the first three noble gases.

When this occurs, it usually allows the elements involved to obtain noble gas configurations. Let us look at magnesium oxide as an example:

- Magnesium forms Mg^{2+}, which has the electronic configuration $1s^22s^22p^6$. This is the same as neon, so magnesium has achieved a noble gas configuration.
- Oxygen forms O^{2-}, which has the electronic configuration $1s^22s^22p^6$. This is also the same as neon, so oxygen has also achieved a noble gas configuration.

Covalent bonding

Covalent bonding occurs in compounds consisting of two *non-metals*. If we imagine a bond forming between atoms, electrons are *shared* between the atoms and are attracted to the nuclei of both bonded atoms. Examples are O_2, H_2, H_2O and C (diamond and graphite).

Covalent bonding will also usually result in the elements involved achieving noble gas configurations. For example, hydrogen has the electronic configuration $1s^1$:

- If two H atoms covalently bond and share electrons, they both effectively have the electronic configuration $1s^2$. This is the same as helium.
- Each hydrogen atom has therefore obtained a noble gas configuration.

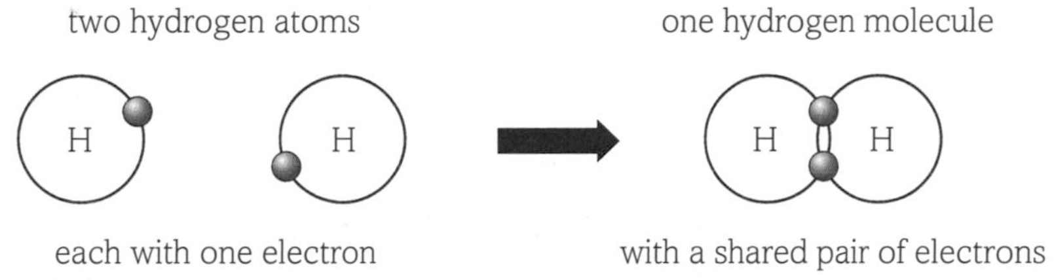

Figure 3 The covalent bonding in hydrogen allows each hydrogen atom to achieve a noble gas configuration.

Metallic bonding

Metallic bonding occurs in *metals*. Electrons are shared between *all* the atoms. Examples include all metals, such as iron, zinc and aluminium, and their alloys, such as brass (copper and zinc) or bronze (copper and tin). See topic 3.1.6 for more on metallic bonding.

Questions

1 Predict the type of bonding in the following compounds:

(a) sodium chloride (b) zinc oxide
(c) hydrogen chloride (d) silver bromide
(e) nitrogen bromide (f) sulfur dioxide.

2 Which of the following compounds have *not* achieved noble gas configuration through bonding? Explain your answer.

(a) H_2 (b) NaI (c) MgO (d) BF_3

Ionic bonding

By the end of this topic, you should be able to demonstrate and apply your knowledge and understanding of:

* ionic bonding as electrostatic attraction between positive and negative ions, and the construction of '*dot-and-cross*' diagrams
* explanation of the solid structures of giant ionic lattices, resulting from oppositely charged ions strongly attracted in all directions, e.g. NaCl

Ionic bonds

Ionic bonds are present in compounds usually consisting of a metal and a non-metal. Imagine an ionic bond being formed between two atoms:

- Electrons are *transferred* from the metal atom to the non-metal atom.
- *Oppositely charged ions* are formed, which are bonded together by electrostatic attraction.
- The metal ion is positive.
- The non-metal ion is negative.

Ionic bonding in sodium oxide, Na_2O

The compound sodium oxide, Na_2O, is formed from atoms of sodium (the metal) and oxygen (the non-metal). We are using it as an example of what happens when ionic bonds form.

A sodium atom has one electron in its outer shell. An oxygen atom has six electrons in its outer shell.

Dot-and-cross diagrams are used to show the origin of electrons in chemical bonding. You use dots to label the electrons of one element and crosses to label the electrons of the other element. Since reactions only involve the outer-shell electrons, you usually only label the electrons in these shells. If there is a third element involved, another symbol is used to represent its electrons.

Figure 1 shows the formation of Na_2O from Na and O atoms. One electron is transferred:

- from each of the two sodium atoms
- to one oxygen atom
- with the formation of two Na^+ ions and one O^{2-} ion.

Notice that both elements involved are able to gain (or be left with) a full outer shell by gaining or losing electrons.

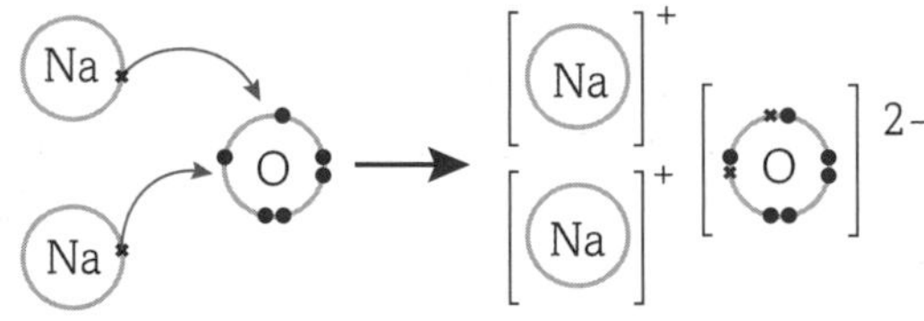

Figure 1 Formation of Na_2O. The substance is bonded together through ionic bonding.

Each sodium atom has lost one electron to obtain the electron configuration of the noble gas neon:

$2Na \rightarrow 2Na^+ + 2e^-$

$1s^22s^22p^63s^1 \rightarrow 1s^22s^22p^6$

An oxygen atom has gained two electrons to obtain the electron configuration of the noble gas neon, too:

$O + 2e^- \rightarrow O^{2-}$

$1s^22s^22p^4 \rightarrow 1s^22s^22p^6$

LEARNING TIP

Dot-and-cross diagrams are usually drawn for the outer shell only, as these are usually the only electrons involved in reactions.
In the sodium oxide example, the sodium shell was shown empty. You can either show this shell empty, to represent the shell that has lost the electron, or full, to represent the full shell below it. It is important to show the charges on ions by adding square brackets around the ion, and placing the charge outside the bracket.
Usually, you will only need to be able to draw the final product, not how it was made.

Giant ionic lattices

In the example above, it looks as if only two Na^+ ions and one O^{2-} ion are involved in the ionic bonding. In reality, the situation for ionic compounds is very different:

- Each ion is surrounded by oppositely charged ions.
- These ions attract each other from all directions, forming a three-dimensioned giant ionic lattice.

All ionic compounds exist as giant ionic lattices in the solid state. When they are in the liquid state (melted), all the ions are free to move around – this is why they are able to conduct electricity.

Sodium chloride

Sodium chloride, NaCl, is an ionic compound and forms a giant ionic lattice:

- Each Na^+ ion is surrounded by six Cl^- ions.
- Each Cl^- ion is surrounded by six Na^+ ions.

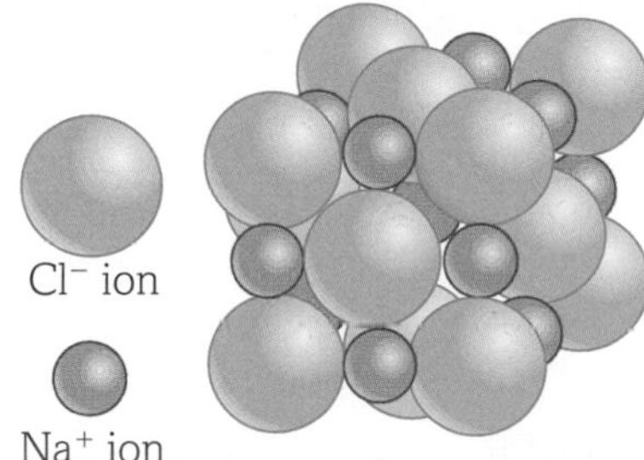

Figure 2 A sodium chloride ionic lattice. Notice how each ion is surrounded by oppositely charged ions.

Sodium chloride's giant ionic lattice structure contains many millions of ions – the actual number depends on the size of the crystal.

Other examples of ionic bonding

Dot-and-cross diagrams are very useful, as they show the number and source of electrons involved in the bonding. Figures 3 and 4 show dot-and-cross diagrams for calcium oxide, CaO, and aluminium fluoride, AlF_3.

- Look at each example and check to see where the electrons have come from.
- Note the charge on each ion.
- See how the dots and crosses help you to work out the charges on the ions.

Ionic bonding in calcium oxide, CaO

The calcium atom has lost two electrons to form a Ca^{2+} ion, with the electron configuration of argon. These two electrons, shown as crosses, now form part of the oxide ion, O^{2-}.

$Ca \rightarrow Ca^{2+} + 2e^-$

$O + 2e^- \rightarrow O^{2-}$

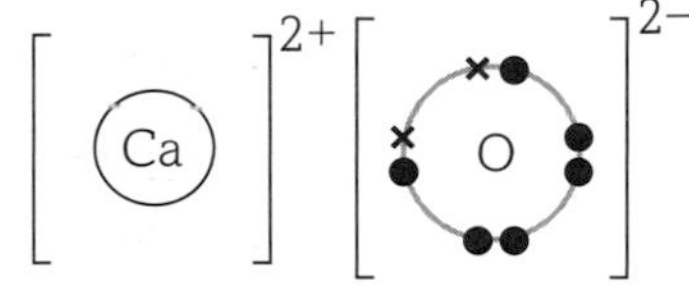

Figure 3 Dot-and-cross diagrams of CaO.

Ionic bonding in aluminium fluoride, AlF_3

The aluminium atom has lost three electrons to form an Al^{3+} ion, with the electron configuration of neon. Each of the three fluorine atoms accepts one of these electrons, forming F^-. Each fluorine atom has achieved the noble gas configuration of neon by accepting an electron to form a 1− ion.

$Al \rightarrow Al^{3+} + 3e^-$

$3F + 3e^- \rightarrow 3F^-$

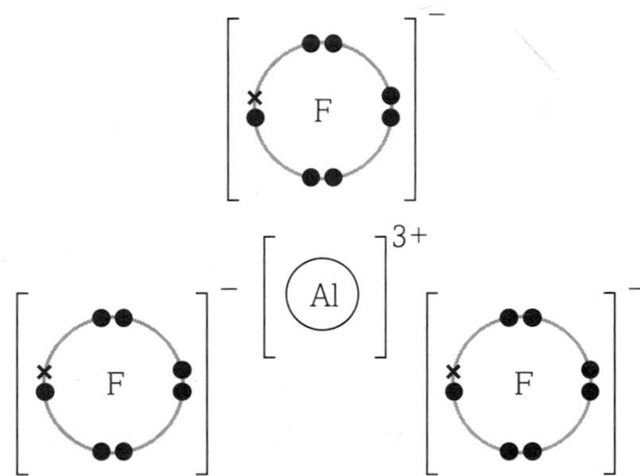

Figure 4 Dot-and-cross diagrams of AlF_3.

Questions

1. Draw dot-and-cross diagrams for:
   (a) MgO
   (b) $CaBr_2$
   (c) Na_3P
   (d) Al_2O_3

2. For each atom and ion below, write down the electron configuration, in terms of sub-shells.
   (a) K and K^+
   (b) Ca and Ca^{2+}
   (c) S and S^{2-}
   (d) Al and Al^{3+}
   (e) N and N^{3-}

 Structures of ionic compounds

By the end of this topic, you should be able to demonstrate and apply your knowledge and understanding of:

* explanation of the solid structures of giant ionic lattices, resulting from oppositely charged ions strongly attracted in all directions, e.g. NaCl
* explanation of the effect of structure and bonding on the physical properties of ionic compounds, including melting and boiling points, solubility and electrical conductivity in solid, liquid and aqueous states

Giant ionic lattices

As we discussed in topic 2.2.4, ionic substances, such as NaCl, always exist as giant lattices in the solid state. This means they can contain millions of ions that have been formed through electron transfer. Their actual size will depend on the amount of ions involved.

Each ion attracts oppositely charged ions from all directions:

- Each ion is surrounded by oppositely charged ions.
- The ions attract each other, forming a *giant ionic lattice.*

Because giant ionic lattices involve huge amounts of electrostatic attraction between ions, they have unique properties.

Properties of ionic compounds

High melting and boiling points

Ionic compounds are solids at room temperature. A large amount of energy is needed to break the *strong* electrostatic bonds that hold the oppositely charged ions together in the solid lattice. For this reason, ionic compounds have high melting and boiling points.

Table 1 gives the melting points of sodium chloride and magnesium oxide. The melting point of MgO is higher than that of NaCl. This is because the charges on the Mg^{2+} and O^{2-} ions are greater than those on Na^+ and Cl^-. The greater the charge, the stronger the electrostatic forces between the ions. This means that more energy is required to break up the ionic lattice during melting. The melting point of magnesium oxide is so high that it is used to line furnaces for brick-making.

Compound	Ions present	Melting point/°C
NaCl	Na^+ and Cl^-	801
MgO	Mg^{2+} and O^{2-}	2852

Table 1 Melting point of NaCl compared with MgO. The different values show the effect of charge sizes.

LEARNING TIP

Remember, it is the strong electrostatic forces between all the ions in a giant ionic lattice that mean ionic substances have high melting and boiling points. All of these forces have to be overcome when melting an ionic substance.

Electrical conductivity

In a solid ionic lattice:

- the ions are held in fixed positions and no ions can move
- the ionic compound does not conduct electricity.

When an ionic compound is *melted* or *dissolved* in water:

- the solid lattice breaks down and the ions are free to move
- the ionic compound is now a conductor of electricity.

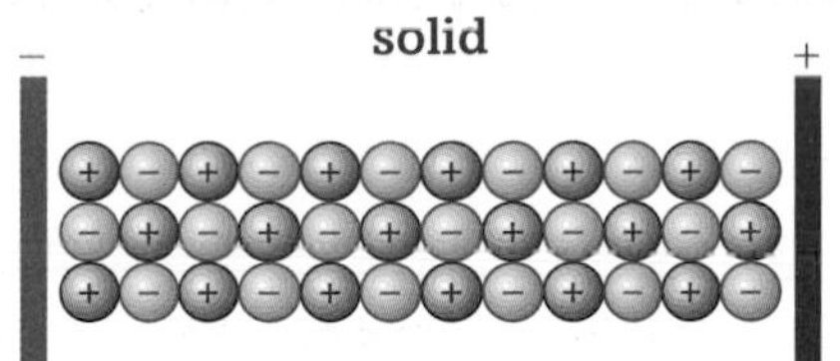

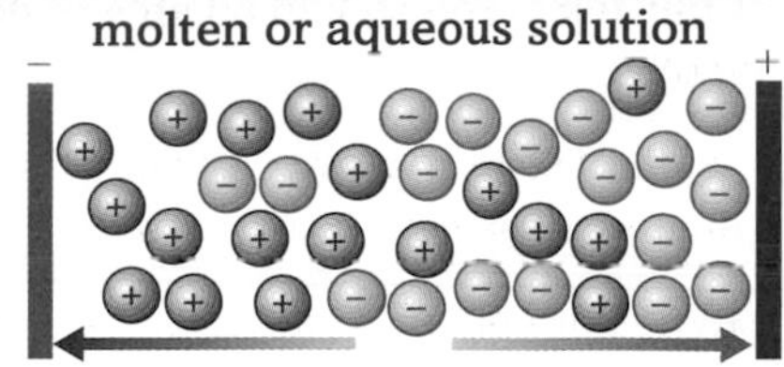

Figure 1 The diagram on the left shows ions in fixed positions. This substance would not be able to conduct electricity. The diagram on the right shows a molten (or dissolved) ionic substance. This would be able to conduct electricity.

Solubility

An ionic lattice dissolves (i.e. is soluble) in *polar* solvents, such as water. Polar solvents contain substances that have polar bonds. A polar bond occurs between atoms that do not share electrons equally, and it results in the atoms having very small charges on them. See topic 2.2.10 for further information on polar bonds.

Polar water molecules break down an ionic lattice by surrounding each ion to form a solution. The slight charges within the polar substance are able to attract the charged ions in the giant ionic lattice. This means the lattice is disrupted and ions are pulled out of it.

Let's look at NaCl as an example. When NaCl is dissolved in water, the giant ionic lattice breaks down:

- Water molecules attract the Na^+ and Cl^- ions.
- The ionic lattice breaks down as it dissolves. Water molecules surround the ions.
- Na^+ attracts $\delta-$ charges on the O atoms of the water molecules.
- Cl^- attracts $\delta+$ charges on the H atoms of the water molecules.

This is illustrated in Figure 2.

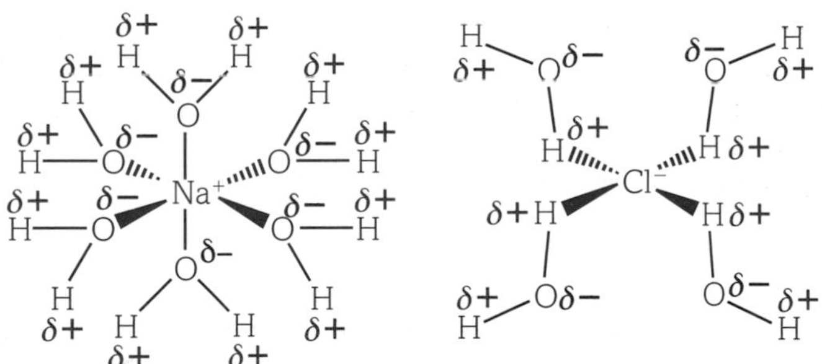

Figure 2 NaCl is soluble in polar solvents such as water. The slight charges on the water molecules break the giant ionic lattice apart. The positive Na^+ ions will be surrounded by the δ^- on the water molecules. The Cl^- ion will be surrounded by the δ^+ on the water molecule.

LEARNING TIP

Remember that when an ionic substance is melted or dissolved in water, it is able to conduct electricity because the ions are able to move and carry the charge.

LEARNING TIP

$\delta-$ means 'slight negative charge' and $\delta+$ means 'slight positive charge'. The symbol δ is known as delta, so sometimes you may hear or see these slight charges referred to as delta positive and delta negative.

Questions

1. NaCl and CCl_4 both contain chlorine but have very different melting points. Explain why NaCl has a high melting point of 801 °C but CCl_4 has a low melting point of −23 °C.
2. Explain why ionic substances are able to conduct electricity when molten or dissolved, but not when solid.
3. Explain why ionic compounds are soluble in water.

6 Covalent bonding

By the end of this topic, you should be able to demonstrate and apply your knowledge and understanding of:

* covalent bond as the strong electrostatic attraction between a shared pair of electrons and the nuclei of the bonded atoms
* construction of *'dot-and-cross'* diagrams of molecules and ions to describe:
  (i) single covalent bonding
  (ii) multiple covalent bonding
* use of the term *average bond enthalpy* as a measurement of covalent bond strength

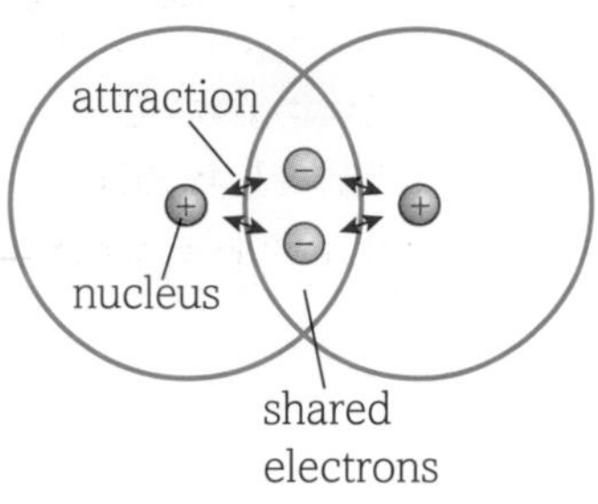

Figure 1 Strong electrostatic attraction between the shared pair of electrons and the nuclei of the bonded atoms.

Covalent bonds

As we saw in topic 2.2.3, covalent bonding occurs in compounds consisting of non-metals bonded to other non-metals. In a **covalent bond**, an electron pair occupies the space between the two atoms' nuclei, as shown in Figure 1. This shared pair of electrons forms a covalent bond:

- The negatively charged shared pair of electrons is attracted to the positive charges of *both* nuclei.
- This attraction overcomes the repulsion between the two positively charged nuclei.
- The resulting attraction is the covalent bond that holds the two atoms together.
- The two electrons are *shared*. This is in contrast to the *transfer* of electrons that results in an ionic bond.

Dot-and-cross diagrams

As we work through this topic, you will notice that covalent bonds are represented using *dot-and-cross* diagrams:

- Dots are used to represent electrons from one atom; crosses are used for the other atom.
- Circles are often used to represent the outer shell.
- A shared pair of electrons is shown using a dot and a cross together.
- The circles representing shells are drawn overlapping, to show that electrons are being shared, not exchanged.
- As with ionic compounds, there is no need to show shells or electrons except for those in the outer shell.

Single covalent bonding

If atoms are bonded by one shared pair of electrons, this is known as a *single bond*. Whereas ionic attractions act in all directions, covalent bonds act in one direction only – solely between the atoms involved in the bond. Some examples of covalently bonded substances are discussed below.

Hydrogen, H_2, is formed from two hydrogen atoms (see Figure 2):

- Each hydrogen atom has one electron in its outer shell.
- Each hydrogen atom contributes one electron to the covalent bond.
- The covalent bond is often written as a line, for example H–H.

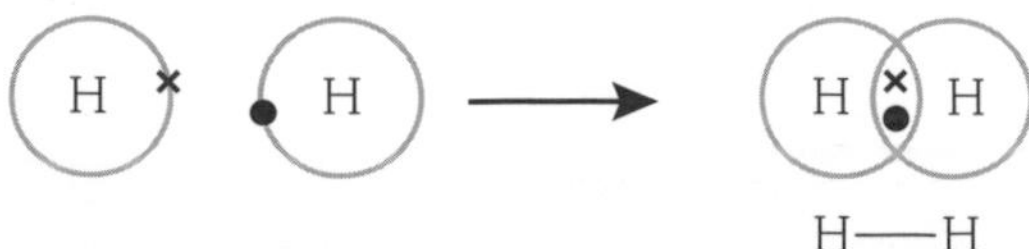

Figure 2 Covalent bond formation in a molecule of H_2.

By sharing electrons and forming a single covalent bond, each hydrogen atom fills its 1s sub-shell and achieves the noble gas configuration of helium.

Some atoms will share electrons with more than one other atom. Each shared pair between different atoms (they can be atoms of the same element) equals one single covalent bond. Examples of this are shown in Figure 3.

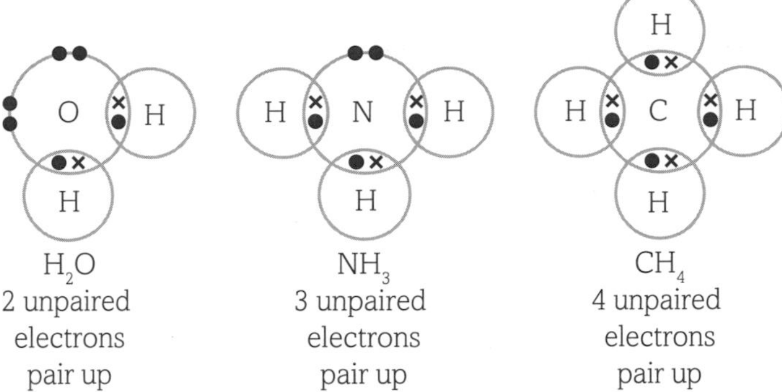

Figure 3 Covalent bonding in H_2O, NH_3 and CH_4. H_2O contains two single covalent bonds, NH_3 involves three single covalent bonds and CH_4 involves four single covalent bonds. Notice that each of the atoms involved has been able to obtain a noble gas configuration by sharing electrons and forming covalent bonds.

DID YOU KNOW?

Some elements will always make the same number of covalent bonds when they bond:

- Carbon will always make four covalent bonds.
- Nitrogen will always make three covalent bonds.
- Oxygen will always make two covalent bonds.
- Hydrogen will always make one covalent bond.

You may have noticed that the number of bonds these elements form corresponds exactly to how many electrons they need in order to obtain a noble gas configuration.

Lone pairs

Look at the dot-and-cross diagram of ammonia, NH_3, in Figure 3. Can you see that there are three shared electrons and one pair that belong only to nitrogen? A pair of electrons like this, that are not used for bonding, is called a lone pair.

Now look at the dot-and-cross diagram of water, H_2O, in Figure 3. Can you see that it has two lone pairs?

A lone pair gives a concentrated region of negative charge around the atom. Lone pairs can influence the chemistry of a molecule in several ways (see topics 2.2.7 and 2.2.9).

LEARNING TIP

A lone pair is an outer-shell pair of electrons that is not involved in chemical bonding.

Multiple covalent bonding

Some non-metallic atoms can share more than one pair of electrons with another atom to form a *multiple bond*.

- Sharing of *two* pairs of electrons forms a *double bond*. For example, oxygen, O_2, bonds in this way. This can be written as O=O.
- Sharing of *three* pairs of electrons forms a *triple bond*. For example, nitrogen, N_2, bonds in this way. This can be written as N≡N.
- Carbon dioxide, CO_2, has two double bonds. This can be written as O=C=O.

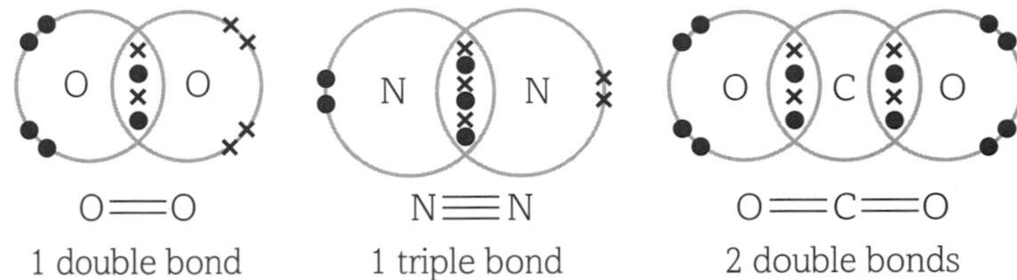

Figure 4 Dot-and-cross diagrams of the double bond in O_2, the triple bond in N_2 and the two double bonds in CO_2.

Average bond enthalpy

Not all covalent bonds are the same strength. Some are much stronger than others, which means that more energy would be required to break them, for example during a reaction. A measure of the average energy of a bond (and therefore the energy that would be needed to break it) is called the *average bond enthalpy*.

Table 1 shows some covalent bonds and their bond enthalpies. They are shown in order of increasing enthalpy and therefore increasing strength. Note that the values are shown as positive figures (the + sign is used). This is because bond breaking is an endothermic change, one that requires energy. Endothermic values are shown with a + sign. The bigger the value, the more energy is needed to break the bond.

Bond	Average bond enthalpy/kJ mol^{-1}
C–H	+413
H–H	+436
O=O	+497
C=C	+612

Table 1 Bond enthalpies for different bonds.

Questions

1. Draw dot-and-cross diagrams for:
   (a) F_2 (b) HF (c) SiF_4 (d) SCl_2
2. Draw dot-and-cross diagrams for:
   (a) CS_2 (b) C_2H_4 (c) HC≡N (d) $H_2C=O$
3. Which bond would be considered stronger, O–H, which has a bond enthalpy of +463 kJ mol^{-1}, or N=O, with a bond enthalpy of +607 kJ mol^{-1}?

2.2 

Dative covalent bonding

By the end of this topic, you should be able to demonstrate and apply your knowledge and understanding of:

* construction of '*dot-and-cross*' diagrams of molecules and ions to describe dative covalent (coordinate) bonding

Dative covalent bonding

In a dative covalent or coordinate bond, *one* of the atoms supplies both the shared electrons to the covalent bond.

- A dative covalent bond can be written as A→B.
- The direction of the arrow shows the direction in which the electron pair has been donated. In the example above, A is donating a pair of electrons to B.

LEARNING TIP

Once formed, dative covalent bonds are equivalent to ordinary covalent bonds.

The ammonium ion, NH_4^+

The ammonium ion, NH_4^+, has three covalent bonds and one dative covalent bond. Figure 1 shows the formation of the NH_4^+ ion from ammonia, NH_3, and H^+. One of the electron pairs around the nitrogen atom in an NH_3 molecule is a *lone pair*. In the formation of an ammonium ion, NH_4^+, this lone pair provides both the bonding electrons when bonding with the H^+ ion, and the resulting NH_4^+ ion has a positive charge of 1+.

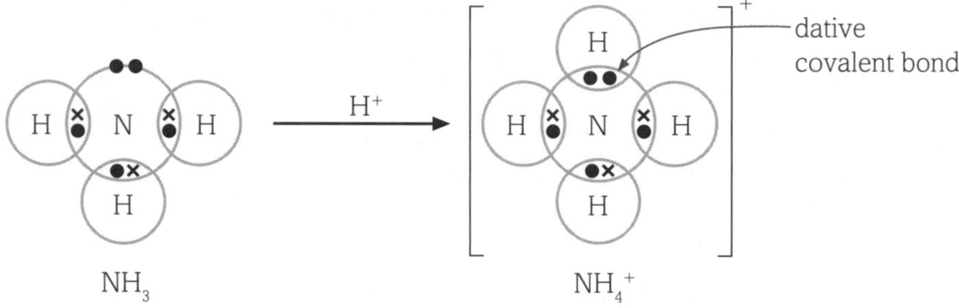

Figure 1 Dative covalent bonding in NH_4^+. The N in NH_3 donates a pair of electrons to form a bond with H.

In a dative covalent bond, one atom provides both bonding electrons from a lone pair of electrons. However, once formed, this *dative* covalent bond is equivalent to all the other covalent bonds. For example, in the NH_4^+ ion, you cannot tell which bond was formed from the N lone pair.

The diagram in Figure 2 shows how the ammonium ion (and all compounds that involve dative covalent bonds) can be represented to show the dative covalent bonding within it.

H
↑
H—N—H
|
H
+

Figure 2 The ammonium ion. Notice the dative covalent bond shown with an arrow. The arrow indicates the origin of the bonded pair.

The oxonium ion, H_3O^+

When an acid is added to water, water molecules form oxonium ions, H_3O^+. For example, hydrogen chloride gas forms hydrochloric acid when added to water:

$$HCl(g) + H_2O(l) \rightarrow H_3O^+(aq) + Cl^-(aq)$$

H_3O^+ ions are responsible for reactions of acids. In equations, the oxonium ion is often simplified as $H^+(aq)$.

In the oxonium ion, one of the lone pairs around the oxygen atom in an H_2O molecule provides both the bonding electrons to form a dative covalent bond. The bonding in the oxonium ion is shown in Figure 3.

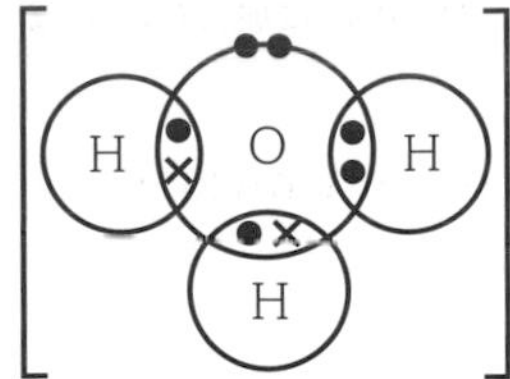

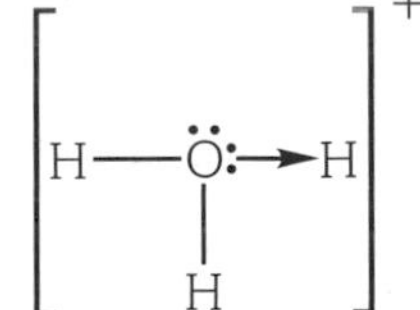

Figure 3 The oxonium ion.

How many covalent bonds can be formed?

When covalent bonds form, unpaired electrons often pair up so that the bonded atoms obtain a noble gas electron configuration, obeying the Octet Rule (also see topic 2.2.3). This is not always possible:

- There may not be enough electrons to reach an octet.
- More than four electrons may pair up in bonding (known as an expansion of the octet).

Not enough electrons for an octet

Within period 2, the elements beryllium, Be, and boron, B, both form compounds with covalent bonds. Neither of these elements has enough unpaired electrons to reach a noble gas electron configuration. However, they *can* pair up any unpaired electrons.

Consider the compound boron trifluoride, BF_3 (see Figure 4). Boron has three electrons in its outer shell. Each fluorine atom has seven electrons in its outer shell.

- Three covalent bonds can be formed.
- Each of boron's three outer electrons is paired, so there are now six electrons in boron's outer shell.
- Each of the three fluorine atoms has eight electrons in its outer shell, attaining an octet. The central boron atom does not achieve an octet.

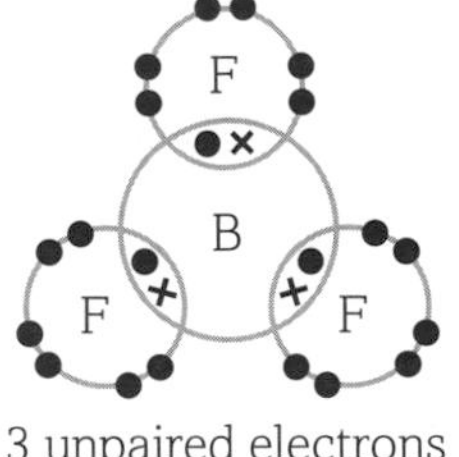

Figure 4 Dot-and-cross diagram of BF_3.

Expansion of the octet

For elements in groups 15–17 of the periodic table, something odd happens from period 3. As we move down the periodic table, more of the outer-shell electrons are able to take part in bonding. In the resulting molecules, one of the bonding atoms may finish up with more than eight electrons in its outer shell. This breaks the Octet Rule and is often called *expansion of the octet*.

- Atoms of non-metals in group 15 can form three or five covalent bonds, depending on how many electrons are used in bonding.
- Atoms of non-metals in group 16 can form two, four or six covalent bonds, depending on how many electrons are used in bonding.
- Atoms of non-metals in group 17 can form one, three, five or seven covalent bonds, depending on how many electrons are used in bonding.

Table 1 lists the elements that can expand their octets.

Group 15	Group 16	Group 17
P	S	Cl
As	Se	Br
	Te	I
		At

Table 1 Elements that expand their octet.

For example, in the compound sulfur hexafluoride, SF_6 (see Figure 5), sulfur has expanded its octet:

- Sulfur has six electrons in its outer shell.
- Six covalent bonds can be formed.
- Each of sulfur's six electrons is paired, so sulfur now has 12 outer electrons. It no longer obeys the Octet Rule – it has expanded the octet.
- Each of the six fluorine atoms has eight electrons in its outer shell, attaining the octet.

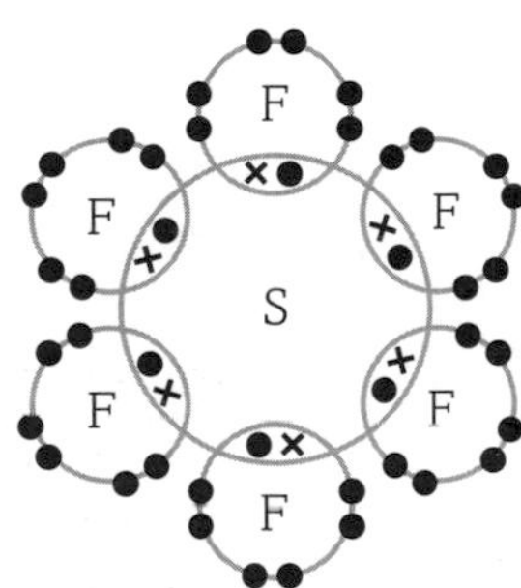

Figure 5 Dot-and-cross diagram of SF_6. Sulfur has expanded its octet and is surrounded by 12 outer electrons. Each fluorine obeys the Octet Rule as it is surrounded by eight outer electrons.

Modifying the Octet Rule

A better rule than the Octet Rule would be:

- unpaired electrons pair up
- the maximum number of electrons that can pair up is equivalent to the number of electrons in the outer shell.

Questions

1. Draw dot-and-cross diagrams for:
   (a) PCl_4^+ (b) H_3O^+ (c) H_2F^+

2. Draw dot-and-cross diagrams for:
   (a) BF_3 (b) PF_5 (c) SO_2 (d) SO_3

Structures of covalent compounds

By the end of this topic, you should be able to demonstrate and apply your knowledge and understanding of:

* explanation of the solid structures of simple molecular lattices, as covalently bonded molecules attracted by intermolecular forces, e.g. I_2, ice
* explanation of the effect of structure and bonding on the physical properties of covalent compounds with simple molecular lattice structures including melting and boiling points, solubility and electrical conductivity
* explanation of physical properties of giant covalent lattices, including melting and boiling points, solubility and electrical conductivity, in terms of structure and bonding

Types of covalent structure

Elements and compounds with covalent bonds can have either of these structures:

- a simple molecular lattice
- a giant covalent lattice.

Simple molecular structures

Simple molecular structures are made up from small, simple molecules, such as Ne, H_2, O_2, N_2 and H_2O. In a solid simple molecular lattice (see Figure 1):

- the atoms within each molecule are held together by strong covalent bonds
- the different molecules are held together by weak intermolecular forces (such as van der Waals' or London forces).

An example of a simple molecular structure is iodine:

- Within each I_2 molecule, I atoms are held together by strong covalent bonds.
- When solid I_2 (an example of a simple molecular lattice) changes state, the weak intermolecular forces between the I_2 molecules break.

The structure of iodine is illustrated in Figure 1.

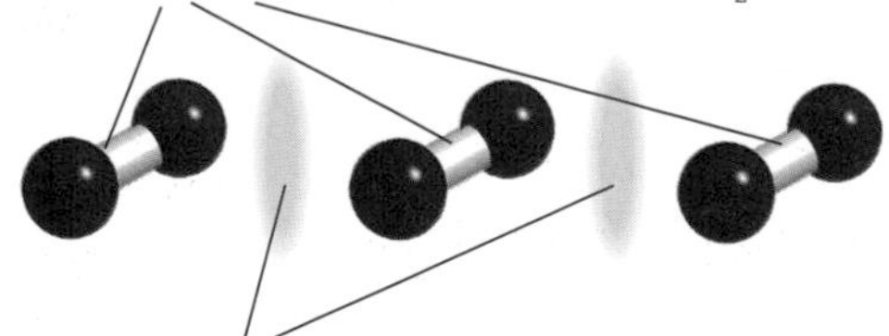

Figure 1 The forces holding different molecules of iodine together and those bonding the atoms together within the molecule.

Properties of simple molecular structures

Melting and boiling point

Simple molecular structures have low melting and boiling points because the intermolecular forces are weak, so a relatively small amount of energy is needed to break them.

Electrical conductivity

Simple molecular structures are *non-conductors* of electricity because there are no charged particles free to move.

Solubility

Simple molecular structures are generally soluble in *non-polar* solvents, such as hexane. Weak London forces are able to form between covalent molecules and these solvents. This helps the molecular lattice to break down and the substance dissolves.

LEARNING TIP

Remember the phrase 'like dissolves like' when thinking about the solubility of simple molecular lattices. Non-polar solvents will only be held together by weak van der Waals' forces (such as London forces), just like simple molecular lattices. See topic 2.2.11 for more on intermolecular forces.

Figure 2 Iodine is shown here dissolved in a non-polar solvent (hexane) and a polar solvent (water). Notice that the hexane top layer is dark purple – the iodine is highly soluble in this solvent. In the bottom water layer, the colour is a very faint brown. The iodine is only sparingly soluble in this solvent.

Giant covalent structures

Diamond, graphite and SiO_2 are all examples of giant covalent lattices.

Properties of giant covalent structures

Melting and boiling point

Giant covalent structures have high melting and boiling points because high temperatures are needed to break the strong covalent bonds within the lattice.

Refer to topic 2.2.11 for more information about the forces affecting physical properties such as boiling points.

Electrical conductivity

Giant covalent structures are non-conductors of electricity because there are no free charged particles, except for graphite (see Figure 3).

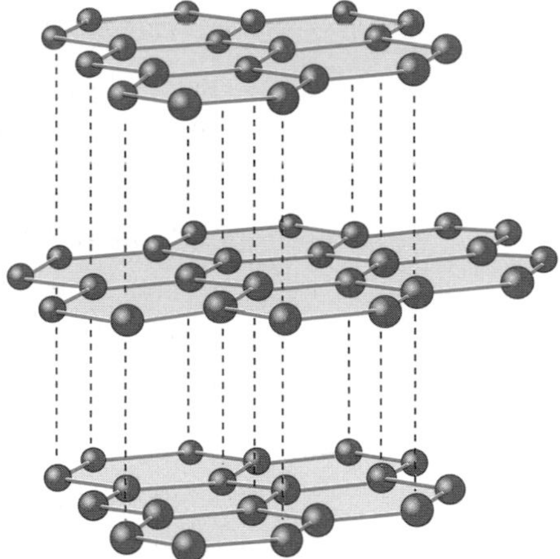

Figure 3 Graphite is the only giant covalent structure able to conduct electricity. Delocalised electrons between the layers are able to move freely, parallel to the layers, when a voltage is applied.

Solubility

Giant covalent structures are insoluble in *both* polar and non-polar solvents because the covalent bonds in the lattice are too strong to be broken by either polar or non-polar solvents.

LEARNING TIP

When a *simple molecular substance* is melted or boiled, weak intermolecular forces *between* molecules need to be broken. This does not require much energy – these substances have *low melting and boiling points.*
When a *giant covalent lattice* is melted or boiled, strong covalent bonds *within* molecules need to be broken. This requires a huge amount of energy – these substances have *very high melting and boiling points.*

Questions

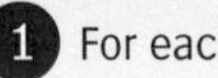

1. For each of the substances below, predict the:
   (a) structure
   (b) melting point
   (c) electrical conductivity
   (d) solubility.
   (i) NaCl (ii) SiO_2 (sand) (iii) Br_2 (iv) C_2H_5OH

2. Suggest what type of structure is found in the following substances, given experimental data on them. (You may need to recall types of structures you have covered previously that are not mentioned in this topic.)

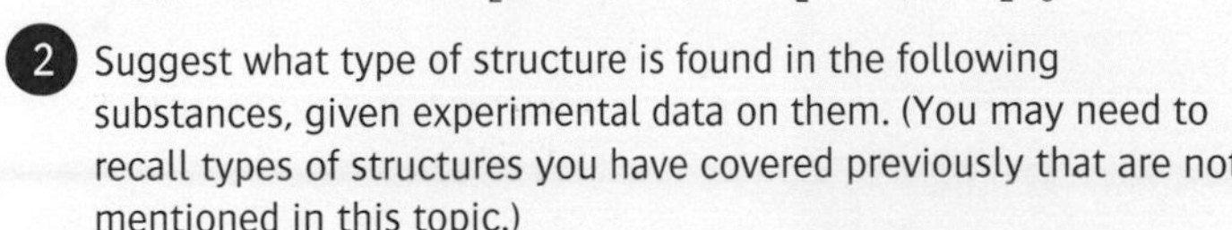

   Substance A: melting point 1725 °C, did not dissolve in any solvent, did not conduct electricity

   Substance B: melting point 114 °C, dissolved in hexane, did not conduct electricity

   Substance C: melting point 801 °C, did not dissolve in hexane but dissolved in water, did not conduct electricity in solid state, aqueous solution did conduct electricity.

Shapes of molecules and ions

By the end of this topic, you should be able to demonstrate and apply your knowledge and understanding of:

* the shapes of, and bond angles in, molecules and ions with up to six electron pairs (including lone pairs) surrounding the central atom as predicted by electron pair repulsion, including the relative repulsive strengths of bonded pairs and lone pairs of electrons
* electron pair repulsion to explain the following shapes of molecules and ions: linear, non-linear, trigonal planar, pyramidal, tetrahedral and octahedral

Electron pair repulsion theory

The shape of a molecule or ion is determined by the number of electron pairs in the outer shell surrounding the central atom. These pairs can be bonding pairs or lone pairs:

- As electrons all have a negative charge, each electron pair repels other electron pairs.
- The shape adopted will be the shape that allows all the pairs of electrons to be as far apart as possible.

Molecules surrounded by bonded pairs or bonding regions

If the central atom in a compound is only surrounded by bonded pairs, all these pairs will repel one another equally. The number of pairs around the central atom dictates the shape of the compound. When we consider shapes of molecules, a double bond (or triple bond) can be thought of as a 'bonding region'. Because the double bond is fixed in place between the central atom and the other bonded atom, it will repel in the same way as an ordinary bonding pair of electrons.

Table 1 outlines the shapes of molecules with up to six bonded pairs or bonding regions around them.

Number of bonded electron pairs around central atom	1	2	3	4	5	6
Name of shape	linear	linear	trigonal planar	tetrahedral	trigonal bipyramid	octahedral
Example molecule	H_2	CO_2	BF_3	CH_4	PCl_5	SF_6
Dot-and-cross diagram	H H	O C O	F B F F	H H C H H	Cl Cl Cl P Cl Cl	F F F S F F F
Shape and bond angles	H–H	O=C=O 180°	F B F F 120°	H C H H H 109.5°	Cl 90° Cl P Cl 120° Cl Cl	F F F S F F F 90°

Table 1 The shapes of molecules with up to six bonded electron pairs around them.

DID YOU KNOW?

Many molecules have 3D shapes and we need a simple way of drawing these shapes on a flat sheet of paper. Chemists use bold wedges and dotted wedges – as in the diagrams of methane, CH_4, below:

- A normal line shows a bond in the plane of the paper.
- A bold wedge shows the bond coming out from the plane of the paper towards you.
- A dotted wedge shows the bond going out from the plane of the paper away from you.

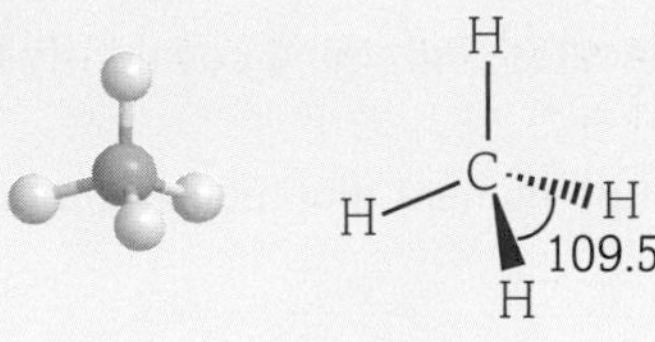

Molecules with lone pairs

A lone pair of electrons is slightly more electron-dense than a bonded pair. Therefore, a lone pair repels more than a bonded pair. The relative strengths of the repulsion are:

lone pair/lone pair > bonded pair/lone pair > bonded pair/bonded pair.

- The diagrams in Figure 1 show how the shapes of molecules are affected by the presence of lone pairs of electrons.
- Each lone pair reduces the bond angle by about 2.5°. This is the result of the extra repulsive effect of each lone pair.

A methane molecule is **tetrahedral** with bond angles of 109.5°.

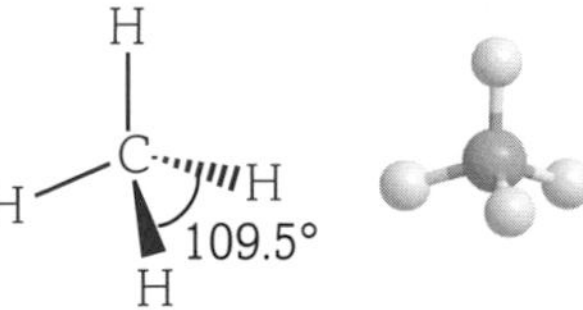

An ammonia molecule is **pyramidal** with bond angles of 107°.

A water molecule is **non-linear** with a bond angle of 104.5°.

Figure 1 The effect of lone pairs on shapes of molecules. Notice how, if one out of four areas of electrons is a lone pair, the tetrahedral shape becomes pyramidal. If two out of four of the areas of electron density are lone pairs, the shape becomes non-linear.

LEARNING TIP

When you are thinking about the shapes of molecules that include lone pairs, firstly think of the shape that the molecule would have if all the pairs were bonded pairs or bonding regions. The actual shape will be based on this, but the bonded pairs will all be a bit closer together with a smaller angle between them. Each lone pair reduces the original angle (if there were no lone pairs) by about 2.5°.

Shapes of ions

The principles discussed above can also be applied to any molecular ion. For example, have a look at the ammonium ion in Figure 2.

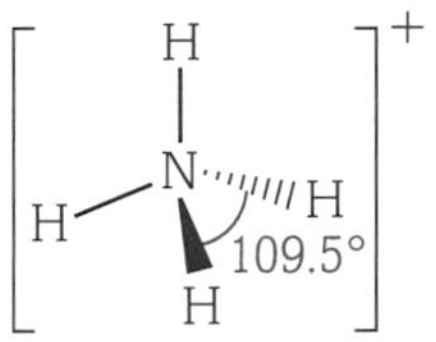

Figure 2 Shape of NH_4^+ ion.

- There are four electron pairs around the central N atom.
- The shape will be tetrahedral.
- The charge is distributed across the whole molecule – shown by square brackets and a + sign.

Questions

1. For each of the following molecules, draw a dot-and-cross diagram and predict the shape and bond angles.
   (a) H_2S (b) $AlCl_3$ (c) SiF_4 (d) PH_3
2. For each of the following ions, draw a dot-and-cross diagram and predict the shape and bond angles.
   (a) NH_4^+ (b) H_3O^+ (c) NH_2^-
3. Each of the following molecules has at least one multiple bond. For each molecule, draw a dot-and cross diagram and predict the shape and bond angles.
   (a) HCN (b) C_2H_4 (c) SO_2 (d) SO_3

2.2 **10**

Electronegativity and bond polarity

By the end of this topic, you should be able to demonstrate and apply your knowledge and understanding of:

* electronegativity as the ability of an atom to attract the bonding electrons in a covalent bond
* interpretation of Pauling electronegativity values
* explanation of:
  (i) a polar bond and permanent dipole within molecules containing covalently bonded atoms with different electronegativities
  (ii) a polar molecule and overall dipole in terms of permanent dipole(s) and molecular shape

What is electronegativity and how is it measured?

In 1932, the US chemist Linus Pauling invented the *Pauling* scale to measure the **electronegativity** of an atom. Electronegativity measures the attraction of a bonded atom for the pair of electrons in a covalent bond. Electronegativity increases towards the top right of the periodic table, with fluorine having the most electronegative atoms.

The periodic table in Figure 1 shows the electronegativities of all the elements. The electronegativity depends on an element's position in the periodic table. The general trend is that electronegativity increases towards fluorine, from all directions (that is, left to right across each period and up each group).

Polar and non-polar bonds

A covalent bond is a shared pair of electrons. In a molecule of hydrogen, H_2, the two bonding atoms are *identical*. Each hydrogen atom has an equal share of the pair of electrons in the bond, resulting in a perfect 100 per cent covalent bond. The nucleus of each bonded atom is equally attracted to the bonding electron pair. We say that a H–H bond is *non-polar* – the electrons in the bond are evenly distributed between the atoms that make up the bond.

If the bonding atoms are *different*, one of the atoms is likely to attract the bonding electrons more. The bonding atom with a greater attraction for the electron pair is said to be more electronegative than the other atom.

Electronegativity increases →

Period (electronegativity increases ↑)

Period																		
1	H 2.20																	He
2	Li 0.98	Be 1.57											B 2.04	C 2.55	N 3.04	O 3.44	F 3.98	Ne
3	Na 0.93	Mg 1.31											Al 1.61	Si 1.90	P 2.19	S 2.58	Cl 3.16	Ar
4	K 0.82	Ca 1.00	Sc 1.36	Ti 1.54	V 1.63	Cr 1.66	Mn 1.55	Fe 1.83	Co 1.88	Ni 1.91	Cu 1.90	Zn 1.65	Ga 1.81	Ge 2.01	As 2.18	Se 2.55	Br 2.96	Kr 3.00
5	Rb 0.82	Sr 0.95	Y 1.22	Zr 1.33	Nb 1.6	Mo 2.16	Tc 1.9	Ru 2.2	Rh 2.28	Pd 2.20	Ag 1.93	Cd 1.69	In 1.78	Sn 1.96	Sb 2.05	Te 2.1	I 2.66	Xe 2.6
6	Cs 0.79	Ba 0.89	*	Hf 1.3	Ta 1.5	W 2.36	Re 1.9	Os 2.2	Ir 2.20	Pt 2.28	Au 2.54	Hg 2.00	Tl 1.62	Pb 2.33	Bi 2.02	Po 2.0	At 2.2	Rn
7	Fr 0.7	Ra 0.9	**	Rf	Db	Sg	Bh	Hs	Mt	Ds	Rg	Uub	Uut	Uuq	Uup	Uuh	Uus	Uuo

Lanthanides *	La 1.1	Ce 1.12	Pr 1.13	Nd 1.14	Pm 1.13	Sm 1.17	Eu 1.2	Gd 1.2	Tb 1.1	Dy 1.22	Ho 1.23	Er 1.24	Tm 1.25	Yb 1.1	Lu 1.27
Actinides **	Ac 1.1	Th 1.3	Pa 1.5	U 1.38	Np 1.36	Pu 1.28	Am 1.13	Cm 1.28	Bk 1.3	Cf 1.3	Es 1.3	Fm 1.3	Md 1.3	No 1.3	Lr 1.3

Figure 1 The variation of Pauling electronegativity values across the periodic table. Electronegativity increases towards F. The most electronegative elements are shown in pink.

In a molecule of hydrogen chloride, HCl, the two bonding atoms are different:

- The Cl atom is more electronegative than the H atom.
- The Cl atom has a greater attraction for the bonding pair of electrons than the H atom.
- The bonding electrons are held closer to the Cl atom than to the H atom.

There is now a small charge difference across the H–Cl bond (see Figure 2). This charge difference is always present. It is called a *permanent dipole* and is shown by:

- a small positive charge on the H atom, written $\delta+$
- a small negative charge on the Cl atom, written $\delta-$.

We now have a *polar covalent bond*: $H^{\delta+}—Cl^{\delta-}$

LEARNING TIP

A permanent dipole is a small charge difference across a bond that results from a difference in the electronegativities of the bonded atoms. A polar covalent bond has a permanent dipole.

δ+ ⟶ δ−
H ˣ• Cl

The bonded electron pair is attracted towards the Cl atom.

Figure 2 HCl contains a permanent dipole. The bond between H and Cl is called a polar bond.

Polar and non-polar molecules

Molecules such as HCl have polar bonds and are polar molecules. A hydrogen chloride molecule is non-symmetrical. There is a charge difference across the H–Cl bond and across the whole HCl molecule.

For molecules that are symmetrical, the dipoles of any bonds within the molecule may cancel out. Figure 3 shows tetrachloromethane, CCl_4. This is a non-polar molecule but it contains polar C–Cl bonds. How is this possible?

- Each C–Cl bond is polar.
- However, a CCl_4 molecule is symmetrical.
- Therefore, the dipoles act in different directions and cancel each other out.

Figure 3 CCl_4. The molecule contains polar C–Cl bonds. It is non-polar because it is symmetrical; the pull of any one electronegative Cl is cancelled out by the pull of the others.

Water and carbon dioxide

Oxygen is one of the most electronegative elements. When it is bonded to other atoms, these bonds will be polar. But this does not mean that all molecules containing oxygen will be polar.

Carbon dioxide and water are shown in Figure 4. All the bonds containing oxygen are polar. In CO_2 there is no overall dipole because the molecule is symmetrical. In H_2O, there is an overall dipole as the molecule is not symmetrical; it is non-linear or bent.

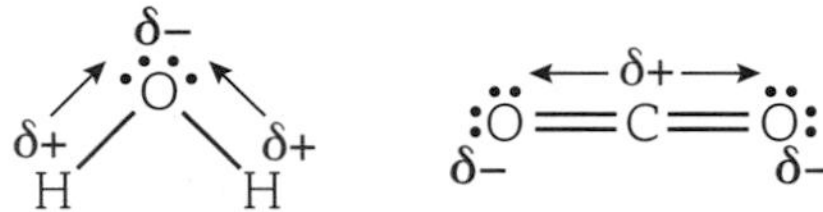

Figure 4 H_2O and CO_2 both contain polar bonds. Only H_2O has an overall dipole as it is not a symmetrical molecule.

DID YOU KNOW?

Bonds between atoms with different electronegativity values result in polar bonds. If the difference between the atoms is only small, they share the electrons unequally and form a polar covalent bond. The greater the difference in electronegativity values, the greater the permanent dipole. If the difference is very great, the most electronegative element may actually be able to effectively take the electron. They then form an ionic bond.

Between the extremes of 100 per cent ionic and 100 per cent covalent bonding, we have a whole range of intermediate bonds with both ionic and covalent contributions.

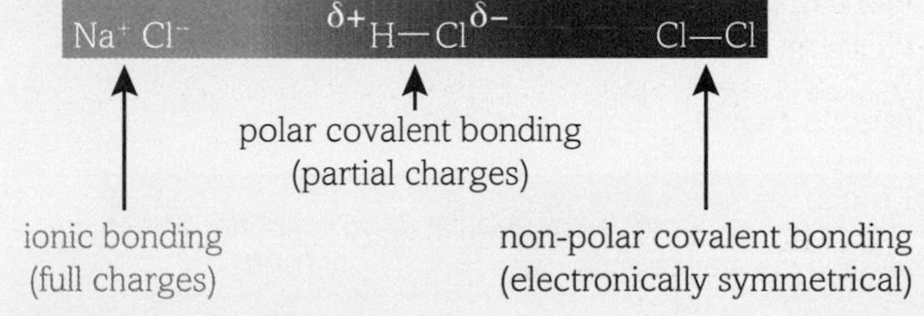

Figure 5 Range of bonding from covalent to ionic.

Questions

1. Draw diagrams of the bonds present in the following molecules. Label any polar covalent bonds and dipoles present on your diagrams.
   (a) Br_2 (b) H_2O (c) O_2
   (d) HBr (e) NH_3 (f) CF_4
2. (a) Predict the shape of the molecules for BF_3 and PF_3.
   (b) Explain why BF_3 is non-polar, whereas PF_3 is polar.
3. Explain why H_2O is polar, whereas CO_2 is non-polar.
4. (a) Predict the shape of the molecule SF_6.
   (b) Explain whether the molecule will be polar or non-polar.

2.2

11 Intermolecular forces

By the end of this topic, you should be able to demonstrate and apply your knowledge and understanding of:

* intermolecular forces based on permanent dipole–dipole interactions and induced dipole–dipole interactions

What are intermolecular forces?

As we have discussed, atoms in compounds are held together by bonds – ionic and covalent bonds. Individual compounds are attracted to one another by forces called intermolecular forces; 'intermolecular' means 'between molecules'.

Intermolecular forces occur due to constant random movements of the electrons within the shells of the atoms in molecules. They do not involve any sharing or transfer of electrons.

There are two main types of intermolecular force. They are:

- hydrogen bonding
- van der Waals' forces.

These forces are much weaker than chemical bonds. Their relative strengths are shown in Table 1.

LEARNING TIP

An intermolecular force is an attractive force between neighbouring molecules.

Bond type	Relative strength
Ionic and covalent bonds	1000
Hydrogen bonds	50
Permanent dipole–dipole forces (any intermolecular force, excluding hydrogen bonding, where a permanent dipole is involved)	10
London (dispersion) forces	1

Table 1 Comparison of the relative strengths of intermolecular forces, compared with ionic and covalent bonding.

Van der Waals' forces

DID YOU KNOW?

The names used in intermolecular bonding can be confusing. The term 'van der Waals' forces' is used to describe several different types of intermolecular bonding:

- dipole–dipole interactions (which include permanent dipole–induced dipole interactions and permanent dipole–permanent dipole interactions)
- London (dispersion) forces (which are also known as induced dipole–dipole interactions).

Permanent dipole–induced dipole interactions

Some molecules have a permanent dipole, due to polar bonds being present. If a molecule has a permanent dipole it has a slightly negative ($\delta-$) and a slightly positive ($\delta+$) end. When it is near to other molecules that are non-polar, it is able to cause electrons in the shells in the nearby molecule to shift slightly (by being repelled by the $\delta-$ end or attracted to the $\delta+$ side). This causes the non-polar molecule to become slightly polar and then an attraction occurs. The molecule with a permanent dipole has *induced* a dipole in the other molecule.

This is known as a *permanent dipole–induced dipole interaction.* It is the first of two interactions called permanent dipole–dipole interactions that are discussed in this topic. It is illustrated in Figure 1.

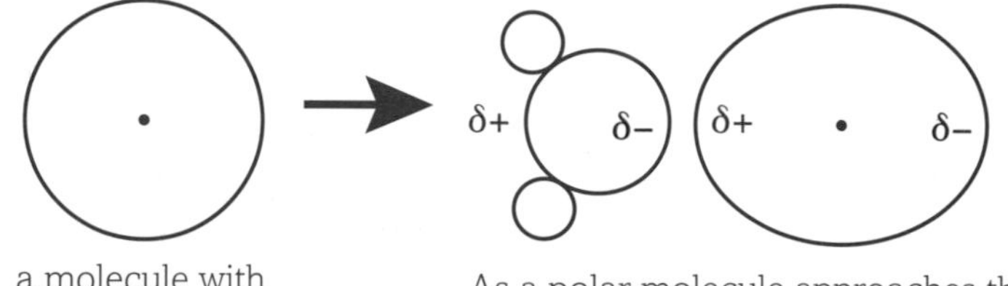

a molecule with no dipole

As a polar molecule approaches the molecule, it induces a dipole. The two molecules are then attracted.

Figure 1 A permanent dipole–induced dipole interaction.

Permanent dipole–permanent dipole interactions

Molecules with permanent dipoles will also be attracted to other molecules with permanent dipoles. They act a bit like little bar magnets, with opposite ends being attracted to one another. This is known as a *permanent dipole–permanent dipole interaction.* It is the second type of interaction called a permanent dipole–dipole interaction. Molecules of HCl are attracted together by such forces, as shown in Figure 2 below.

$$\overset{\delta+}{\mathrm{H}} \longrightarrow \overset{\delta-}{\mathrm{Cl}} \cdots\cdots \overset{\delta+}{\mathrm{H}} \longrightarrow \overset{\delta-}{\mathrm{Cl}}$$

Permanent dipole–dipole interaction between a δ+ atom of one molecule and a δ− atom of another molecule.

Figure 2 A permanent dipole–permanent dipole interaction.

These two types of intermolecular force are usually referred to as permanent dipole–dipole forces, as they both involve at least one permanent dipole.

DID YOU KNOW?

The microscopic hairs on a gecko's feet create millions of van der Waals' forces. It is these forces that allow geckos to stick to virtually any surface.

London (dispersion) forces

Not all molecules contain dipoles; some are non-polar. Non-polar molecules may still be attracted to one another, although such intermolecular forces are very weak:

- London (dispersion) forces are caused by the constant random movement of electrons in atoms' shells. This movement unbalances the distribution of charge within the electron shells – rather like the electron density in the shells 'wobbling' from side to side.
- At any moment, there will be an *instantaneous* dipole across the molecule.
- The instantaneous dipole *induces* a dipole in neighbouring molecules, which in turn induce further dipoles on their neighbouring molecules.
- The small induced dipoles attract one another, causing weak intermolecular forces known as London (dispersion) forces or instantaneous dipole–induced dipole forces.

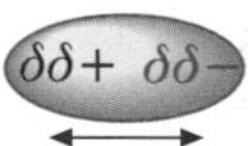

At any moment, oscillations produce an instantaneous dipole

Instantaneous dipole induces dipoles in neighbouring molecules.

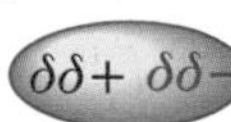

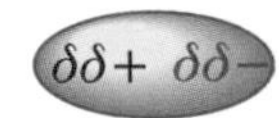

Induced dipoles attract one another.

Figure 3 London (dispersion) forces arise due to natural oscillations in electron shells.

The size of London (dispersion) forces increases with increasing numbers of electrons. The greater the number of electrons, the larger the induced dipoles and the greater the attractive forces between molecules.

LEARNING TIP

A permanent dipole–dipole interaction is a weak attractive force between permanent dipoles and permanent dipoles or induced dipoles in neighbouring polar molecules.
London (dispersion) forces are attractive forces between induced dipoles in neighbouring molecules.

The effect of London forces on boiling points

When a substance is boiled, heat energy is being used to overcome intermolecular forces between molecules. The bonding within molecules stays intact; the molecules just become separated and are no longer attracted to one another. In the case of non-polar molecules, there are only London forces to be overcome. Because these forces are so weak, such substances have very low boiling points. The boiling points of the noble gases are shown in Table 2.

Noble gas	Boiling point/°C	Number of electrons
He	−269	2
Ne	−246	10
Ar	−186	18
Kr	−153	36
Xe	−108	54
Rn	−62	86

Table 2 The boiling points of the noble gases illustrate that London forces increase in size as more electrons are present.

As the number of electrons increases, so does the strength of the van der Waals' forces. In this case they would be London (dispersion) forces, as no dipoles are present in any molecules before they interact. The boiling point increases as we move down the noble gas group. If there were no van der Waals' forces (in this case, London forces only), it would be impossible to liquefy the noble gases.

Questions

1. Explain how the different types of van der Waals' forces arise.
2. The boiling points of the group 17 elements are shown below. Each element exists as diatomic molecules:

   F_2, −188 °C; Cl_2, −35 °C; Br_2, 59 °C; I_2, 184 °C

   Explain this trend, in terms of intermolecular forces.

2.2 (12) Hydrogen bonding

By the end of this topic, you should be able to demonstrate and apply your knowledge and understanding of:

* hydrogen bonding as intermolecular bonding between molecules containing N, O or F and the H atom of –NH, –OH or HF
* explanation of anomalous properties of H_2O resulting from hydrogen bonding, e.g.:
  (i) the density of ice compared with water
  (ii) its relatively high melting and boiling points

Hydrogen bonding

Oxygen, nitrogen and fluorine are all highly electronegative elements. Molecules containing O–H, N–H and F–H bonds are polar with permanent dipoles. These dipoles are particularly strong. The permanent dipole–dipole interaction between molecules containing O–H, N–H and F–H bonds is a type of intermolecular bond and is given a special name: a *hydrogen bond*.

LEARNING TIP

A hydrogen bond is a strong permanent dipole–permanent dipole attraction between:

- an electron-deficient hydrogen atom (O–$H^{\delta+}$, N–$H^{\delta+}$ or F–$H^{\delta+}$) on one molecule; and
- a lone pair of electrons on a highly electronegative atom (O, N or F) on a different molecule.

In a hydrogen bond the electron-deficient $H^{\delta+}$ on one molecule attracts a lone pair of electrons on a $O^{\delta-}$, $N^{\delta-}$ or $F^{\delta-}$ bonded within a different molecule. This is illustrated in Figure 1 and Figure 2.

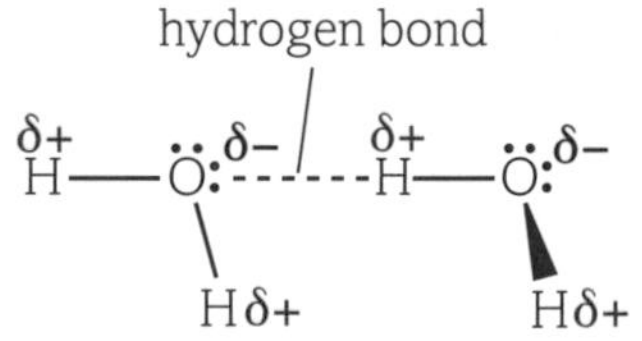

hydrogen bond

δ+ H—Ö:δ– - - - - - δ+ H—Ö:δ–

Hδ+ Hδ+

Figure 1 Hydrogen bonding between two molecules of water. Water is able to form hydrogen bonds because the O has lone pairs on it. These can form a hydrogen bond with the H on another water molecule, as it is electron-deficient.

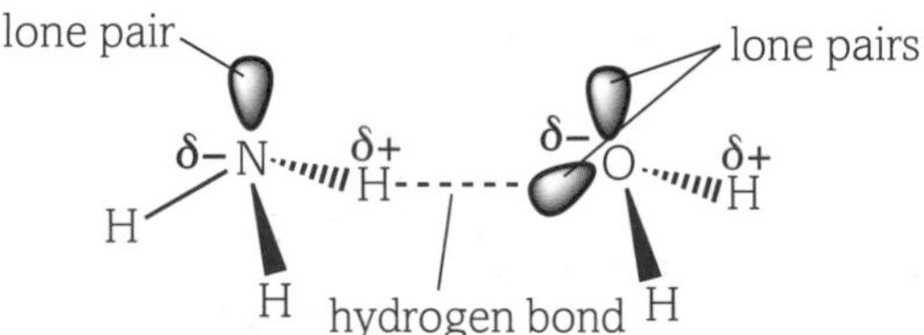

Figure 2 Hydrogen bonding between a molecule of water and a molecule of ammonia. The lone pairs on the electronegative O atom are attracting the H on the molecule of ammonia. This H atom is very electron-deficient because it is attached to N, which is highly electronegative and has lone pairs on it.

Hydrogen bonding is found in lots of different situations, including in proteins and between base pairs within DNA strands.

LEARNING TIP

It is essential that a hydrogen bond is shown as a dotted line starting at $H^{\delta+}$ and going to a lone pair on $O^{\delta-}$, $N^{\delta-}$ or $F^{\delta-}$. To spot where hydrogen bonding will occur, look for molecules with: O–H; N–H; or F–H. Common examples are H_2O, NH_3, CH_3OH and C_2H_5OH.

The effect of hydrogen bonding on the properties of water

A hydrogen bond in water has only about 5 per cent of the strength of the O–H covalent bonds. However, hydrogen bonding is strong enough to have significant effects on physical properties. This results in some unexpected properties for water.

Ice is less dense than water

In almost all materials, the solid is denser than the liquid. Water is the exception, with ice (solid H_2O) being less dense than water (liquid H_2O). Why is this?

- When ice forms, water molecules arrange themselves into an orderly pattern and hydrogen bonds form between the molecules (these will occur in the liquid phase but will not occur as often, as molecules can move past each other and hence overcome these bonds).
- Ice has an open lattice with hydrogen bonds holding the water molecules apart (see Figure 3).
- When ice melts, the rigid hydrogen bonds collapse, allowing the H_2O molecules to move closer together.
- So, ice is less dense than water. This is why ice floats on water.

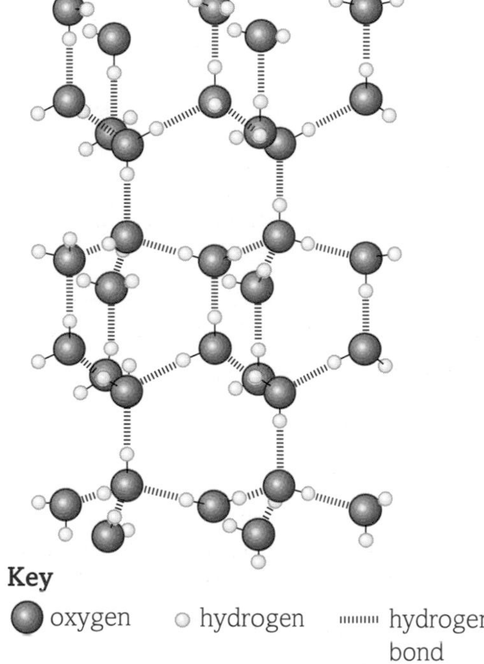

Figure 3 The ice lattice is an open network of H_2O molecules. Notice how the hydrogen bonds hold the water molecules apart. It is because the hydrogen bonds make ice take up more space that it is less dense than water. Notice also the shape of the lattice. Each oxygen atom has four bonds: two covalent bonds and two hydrogen bonds. The hydrogen bonds are slightly longer. The open structure is made up of rings of six oxygen atoms. Snowflakes are based on six-sided shapes. There is an almost infinite number of possible six-sided arrangements, so that every snowflake has a unique shape.

Water has higher than expected melting and boiling points

There are relatively strong hydrogen bonds between H_2O molecules.

- The hydrogen bonds are much stronger than other intermolecular forces.
- The extra strength of these forces has to be overcome in order to melt or boil H_2O. This results in H_2O having higher melting and boiling points than would be expected if hydrogen bonds were not present.
- Other hydrides of group 16 elements (for example SiH_2) show the same structure as water but do not exhibit hydrogen bonding. The differences in their melting and boiling points illustrate the effect hydrogen bonding has. This is shown in Figure 4.

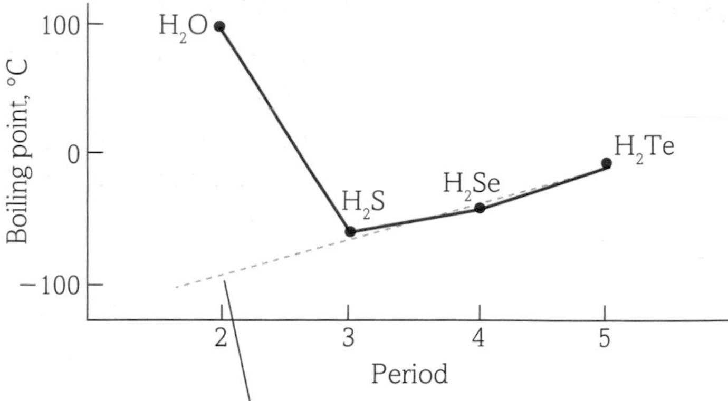

Figure 4 Water has an unusually high boiling point compared with other group 16 hydrides. This is because water is able to form hydrogen bonds, which are much stronger than other intermolecular forces.

Other properties of water

The extra intermolecular bonding from hydrogen bonds also explains the relatively high surface tension and viscosity of water. When small insects walk on water, they are walking across a raft of hydrogen bonds – see Figure 5.

Figure 5 The high surface tension of water caused by the presence of hydrogen bonding keeps this pond skater on the surface.

Questions

1. Which of the following molecules have hydrogen bonding?
   H_2S; CH_4; HF; CH_3OH; PH_3; NO_2; CH_3NH_2
2. Draw diagrams showing hydrogen bonding between:
   (a) two molecules of water
   (b) two molecules of ammonia
   (c) two molecules of ethanol, C_2H_5OH
   (d) one molecule of water and one molecule of ethanol
   (e) two molecules of hydrogen fluoride, HF.

BORING BORON?

This article is taken from the *New Scientist,* which is a magazine and an online news site. Read the article and then answer the questions below.

FIRST BORON BUCKYBALLS ROLL OUT OF THE LAB

Score one for boron. For the first time, a version of the famous football-shaped buckyball has been created from boron.

Discovered in 1985, buckyballs are made from 60 carbon atoms linked together to form hollow spheres. The molecular cages are very stable and can withstand high temperatures and pressures, so researchers have suggested they might store hydrogen at high densities, perhaps making it a viable fuel source. At normal pressures, too much of the lightweight gas can escape from ordinary canisters, and compressing it requires bulky storage tanks.

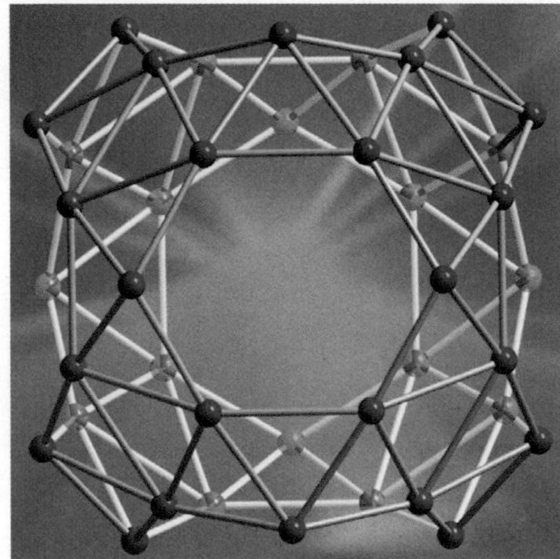

Figure 1 Meet borospherene

Meet borospherene

Boron sits next to carbon in the periodic table, so a boron ball may also display useful properties. But it wasn't clear whether boron could form such structures.

Now Lai-Sheng Wang at Brown University in Providence, Rhode Island, and his colleagues have made a cage-like molecule with 40 boron atoms by vaporising a chunk of boron with a laser then freezing it with helium, creating boron clusters. The team analysed the energy spectra of these clusters and compared them with computer models of 10,000 possible arrangements of boron atoms. The matching configuration revealed they had created the boron ball.

Unlike carbon buckyballs, in which the faces are made of hexagons and pentagons, the boron buckyball is made from triangles, hexagons and heptagons. As a result, it is less spherical but still an enclosed structure. Wang has dubbed the molecule "borospherene". The team is now hunting for a boron analogue of graphene – a strong sheet of carbon just one atom thick that is often touted as a "wonder" material because of its unique electrical properties.

Mark Fox at Durham University, UK likes the name – and is excited at the prospect of finding a boron version of graphene. Buckyballs led to the discovery of graphene, he says, and history may repeat itself with boron.

- New Scientist: http://www.newscientist.com/article/dn25891-first-boron-buckyballs-roll-out-of-the-lab.html; journal reference: *Nature Chemistry,* DOI: 10.1038/nchem.1999

Let us start by considering the nature of the writing in the article.

1. **This article has been written with a broad audience in mind. What aspects of the article suggest that the piece was not written for scientists?**

Questions 2–4 are about the chemistry in, or connected to, this article. Some of the questions will build on ideas that you covered at GCSE and others will be covered later in the course. The timeline below tells you about other chapters in this book that are relevant to this question: Do not worry if you are not ready to give answers to these questions at this stage. You may like to read through the activity at this initial stage and return to the questions once you have covered other topics later in the book.

2. a. Name the type of bond present in borospherene.

b. The melting point of a different form of boron is 2300 °C. What does this suggest about the structure formed in this case?

3. An important reducing agent in chemistry is the compound sodium borohydride. Describe the bonding in $NaBH_4$ and use your knowledge of electron pair repulsion theory to predict and explain the shape of the BH_4^- ion.

4. Assign an oxidation number to boron in each of the following compounds:

a. $Na_2B_4O_7$

b. $NaBH_4$

c. BF_3

Activity

Boron is currently being considered for its potential medical uses in a number of areas. Carry out a web search and find out what you can about its potential uses. Aim to deliver a presentation to your peers (no more than five slides) summarising as many aspects of research as you can. Be prepared to defend your presentation in a 5 min Q&A session.

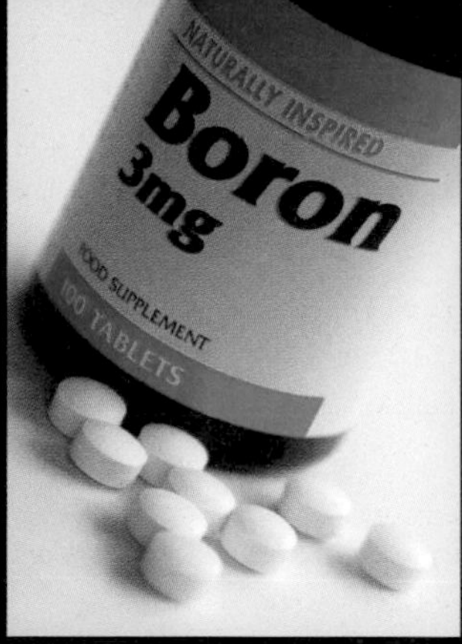

Figure 2 Would you take a boron supplement?

Be constructively critical about the nature of the source material on the web. What constitutes a reliable resource?

3.2 4.1 4.2

2.2 Practice questions

1. Which of the following statements about the S^{2-} ion is true?
   A. It has an electronic configuration of $1s^22s^22p^63s^23p^4$.
   B. It has the same electronic configuration as a neon atom.
   C. It has the same electronic configuration as an argon atom.
   D. It has the same outer shell electron structure as an oxygen atom. [1]

2. Which of the following letter options gives the best description of the compound potassium chloride (KCl)?

	Melting point	**Solubility in water**	**Conductivity of molten compound**
A	below 200°C	very soluble	does not conduct
B	below 200°C	insoluble	is a poor conductor
C	below 200°C	very soluble	is a good electrical conductor
D	above 200°C	very soluble	is a good conductor

[1]

3. How many electrons are there in the outer shell of nitrogen in the ammonium (NH_4^+) ion?
   A. One less than N in an NH_3 molecule.
   B. One more than N in an NH_3 molecule.
   C. The same as N in an NH_3 molecule.
   D. The same as N in an NH_2^- ion. [1]

4. The melting point of SiO_2 is over 2000 °C higher than that for SO_2. What is the best explanation for this?
   A. There is a greater difference in electronegativity between sulfur and oxygen than between silicon and oxygen.
   B. Silicon dioxide forms a giant covalent lattice and sulfur dioxide forms a simple molecular structure.
   C. There is a higher degree of ionic character in silicon dioxide than sulfur dioxide.
   D. Sulfur has a higher relative atomic mass than silicon. [1]

[Total: 4]

5. Linus Pauling was a Nobel Prize winning chemist who devised a scale of electronegativity.

   Some Pauling electronegativity values are shown in the table.

Element	**Electronegativity**
B	2.0
Br	2.8
N	3.0
F	4.0

(a) What is meant by the term *electronegativity*? [2]

(b) Show, using δ^+ and δ^- symbols, the permanent diploes on each of the following bonds.
N–F N–Br [1]

(c) Boron trifluoride, BF_3, ammonia, NH_3, and sulfur hexafluoride, SF_6, are all covalent compounds. The shapes of their molecules are different.
   (i) State the shape of a molecule of SF_6. [1]
   (ii) Using outer electron shells only, draw dot-and-cross diagrams for molecules of BF_3 and NH_3.
   Use your diagrams to explain why a molecule of BF_3 has bond angles of 120° and NH_3 has bond angles of 107°. [5]
   (iii) Molecules of BF_3 contain polar bonds, but the molecules are non-polar. Suggest an explanation for this difference. [2]

[Total: 11]

6. A graph showing the boiling points of group 17 halides is shown below.

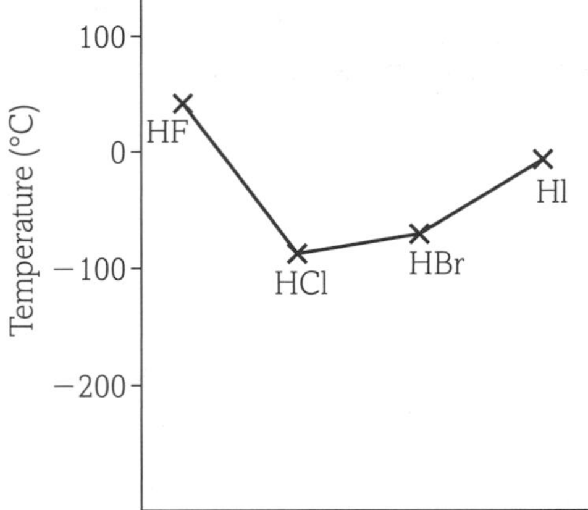

Use your knowledge of the different types of intermolecular force to explain this pattern. [5]

[Total: 5]

7. The following question concerns the s,p,d electronic structures of elements and their ions from hydrogen $Z = 1$ to krypton $Z = 36$.
   (a) Give the symbol of an atom with a single electron in each p-orbital. [1]
   (b) Give the formula of an ion with a single electron in each d-orbital. [1]
   (c) Give the formula of the halide ion with the same electronic configuration as Ar. [1]
   (d) Give the formula of the period 4 ion that forms an ionic chloride of the type MCl_2. [1]
   (e) Give the full s, p, d notation of the Se^{2-} ion. [1]

[Total: 5]

8. Draw dot and cross diagrams for each of the following:
   (a) H_2S [2]
   (b) K_2S [2]
   (c) NH_4Cl [2]
   (d) SF_6. [2]

   For each example indicate the type of bonding present, and where appropriate, any dative covalent bond. If the compound is molecular indicate and explain any bond angles.

[Total: 8]

9. Use your knowledge of bonding to explain the following *apparent* anomalies:
   (a) Carbon dioxide CO_2 is a gas at room temperature whereas SiO_2 is a solid with a melting point of 1610 °C but both form covalent bonds. [4]
   (b) Both bromine Br_2 and water H_2O are liquids at room temperature, but bromine with a molecular mass of 160 g mol^{-1} boils at 58 °C whereas water with a molecular mass of 18 g mol^{-1} boils at 100 °C. [4]

[Total: 8]

MODULE 3 The periodic table and energy

CHAPTER 3.1

THE PERIODIC TABLE

Introduction

The periodic table has not always looked as it does now. It has been developed over many years and has been influenced by the work of many scientists. The periodic table is a hugely important tool for chemists – it contains information about every element.

The organisation of the periodic table into columns (groups) and rows (periods) helps chemists to predict the physical properties (like melting points and bonding types) of elements – it also informs us about how the elements behave chemically.

In this chapter you will look in depth at the metallic group 2 elements (the alkaline earth metals), and the non-metallic group 17 elements (the halogens). These elements are used widely in the world around us, from building materials to swimming-pool water. You will learn how the elements undergo reactions and how this changes on moving down each group.

This chapter will also look at qualitative chemical tests that can be used to identify positive ions and negative ions in unknown substances.

All the maths you need

To unlock the puzzles of this chapter you need the following maths:

- The units of measurement (*e.g. °C*)
- Using an appropriate range of standard figures (*e.g. quoting ionisation energies*)
- Plotting and interpreting graphs (*e.g. periodic trends*)
- Visualising and representing 2D and 3D shapes (*e.g. describing the changes in giant structures across a period*)

What have I studied before?

- How the periodic table is organised into groups and periods
- The information given with each element symbol
- Where metals and non-metals are found on the periodic table
- How the reactivity and the properties of elements can change moving down a group
- Chemical tests, such as using limewater for carbon dioxide and a lit splint for hydrogen

What will I study later?

- The properties and behaviour of the transition elements in the d-block of the periodic table (AL)
- Further tests for ions (AL)
- The role of halogens in reactions with organic molecules (AL)

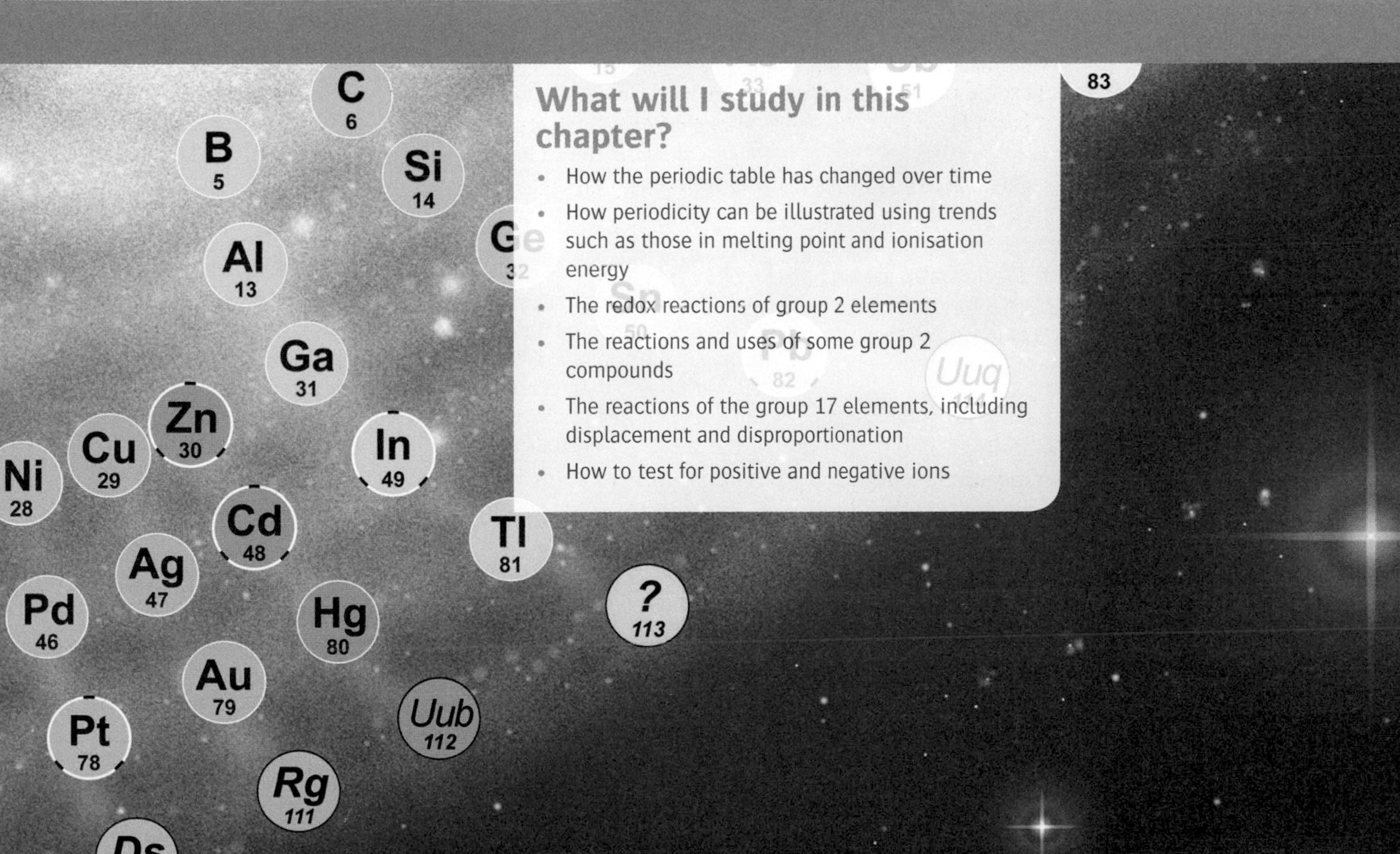

What will I study in this chapter?

- How the periodic table has changed over time
- How periodicity can be illustrated using trends such as those in melting point and ionisation energy
- The redox reactions of group 2 elements
- The reactions and uses of some group 2 compounds
- The reactions of the group 17 elements, including displacement and disproportionation
- How to test for positive and negative ions

3.1 ① The development of the periodic table

By the end of this topic, you should be able to demonstrate and apply your knowledge and understanding of:

- the periodic table as the arrangement of elements:
  - (i) by increasing atomic (proton) number
  - (ii) in periods showing repeating trends in physical and chemical properties (periodicity)
  - (iii) in groups having similar chemical properties

Before the periodic table

People interested in the world around them often considered what the world was made up of. Ancient philosophers, such as Aristotle (384–322 BCE), believed the world was made up of four elements – earth, water, air and fire. It could be argued that these four elements are similar to what we now call the states of matter – solid, liquid and gas, with fire representing more unusual phenomena such as 'plasma'.

As scientists and ordinary people began to extract and use materials from the world around them, a more in-depth understanding of substances evolved. Some elements, such as gold, were easy to find; others had to be extracted from their ores. This deepened the knowledge about the way substances behaved.

Over time, understanding about the world around us has led to the development of what we now call the periodic table. This lists the known elements and groups them according to their behaviour and electronic structure. It has not always looked as it does now – there have been changes along the way, with mistakes and disagreements!

Antoine-Laurent de Lavoisier

In 1789, the French chemist Antoine-Laurent de Lavoisier produced what is now considered to be the first 'modern' chemical textbook. In this he compiled the first extensive list of elements, which he described as 'substances that could not be broken down further'. He also devised a theory about the formation of compounds from elements.

Lavoisier's list of elements included oxygen, nitrogen, hydrogen, phosphorus, mercury, zinc and sulfur. It also distinguished between metals and non-metals. Unfortunately it also included some compounds and mixtures, along with terms such as 'light' and 'caloric' (heat), which he believed to be material substances.

Jöns Jakob Berzelius

In 1828, the Swedish chemist Jöns Jakob Berzelius published a table of atomic weights, and determined the composition by mass of many compounds. Berzelius was also responsible for introducing letter-based symbols for elements – previously, signs had been used.

Johann Wolfgang Döbereiner

Döbereiner noticed that certain groups of three elements (he called them triads) ordered by atomic weight would have a middle element with a weight and properties (such as density) that were roughly an average of the other two elements. Some of the triads he considered were:

- calcium, strontium and barium
- chlorine, bromine and iodine
- lithium, sodium and potassium.

John Newlands

The English chemist John Newlands devised a periodic table that had the elements arranged in order of their relative atomic weights (now called 'relative atomic masses'). In 1865 he suggested that, rather than being in triads, elements show similar properties to the element eight places after it in the table. He named this the 'law of octaves'. Newlands' version of the periodic table is shown in Figure 1.

DID YOU KNOW?

Despite making a major contribution to the development of the periodic table, during the French Revolution Lavoisier was branded a traitor and was beheaded.

DID YOU KNOW?

Even though you still learn about the contribution that Newlands made to the development of the periodic table, he was ridiculed by many people in the scientific community, primarily for the use of the musical term 'octave' to describe the behaviour of elements. It was suggested unkindly that Newlands would have found similarities just by ordering the elements alphabetically.

Dmitri Mendeleev

The modern periodic table is based on the one published by Mendeleev in 1869 (Figure 2). His table showed the elements ordered by atomic weights, similar to Newlands', and was ordered periodically.

- Elements with similar properties were arranged in vertical columns.
- Gaps were left where no element fitted the repeating patterns, and the properties of the missing elements were predicted. These missing elements have since been found and matched the predictions Mendeleev made – notably gallium, Ga; germanium, Ge; and scandium, Sc.

Newlands' periodic table contains seven octaves. Many elements were not known about at the time. Newlands left no gaps for undiscovered elements, and when they were found this seemed to undermine his suggestions. Notice that none of the noble gases are included – they had not been discovered.

The coloured elements are some of those that showed similarities.

Notice how Newlands has ordered periods down the table and groups across the table – opposite to the modern periodic table.

H 1	F 8	Cl 15	Co & Ni 22	Br 29	Pd 36	I 42	Pt & Ir 50
Li 2	Na 9	K 16	Cu 23	Rb 30	Ag 37	Cs 44	Tl 53
Gl 3	Mg 10	Ca 17	Zn 25	Sr 31	Cd 38	Ba & V 45	Pb 54
Bo 4	Al 11	Cr 18	Y 24	Ce & La 33	U 40	Ta 46	Th 56
C 5	Si 12	Ti 19	In 26	Zr 32	Sn 39	W 47	Hg 52
N 6	P 13	Mn 20	As 27	Di & Mo 34	Sb 41	Nb 48	Bi 55
O 7	S 14	Fe 21	Se 28	Ro & Ru 35	Te 43	Au 49	Os 51

The numbers with each symbol are the atomic weights known at the time. The elements have been placed in an order that means that these numbers increase sequentially, although Newlands did change the position of some so that elements with similar properties were in the same row.

Figure 1 Newlands' arrangement of elements in octaves.

- The order of elements was rearranged where their properties did not fit. For example, tellurium had a higher atomic weight than iodine, but Mendeleev reversed them to make the properties fit with the rest of the table.
- The arrangement of the elements, or groups of elements, in order of their atomic weights corresponded to their so-called valencies, as well as to their distinctive chemical properties (to some extent). This is apparent in series such as Li, Be, B, C, N, O and F.

			K = 39	Rb = 85	Cs = 133	—	—
			Ca = 40	Sr = 87	Ba = 137	—	—
			—	?Yt = 88?	?Di = 138?	Er = 178?	—
			Ti = 48?	Zr = 90	Ce = 140?	?La = 180?	Th = 231
			V = 51	Nb = 94	—	Ta = 182	—
			Cr = 52	Mo = 96	—	W = 184	U = 240
			Mn = 55	—	—	—	—
			Fe = 56	Ru = 104	—	Os = 195?	—
			Co = 59	Rh = 104	—	Ir = 197	—
Typische Elemente			Ni = 59	Pd = 106	—	Pt = 198?	—
H = 1	Li = 7	Na = 23	Cu = 63	Ag = 108	—	Au = 199?	—
	Be = 9,4	Mg = 24	Zn = 65	Cd = 112	—	Hg = 200	—
	B = 11	Al = 27,3	—	In = 113	—	Tl = 204	—
	C = 12	Si = 28	—	Sn = 118	—	Pb = 207	—
	N = 14	P = 31	As = 75	Sb = 122	—	Bi = 208	—
	O = 16	S = 32	Se = 78	Te = 125?	—	—	—
	F = 19	Cl = 35,5	Br = 80	J = 127	—	—	—

Figure 2 Mendeleev's periodic table of 1869. This early form of the table showed similar chemical elements arranged horizontally. Chemical symbols and atomic weights were used.

DID YOU KNOW?

Another scientist, Julius Lothar Meyer, was carrying out work very similar to that of Mendeleev at the same time as Mendeleev. He was unaware of Mendeleev's work and he did not publish his results until a year later. If he had published earlier, perhaps it would be his name that is associated with being the 'Father of the modern periodic table'.

DID YOU KNOW?

Discovery of the noble gases

In 1868, the noble gas helium was first detected by studying light emitted from the Sun. Helium gets its name from 'Helios', the Greek word for the Sun.

In the 1890s, argon and helium were discovered on Earth as new elements, which were chemically inert.

In 1898, Sir William Ramsay suggested that argon be placed with helium in the periodic table, as part of a group 18. Argon was placed between chlorine and potassium – despite argon having a higher atomic weight than potassium. Ramsay accurately predicted the future discovery and properties of the noble gas neon.

Henry Moseley and Glenn Seaborg

In 1913, Henry Moseley determined the atomic number for all the known elements.

- Moseley modified Mendeleev's periodic law to read that the properties of the elements vary periodically with their atomic *numbers*, rather than atomic weight.
- Moseley's modified periodic law put the elements tellurium and iodine in the correct order, as it did for argon and potassium, and for cobalt and nickel.

In the middle of the twentieth century, the US scientist Glenn Seaborg discovered the transuranic elements from 94 (plutonium) to 102 (nobelium). He also remodelled the periodic table by placing the actinide series below the lanthanide series at the bottom of the table.

Element 106 has been named seaborgium (Sg) in honour of his work.

Questions

1. Look at Newlands' periodic table (Figure 1).
   (a) Identify the elements that Newlands placed out of order, according to their atomic weights.
   (b) What term is given to the sequence of these elements in the modern periodic table?
   (c) Which names in the modern periodic table correspond to the rows and columns in Newlands' periodic table?
2. Which elements did Mendeleev leave spaces for in his periodic table?
3. (a) Look at a modern periodic table and find three pairs of elements that have been placed out of order in terms of increasing atomic mass.
   (b) What property of an atom is the 'real' order of elements in the modern periodic table based on?

3.1

2 The modern periodic table

By the end of this topic, you should be able to demonstrate and apply your knowledge and understanding of:

- the periodic table as the arrangement of elements:
  - (i) by increasing atomic (proton) number
  - (ii) in periods showing repeating trends in physical and chemical properties (periodicity)
  - (iii) in groups having similar chemical properties
- the periodic trend in electron configurations across periods 2 and 3

Arranging the elements

In the modern periodic table, elements are arranged in order of increasing atomic numbers, starting with hydrogen, the element with the smallest atomic number.

LEARNING TIP

Remember, the atomic number is the number of protons in the nucleus.

Periods

Each horizontal row in the table is called a period. The numbers increase from left to right across each period and with each successive period following the same pattern.

The elements show trends (gradual changes) in properties across a period in the table.

- These trends are repeated across each period.
- The repeating pattern of trends is called periodicity.

Groups

Each vertical column is called a group and contains elements with similar properties. The periodic table is the chemists' way of ordering the elements to show patterns in chemical and physical properties. Figure 1 shows the key areas in the periodic table.

The elements in dark green separate metals (to the left) from non-metals (to the right).

These non-metal elements, such as silicon and germanium, are called semi-metals or metalloids. Semi-metals have properties between those of a metal and those of a non-metal.

Periodicity

Periodicity is the trend in properties that is repeated across each period. Using periodicity, predictions can be made about the likely properties of an element and its compounds.

Group

Key: atomic (proton) number / **atomic symbol** / name / relative atomic mass

Period	(1) 1	(2) 2	3	4	5	6	7	8	9	10	11	12	(3) 13	(4) 14	(5) 15	(6) 16	(7) 17	(0) 18
1	1 **H** hydrogen 1.0																	2 **He** helium 4.0
2	3 **Li** lithium 6.9	4 **Be** beryllium 9.0											5 **B** boron 10.8	6 **C** carbon 12.0	7 **N** nitrogen 14.0	8 **O** oxygen 16.0	9 **F** fluorine 19.0	10 **Ne** neon 20.2
3	11 **Na** sodium 23.0	12 **Mg** magnesium 24.3											13 **Al** aluminium 27.0	14 **Si** silicon 28.1	15 **P** phosphorus 31.0	16 **S** sulfur 32.1	17 **Cl** chlorine 35.5	18 **Ar** argon 39.9
4	19 **K** potassium 39.1	20 **Ca** calcium 40.1	21 **Sc** scandium 45.0	22 **Ti** titanium 47.9	23 **V** vanadium 50.9	24 **Cr** chromium 52.0	25 **Mn** manganese 54.9	26 **Fe** iron 55.8	27 **Co** cobalt 58.9	28 **Ni** nickel 58.7	29 **Cu** copper 63.5	30 **Zn** zinc 65.4	31 **Ga** gallium 69.7	32 **Ge** germanium 72.6	33 **As** arsenic 74.9	34 **Se** selenium 79.0	35 **Br** bromine 79.9	36 **Kr** krypton 83.8
5	37 **Rb** rubidium 85.5	38 **Sr** strontium 87.6	39 **Y** yttrium 88.9	40 **Zr** zirconium 91.2	41 **Nb** niobium 92.9	42 **Mo** molybdenum 95.9	43 **Tc** technetium (98)	44 **Ru** ruthenium 101.1	45 **Rh** rhodium 102.9	46 **Pd** palladium 106.4	47 **Ag** silver 107.9	48 **Cd** cadmium 112.4	49 **In** indium 114.8	50 **Sn** tin 118.7	51 **Sb** antimony 121.8	52 **Te** tellurium 127.6	53 **I** iodine 126.9	54 **Xe** xenon 131.3
6	55 **Cs** caesium 132.9	56 **Ba** barium 137.3	57 **La*** lanthanum 138.9	72 **Hf** hafnium 178.5	73 **Ta** tantalum 180.9	74 **W** tungsten 183.8	75 **Re** rhenium 186.2	76 **Os** osmium 190.2	77 **Ir** iridium 192.2	78 **Pt** platinum 195.1	79 **Au** gold 197.0	80 **Hg** mercury 200.6	81 **Tl** thallium 204.4	82 **Pb** lead 207.2	83 **Bi** bismuth 209.0	84 **Po** polonium (209)	85 **At** astatine (210)	86 **Rn** radon (222)
7	87 **Fr** francium (223)	88 **Ra** radium (226)	89 **Ac*** actinium (227)	104 **Rf** rutherfordium (267)	105 **Db** dubnium (268)	106 **Sg** seaborgium (269)	107 **Bh** bohrium (270)	108 **Hs** hassium (269)	109 **Mt** meitnerium (278)	110 **Ds** darmstadtium (281)	111 **Rg** roentgenium (280)	112 **Cn** copernicium (285)		114 **Fl** flerovium (289)		116 **Lv** livermorium (293)		
			•	58 **Ce** cerium 140.1	59 **Pr** praseodymium 140.9	60 **Nd** neodymium 144.2	61 **Pm** promethium (145)	62 **Sm** samarium 150.4	63 **Eu** europium 152.0	64 **Gd** gadolinium 157.3	65 **Tb** terbium 158.9	66 **Dy** dysprosium 162.5	67 **Ho** holmium 164.9	68 **Er** erbium 167.3	69 **Tm** thulium 168.9	70 **Yb** ytterbium 173.1	71 **Lu** lutetium 175.0	
			•	90 **Th** thorium 232.0	91 **Pa** protactinium 231.0	92 **U** uranium 238.0	93 **Np** neptunium (237)	94 **Pu** plutonium (244)	95 **Am** americium (243)	96 **Cm** curium (247)	97 **Bk** berkelium (247)	98 **Cf** californium (251)	99 **Es** einsteinium (252)	100 **Fm** fermium (257)	101 **Md** mendelevium (256)	102 **No** nobelium (259)	103 **Lr** lawrencium (262)	

Figure 1 The periodic table – vertical columns are called groups; horizontal rows are called periods.

Trends down a group may also affect periodic trends.

- Across each period, elements change from metals to non-metals.
- As you move down the periods, this change moves further to the right.
- For example, at the top of group 14 carbon is a non-metal, but at the bottom of the group tin and lead are metals (Figure 2).

So, trends in properties are seen vertically down the groups as well as horizontally across a period.

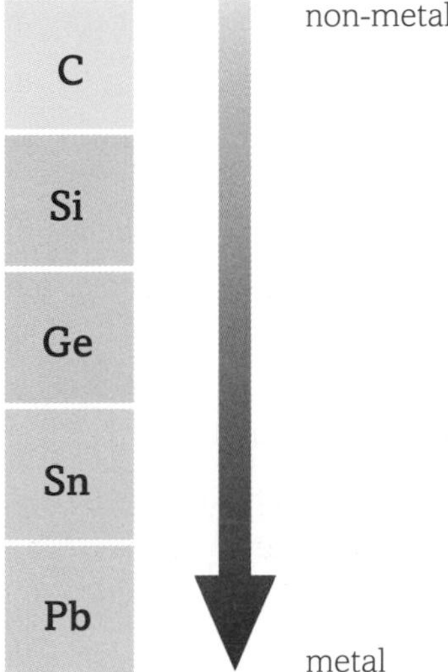

Figure 2 In group 14, the classification of elements changes from 'non-metal' to 'metal' down the group.

Variation in electron structure

Chemical reactions involve electrons in the outer shell of atoms. Any similarity in electron configuration will be reflected in the similarity of chemical reactions.

Similar elements are placed in vertical groups – this is a key principle of the periodic table.

- Elements in a group have atoms with the same number of electrons in their outer shells – this explains their similar chemical behaviour.
- This repeating pattern of similarity is caused by the underlying repeating pattern of electron configuration.

This is shown in Table 1 for elements in period 2 and period 3.

Li [He] $2s^1$	Be [He] $2s^2$	B [He] $2s^22p^1$	C [He] $2s^22p^2$	N [He] $2s^22p^3$	O [He] $2s^22p^4$	F [He] $2s^22p^5$	Ne [He] $2s^22p^6$
Na [Ne] $3s^1$	Mg [Ne] $3s^2$	Al [Ne] $3s^23p^1$	Si [Ne] $3s^23p^2$	P [Ne] $3s^23p^3$	S [Ne] $3s^23p^4$	Cl [Ne] $3s^23p^5$	Ar [Ne] $3s^23p^6$

Table 1 Patterns in electron structures.

LEARNING TIP

A shorthand way of writing electron configurations is to base them on the previous noble gas.
This means that you can concentrate on the outer-shell electrons – those that are responsible for reactions.
For example:
Na $1s^22s^22p^63s^1$ becomes [Ne] $3s^1$

All the elements in a vertical group have:

- the same number of electrons in the outer shell
- the same type of orbitals.

So, the elements in a group react in a similar way because they have similar electron configurations.

LEARNING TIP

Similar outer-shell electron configurations give rise to similar reactions. For example, atoms of group 1 elements all lose one electron to form 1+ ions that have a noble gas electron configuration.

$Li \rightarrow Li^+ + e^-$
[He] $2s^1 \rightarrow$ [He]
$Na \rightarrow Na^+ + e^-$
[Ne] $3s^1 \rightarrow$ [Ne]
$K \rightarrow K^+ + e^-$
[Ar] $4s^1 \rightarrow$ [Ar]
$Rb \rightarrow Rb^+ + e^-$
[Kr] $5s^1 \rightarrow$ [Kr]
$Cs \rightarrow Cs^+ + e^-$
[Xe] $6s^1 \rightarrow$ [Xe]
$Fr \rightarrow Fr^+ + e^-$
[Rn] $7s^1 \rightarrow$ [Rn]

Questions

1 Suggest why the table of elements is called the 'periodic table'.

2 Explain how electronic configurations change:
(a) across a period
(b) down a group.

3 What is the similarity in the electron configurations for atoms and ions of:
(a) the group 2 elements, Be–Ba
(b) the group 17 elements, F–I?

Electrons and the periodic table

By the end of this topic, you should be able to demonstrate and apply your knowledge and understanding of:

* classification of elements into s-, p- and d-blocks

Electron shells overlap

In topic 2.2.2, we discussed that:

- the higher the value of the principal quantum number, *n*, the higher the energy level and the further the shell is from the nucleus
- as we move from one shell to the next, a new type of sub-shell is present
- within a shell, the sub-shell energies increase in the order s < p < d < f.

You need to take care when you go beyond the 3p sub-shell. Remember that the orbitals are filled in order of energy levels, but:

- the 4s energy level is lower than the 3d energy level
- the 4s orbital fills before the 3d orbital
- the 4s orbital would be emptied before the 3d orbital during ionisation.

So the 4th shell starts to fill before the 3rd shell has completely filled. This overlap is shown in Figure 1.

LEARNING TIP

You do not need to know about f sub-shells and the f-block in the periodic table. However, for completeness we have referred to them throughout the topics on electronic configurations and shells.

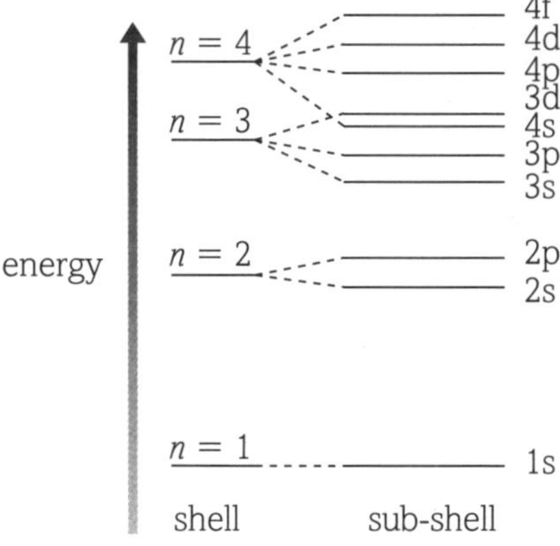

Figure 1 Overlap of 4s and 3d sub-shells.

The diagram in Figure 2 shows how sub-shell orbitals are filled for a potassium atom. The electron configuration is $1s^2 2s^2 2p^6 3s^2 3p^6 4s^1$.

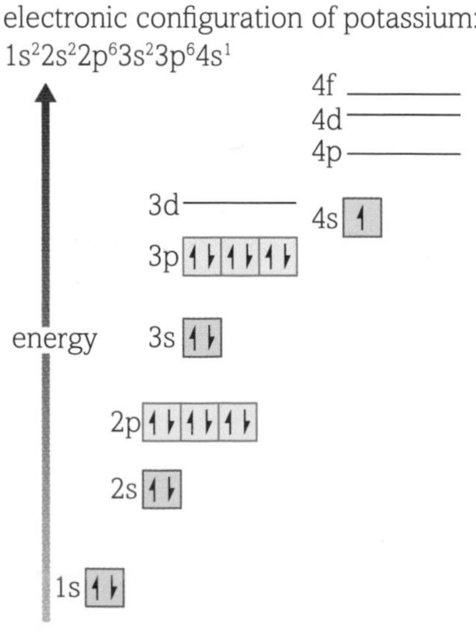

electronic configuration of potassium:
$1s^2 2s^2 2p^6 3s^2 3p^6 4s^1$

energy
4f
4d
4p
3d
4s
3p
3s
2p
2s
1s

Figure 2 Filling the orbitals in potassium.

Sub-shells and the periodic table

The periodic table is structured in blocks that are linked to sub-shells. The pattern mirrors the sub-shells that are being filled. You can see the pattern by dividing the periodic table into blocks, as shown in Figure 3.

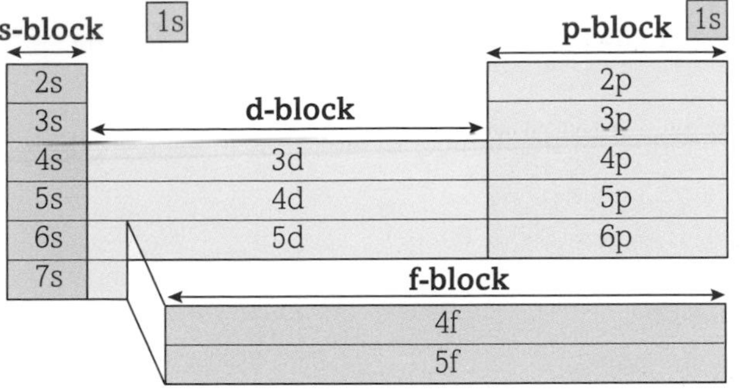

Figure 3 The periodic table is divided into sub-shells, which correspond to the position of the atoms' outermost electrons.

Using the periodic table to determine electron configurations

By following the pattern shown in Figure 3, you can work out the electron configuration for any element using the periodic table. Figure 4 shows how to work out the electron configuration of oxygen.

- Oxygen is the 4th element in the 2p block (the second period, or row, in the p block), i.e. $2p^4$.
- Oxygen has the electron configuration $1s^22s^22p^4$.

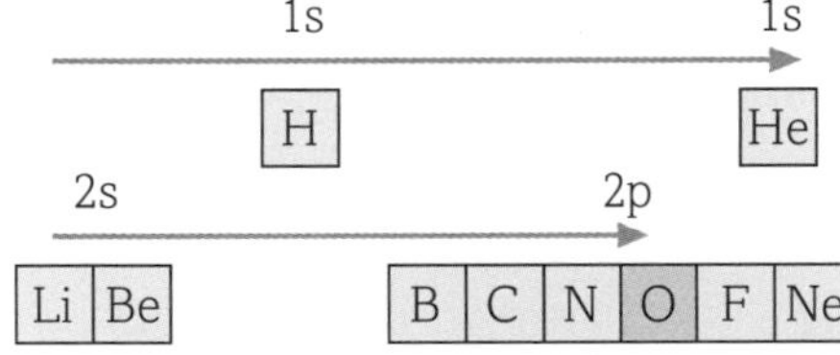

- Oxygen is the 4th element in the 2p block, i.e. $2p^4$
- Oxygen has the electron configuration: $1s^22s^22p^4$

Figure 4 Working out the electronic configuration of oxygen.

LEARNING TIP

You need to remember 4s: first in; first out.
The correct way to show the electron configuration is in shell order. But you might find it more logical to write down the energy level order. So, Ca *should* be shown as $1s^22s^22p^63s^23p^63d^64s^2$, rather than $1s^22s^22p^63s^23p^64s^23d^6$.
Either representation is acceptable.

Shortening an electron configuration

For atoms with many electrons, the electron configuration notation can become lengthy. It is often abbreviated by basing the inner-shell configuration on the noble gas that comes before the element in the periodic table. This shortening also allows you to concentrate on the important outer-shell portion of the electron configuration. It is these electrons that are responsible for the behaviour of the element.

The electronic configurations for some elements in group 1 are shown below in both their full and abbreviated forms. The noble gas that comes before lithium in the periodic table is helium, before sodium is neon, and before potassium is argon.

Li: $1s^22s^1$ or $[He]2s^1$

Na: $1s^22s^22p^63s^1$ or $[Ne]\,3s^1$

K: $1s^22s^22p^63s^23p^64s^1$ or $[Ar]\,4s^1$

Using this notation allows you to see quickly which sub-shell the outer electrons are in and where the element is found in the periodic table.

Questions

1. Which block in the periodic table contains the following elements? (The atomic number of each element is given in brackets.)
   (a) Sr (38) (b) Ga (31) (c) Re (75) (d) Sm (62)
2. Write the electron configuration in terms of sub-shells for these atoms:
   (a) Al (b) Si (c) S (d) Sc (e) Co (f) Zn
3. Write the shorthand electron configuration using the nearest noble gas for:
   (a) Mg (b) Ba (c) Cl

3.1  Evidence for electron shells

Evidence for electron shells

By the end of this topic, you should be able to demonstrate and apply your knowledge and understanding of:

* first ionisation energy (removal of 1 mol of electrons from 1 mol of gaseous atoms) and successive ionisation energies
* prediction from successive ionisation energies of the number of electrons in each shell of an atom and the group of an element

What is ionisation?

Ionisation occurs when atoms lose or gain electrons. For example, in plasma television screens a mixture of positive ions and negative ions is created.

- $Ne(g) \rightarrow Ne^+(g) + e^-$
- $Xe(g) \rightarrow Xe^+(g) + e^-$

The mixture of these ions is called a *plasma*. Plasma televisions can emit light because this plasma is present.

This topic will look specifically at the loss of electrons to form positive ions.

Ionisation energy

The energy needed to form positive ions is known as the ionisation energy. The **first ionisation energy** (1st I.E.) is a measure of how easily an atom loses an electron to form a 1+ ion.

Look at what happens when a sodium atom loses its first electron:

$Na(g) \rightarrow Na^+(g) + e^-$ 1st I.E. = +496 kJ mol^{-1}

Because the equation represents the first ionisation, it means that *one mole of gaseous atoms loses 1 mole of electrons to form 1 mole of gaseous 1+ ions.*

KEY DEFINITION

The **first ionisation energy** of an element is the energy required to remove one electron from each atom in one mole of the gaseous element to form one mole of gaseous 1+ ions.

LEARNING TIP

Remember, ionisation energy is always calculated in the gaseous state and always involves one mole of the element losing one mole of electrons to form one mole of gaseous ions.

Factors affecting ionisation energy

Negative electrons are held in their shells by their attraction to the positive nucleus. To form a positive ion, energy must be supplied to an electron to overcome this attraction. Electrons in the outer shell are removed first because they experience the smallest nuclear attraction. The outer-shell electrons are furthest away from the nucleus and require the least ionisation energy.

The nuclear attraction experienced by an electron depends on three factors:

- atomic radius
- nuclear charge
- electron shielding or screening.

Atomic radius

The larger the atomic radius, the smaller is the nuclear attraction experienced by the outer electrons. This is because the positive charge of the nucleus is essentially further away from the outermost electrons.

Nuclear charge

The higher the nuclear charge, the larger is the attractive force on the outer electrons.

Electron shielding or screening

Inner shells of electrons repel the outer-shell electrons because they are all negative.

- This repelling effect is called electron shielding or screening.
- The more inner shells there are, the larger is the shielding effect and the smaller is the nuclear attraction experienced by the outer electrons.

Successive ionisation energies

Successive ionisation energies are a measure of the amount of energy required to remove each electron in turn.

For example, the *second ionisation energy* is a measure of how easily a 1+ ion loses an electron to form a 2+ ion.

An element has as many ionisation energies as it has electrons. Lithium has three electrons and therefore three successive ionisation energies. Equations to represent the three ionisation energies of lithium are shown in Table 1.

$Li(g) \rightarrow Li^+(g) + e^-$ 1st I.E. = +520 kJ mol^{-1}
$Li^+(g) \rightarrow Li^{2+}(g) + e^-$ 2nd I.E. = +7298 kJ mol^{-1}
$Li^{2+}(g) \rightarrow Li^{3+}(g) + e^-$ 3rd I.E. = +11 815 kJ mol^{-1}

Table 1 Successive ionisation energies.

Each successive ionisation energy is higher than the one before.

- As each electron is removed, there is less repulsion between the remaining electrons and each shell will be drawn in slightly closer to the nucleus.
- The positive nuclear charge will outweigh the negative charge every time an electron is removed.
- As the distance of each electron from the nucleus decreases slightly, the nuclear attraction increases. More energy is needed to remove each successive electron.

LEARNING TIP

Remember, the charge on the ion will tell you which successive ionisation has occurred. For example, the third ionisation of lithium will give a 3+ ion:

$Li^{2+}(g) \longrightarrow Li^{3+}(g) + e^-$

DID YOU KNOW?

Evidence for shells

The Bohr model of the atom describes a positive nucleus, with negatively charged electrons orbiting round it in defined energy levels. Successive ionisation energies support this model.

If you look at the successive ionisation energies for an element you may notice that they do not always increase regularly – there are jumps. These jumps tell us a lot about the electron structure of a substance. See Figure 1 for the ionisation energies of lithium.

When the electrons in the outer shell have all been removed, electrons from the next shell are removed if more ionisation occurs. This takes much more energy because of the closer proximity of the nucleus and the smaller atomic radius, so a 'jump' in ionisation energy is observed. By looking at where these jumps occur it is possible to deduce the electronic structure.

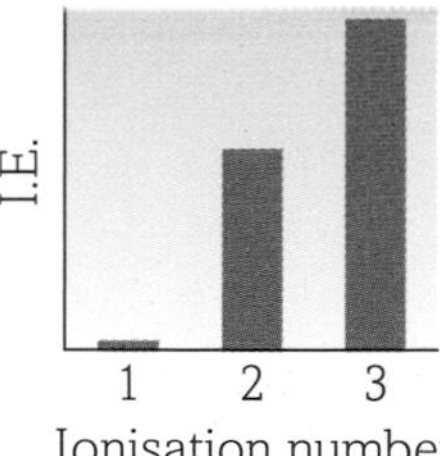

Figure 1 Three successive ionisation energies of lithium – the big jump between ionisations 1 and 2 shows that the second electron is being removed from the next shell, which experiences a higher nuclear attraction.

The graph shown in Figure 2 shows how the successive ionisation energies of nitrogen provide evidence for its shells – the shell number is shown by the symbol n.

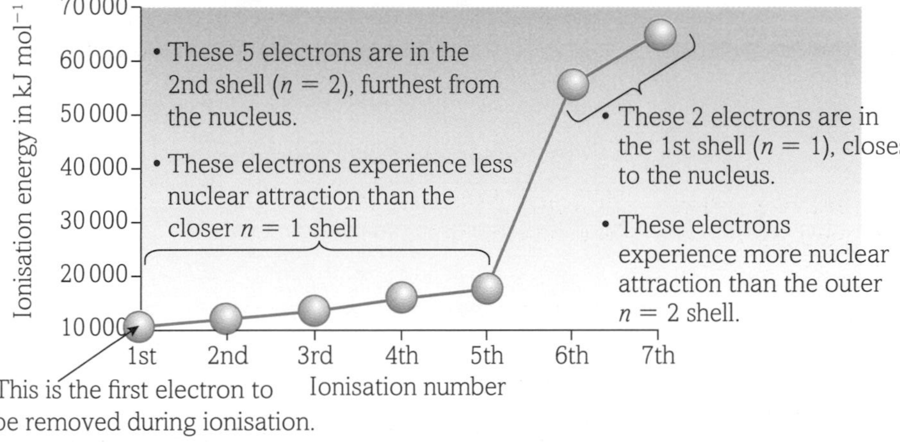

Figure 2 Successive ionisation energies of nitrogen. Note the large increase in ionisation energy as the sixth electron is removed – the successive ionisation energies are reflecting the leap from one shell to another getting closer to the nucleus. If the electrons were all in the same shell, there would be no sharp increase.

Questions

1. Write equations to represent the 1st and 4th ionisation energies of chlorine.
2. Sketch the graph you would expect for the successive ionisation energies of aluminium. Explain where you have placed any significant jumps in ionisation energies.
3. An element in period 3 (Na–Ar) has the following successive ionisation energies, in kJ mol^{-1}:
   789; 1577; 3232; 4356; 16 091; 19 785; 23 787; 29 253.
   Plot a graph of these values and identify the element – explain your reasons.

5 Periodicity: ionisation energies and atomic radii

By the end of this topic, you should be able to demonstrate and apply your knowledge and understanding of:

- first ionisation energy (removal of 1 mol of electrons from 1 mol of gaseous atoms) and successive ionisation energy
- explanation of the trend in first ionisation energies across periods 2 and 3, and down a group, in terms of attraction, nuclear charge and atomic radius

Variation in first ionisation energies and atomic radii

As we discussed in topic 3.1.4, the factors that affect ionisation energies are:

- nuclear charge
- distance (of the electron being removed) from the nucleus
- electron shielding.

LEARNING TIP

When we think about first ionisation energy, the distance of the electron from the nucleus is effectively the atomic radius.
It would be incorrect to describe the distances between successive electrons and the nucleus as atomic radii, because after the first ionisation energy, electrons are being removed from ions. This is why the term 'distance' is used here instead of atomic radius.

The first ionisation energies for the first 20 elements in the periodic table, H–Ca, are shown in Figure 1.

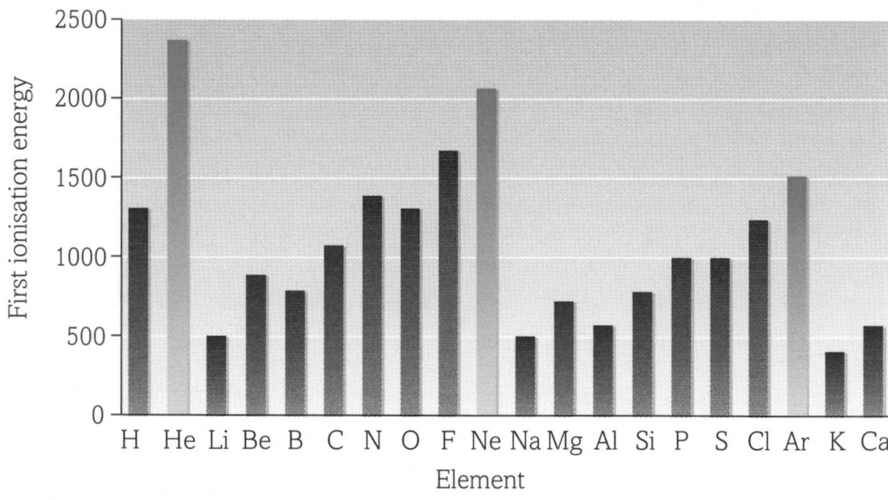

Figure 1 First ionisation energies of the first 20 elements – note how there is a trend across both periods.

The ionisation energy for each noble gas, at the end of each period, is shown in green in Figure 1. You will notice that the ionisation energy of the noble gas is the highest for each period. These atoms have a full outer shell of electrons and a high positive attraction from the nucleus, so ionisation energy values are large.

Trends across a period

Looking at Figure 1, you can see that the ionisation energy values show a general increase across each period.

Across each period:

- the number of protons in the nucleus increases, so there is a higher attraction on the electrons
- electrons are added to the same shell, so the outer shell is drawn inwards slightly
- there is the same number of inner shells, so electron shielding will hardly change.

Moving across a period, the attraction between the nucleus and the outer electrons increases, so more energy is needed to remove an electron. This means that the first ionisation energy increases across a period.

Decreases between groups 2 to 13 and 15 to 16

Look again at Figure 1. There is a small decrease in the first ionisation energy between the group 2 and 13 elements (Be → B, Mg → Al). This is because the group 13 elements have their outermost electron in a p-orbital, whereas group 2 elements have theirs in an s-orbital; and p-orbitals have a slightly higher energy than s-orbitals and so are marginally further away from the nucleus. This means that electrons in these orbitals are slightly easier to remove and so the elements have lower ionisation energies.

A similar decrease occurs between group 15 and 16 (N → O, P → S). As you move from group 13 towards group 18, outer electrons are found in p-orbitals. In groups 13, 14 and 15, each of the p-orbitals contains only a single electron. In group 16, however, the outermost electron is now spin-paired in the p_x orbital. Electrons that are spin-paired experience some repulsion – this makes the first outer electron slightly easier to remove, so a slightly lower first ionisation energy is observed.

LEARNING TIP

Moving across a period, increased nuclear charge is the factor that has the biggest effect on ionisation energy.

There is also a decrease in atomic radius moving across a period, because the increased nuclear charge pulls the electrons in towards it (see Figure 2).

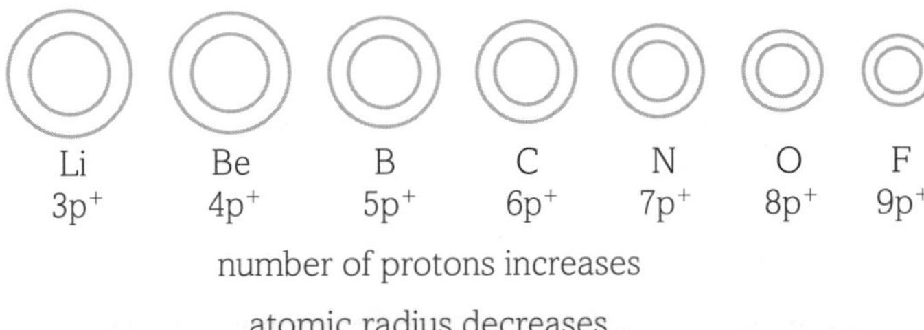

Figure 2 The trend in first ionisation energies and atomic radii across a period.

There is a sharp decrease in first ionisation energy between the end of one period and the start of the next:

- from He at the end of period 1 to Li at the start of period 2
- from Ne at the end of period 2 to Na at the start of period 3
- from Ar at the end of period 3 to K at the start of period 4.

This reflects the addition of a new shell, further from the nucleus, which leads to:

- an increase in the distance of the outermost shell from the nucleus
- an increase in electron shielding of the outermost shell by inner shells.

Trends down a group

Moving down a group, first ionisation energies decrease. Again, looking at Figure 1, you can see that the first ionisation energies decrease moving down each group. This is clearest for the three noble gases – He, Ne and Ar. First ionisation energies decrease down every group:

- the number of shells increases, so the distance of the outer electrons from the nucleus increases; so, there is a weaker force of attraction on the outer electrons
- there are more inner shells, so the shielding effect on the outer electrons from the nuclear charge increases; so, again there is weaker attraction.

The number of protons in the nucleus also increases moving down a group, but the resulting increased attraction is far outweighed by the increases in distance and shielding.

Taking all these factors into account, the attraction between the nucleus and the outer electrons decreases moving down a group, so less energy is needed to remove an electron. This means that the first ionisation energy decreases down a group. However, the atomic radius increases moving down a group because weaker attraction means that the electrons are not pulled as close to the nucleus.

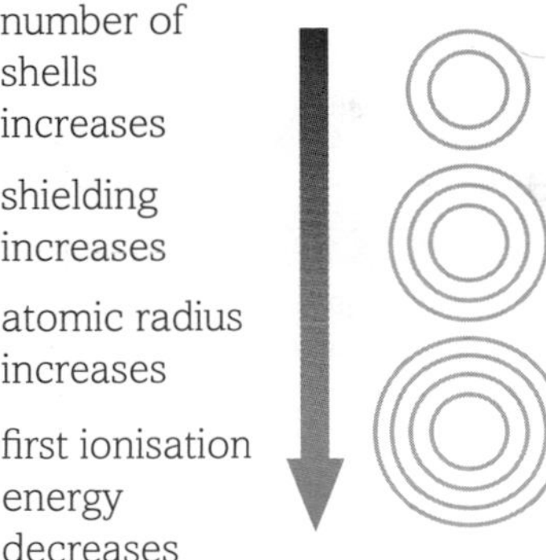

Figure 3 The trends in first ionisation energy and atomic radius down a group.

LEARNING TIP

Moving down a group, increasing distance and shielding are the most important factors influencing ionisation energy.

Questions

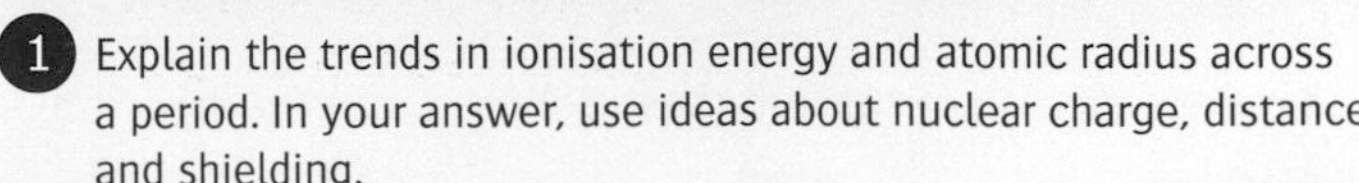
1. Explain the trends in ionisation energy and atomic radius across a period. In your answer, use ideas about nuclear charge, distance and shielding.

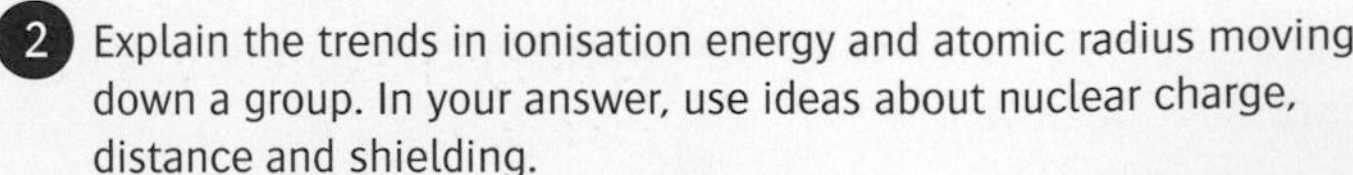
2. Explain the trends in ionisation energy and atomic radius moving down a group. In your answer, use ideas about nuclear charge, distance and shielding.

Metallic bonding and structure

By the end of this topic, you should be able to demonstrate and apply your knowledge and understanding of:

* explanation of metallic bonding as strong electrostatic attraction between cations (positive ions) and delocalised electrons
* explanation of a giant metallic lattice structure, e.g. all metals
* explanation of physical properties of giant metallic and giant covalent lattices, including melting and boiling points, solubility and electrical conductivity in terms of structure and bonding

Introduction

Metallic bonding

The atoms in a solid metal (Figure 1) are held together by metallic bonding.

Figure 1 Copper, mercury and magnesium are common metals. The atoms in these metals, like in all metals, are held together by metallic bonding.

In metallic bonding, the atoms are ionised.

- Positive ions (cations) occupy fixed positions in a lattice.
- The outer-shell electrons are *delocalised* – they are shared between all the atoms in the metallic structure.

The metal is held together by the attractions between all the positive ions and all the negative electrons – especially the delocalised electrons.

Giant metallic lattice structure

In a giant metallic lattice:

- the delocalised electrons are spread throughout the structure
- these electrons can move within the structure
- it is impossible to tell which electron originated from which particular positive ion
- over the whole structure, the charges must balance.

A giant metallic lattice is often described as a lattice of positive ions fixed in position and surrounded by a 'sea' of electrons (Figure 2). These electrons are delocalised and can move. This model helps us to understand the properties of metals.

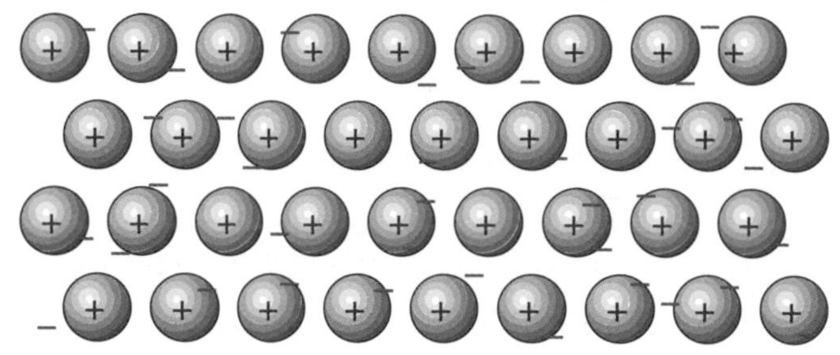

Figure 2 The 'sea of electrons'. Here, there are 36 positive ions balanced by 36 'free' electrons. All the positive ions are attracted to all the delocalised electrons.

Properties of giant metallic lattices

Giant metallic lattices have unique properties.

High melting point and boiling point

Most metals have high melting and boiling points.

- The electrons are free to move throughout the structure, but the positive ions remain where they are.
- The attraction between the positive ions and the negative delocalised electrons is very strong.
- A high temperature is needed to overcome the metallic bonds and dislodge the ions from their rigid positions in the lattice.

Good electrical conductivity

Metals are good conductors of electricity.

- The delocalised electrons can move freely, anywhere within the metallic lattice.
- This allows metals to conduct electricity, even in the solid state (Figure 3).

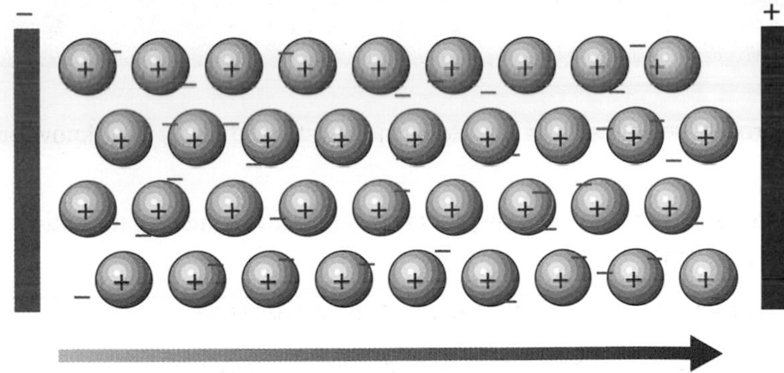

Figure 3 Electrical conductivity in a metal. The electrons are free to flow between the positive ions – the positive ions do not move.

LEARNING TIP

Remember, the positive ions do not move. It is the delocalised electrons that move, and it is these that can carry charge and therefore conduct electricity.

Malleability and ductility

Metals are both ductile and malleable.

- Ductile – can be drawn out or stretched. Ductility permits metals to be drawn into wires.
- Malleable – can be hammered into different shapes. Many metals can be pressed into shapes or hammered into thin sheets.

The delocalised electrons are largely responsible for these properties. Because they can move, the metallic structure has a degree of 'give', which allows atoms or layers to slide past each other.

LEARNING TIP

You should be able to compare the different ways that giant structures (giant metallic lattices and giant covalent lattices) are formed.

Covalent bonds involve *localised* electrons – they cannot move freely; they are fixed between the bonded atoms.

Metallic bonds involve *delocalised* electrons – they can move freely throughout the lattice and are effectively shared by all the positive ions in the lattice.

That's why both structure types have high melting and boiling points, and why only metals can conduct electricity.

Questions

1. Describe how the atoms in metals are bonded together.
2. Explain why metals can conduct electricity.
3. Explain why metals have high melting points and high boiling points.
4. Explain how metallic bonding is different from covalent bonding.

Periodicity: melting points

By the end of this topic, you should be able to demonstrate and apply your knowledge and understanding of:

* explanation of the variation in melting points across periods 2 and 3 in terms of structure and bonding
* explanation of the solid giant covalent lattices of carbon (diamond, graphite and graphene) and silicon as networks of atoms bonded by strong covalent bonds

From metals to non-metals

As you move across periods 2 and 3, the elements change from:

- metal elements to non-metal elements
- solids to gases.

The elements in period 3 clearly illustrate these changes, as shown in Figure 1.

Figure 1 The elements in period 3 clearly change from being solid metals to gaseous non-metals.

Na, Mg and Al are clearly metals, but Si is much harder to classify.

- Si has the shiny appearance of a metal, but is brittle.
- Si conducts electricity, but very poorly.

Silicon is an 'in-between' element, usually classified as a semi-metal or metalloid.

Trends in melting points

The melting points of elements change moving across periods 2 and 3 (see Figure 3). These changes tell us about the structures of the elements. The trend in melting points for both periods follows the same pattern.

- Between group 1 and group 14, melting points increase steadily. This is because the elements have giant structures. For each successive group, if an element has a giant metallic lattice, the nuclear charge increases, as does the number of electrons in the outer shell – this causes a stronger attraction. If the element has a giant covalent lattice, each successive group has more electrons with which to form covalent bonds.
- Between group 14 and group 15, there is a sharp decrease in melting point. This is because the elements have simple molecular structures – each individual molecule is attracted to others by relatively weak intermolecular forces.
- Between group 15 and group 18, the melting points remain relatively low – the elements have simple molecular structures.

LEARNING TIP

The boiling points of elements in periods 2 and 3 show a similar pattern to that in melting points moving across the periods. They:

- increase up to group 14
- decrease between groups 14 and 15
- stay low between groups 15 and 18.

The effect on metallic bonding from extra electrons and greater nuclear charge moving across period 2 is shown in Figure 2.

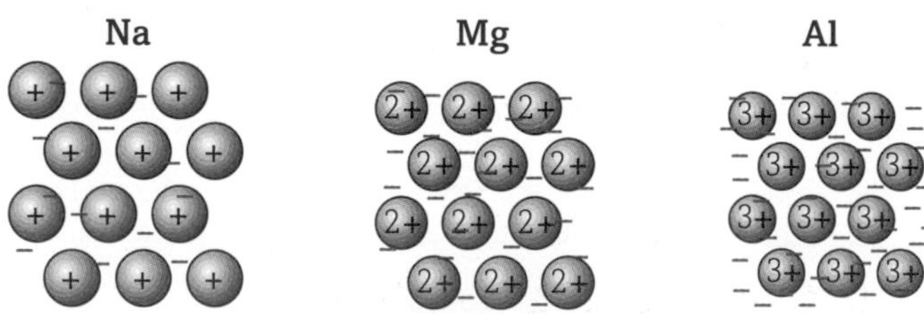

Figure 2 How melting point and boiling point is affected by extra electrons and higher nuclear charge from Na to Al.

Period 2	Li	Be	B	C	N_2	O_2	F_2	Ne
Period 3	Na	Mg	Al	Si	P_4	S_8	Cl_2	Ar
Structure	giant metallic			giant covalent	simple molecular			
Forces	strong forces between positive ions and negative delocalised electrons			strong forces between atoms	weak intermolecular forces between molecules			
Bonding	metallic bonding			covalent	covalent bonding within molecules intermolecular bonding between molecules			

Table 1 Structure and bonding in the periodic table.

Structure and bonding

Table 1 summarises how the bonding and structure changes across periods 2 and 3.

Many forms of carbon and silicon utilise giant covalent bonding, giving them very high melting points. Diamond forms a lattice where each carbon atom forms four other carbon atoms around it. Diamond is one of the hardest materials known.

Graphene forms a two-dimensional giant lattice, one carbon atom thick, of interlocking hexagonal carbon rings. It is extremely strong, very light and can conduct electricity. It has many uses in the field of nanotechnology. See topic 2.2.8 for more on covalent bonding.

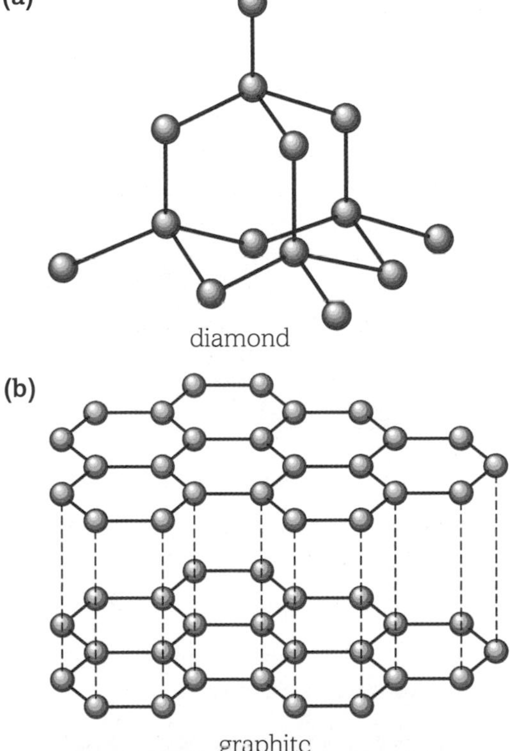

(c)

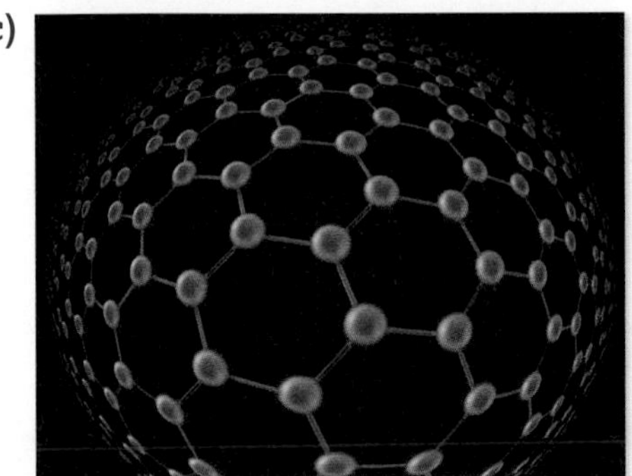

Figure 3 Giant covalent structures of carbon. (a) Diamond has a tetrahedral structure where each carbon is bonded to four other carbon atoms around it. This structure makes it extremely hard with a high melting point. (b) Graphite has a layered structure with delocalised electrons between layers, meaning it is able to conduct electricity. (c) Graphene forms interlocking hexagonal rings that make up a lattice only one atom thick. This makes graphene extremely strong yet extremely light, with many versatile uses.

DID YOU KNOW?

The element silicon has a giant covalent lattice structure similar to that of diamond. However, most silicon is found as silicon dioxide SiO2. Silicon dioxide also has a giant covalent lattice structure, shown in Figure 4.

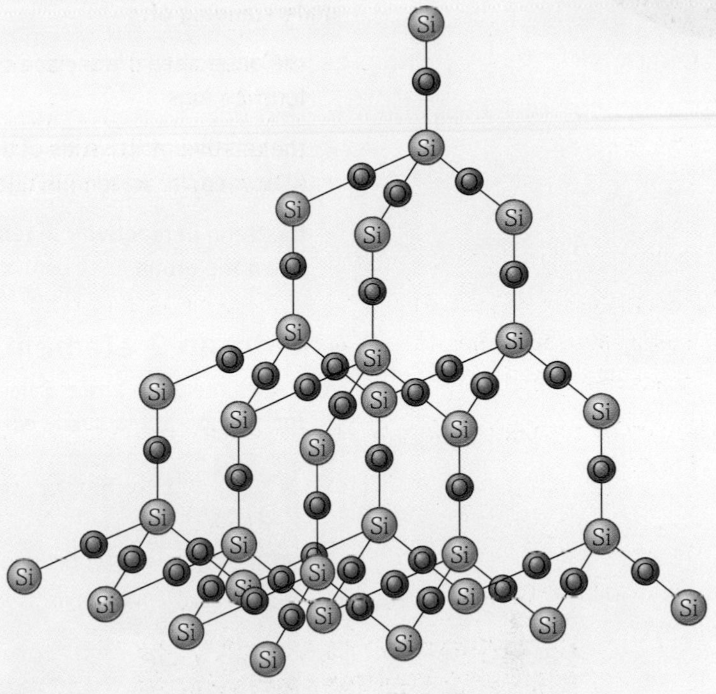

Figure 4 Silicon dioxide has a giant covalent lattice structure.

Questions

1. Explain why the melting point of carbon is much higher than that of nitrogen.
2. Explain why the melting point of aluminium is higher than that of magnesium.
3. Explain the changes of state from solid to gas moving across periods 2 and 3.
4. The graph in Figure 3 shows the changes in the melting points of elements in periods 2 and 3. Identify elements X, Y and Z. Explain how you identified these elements.

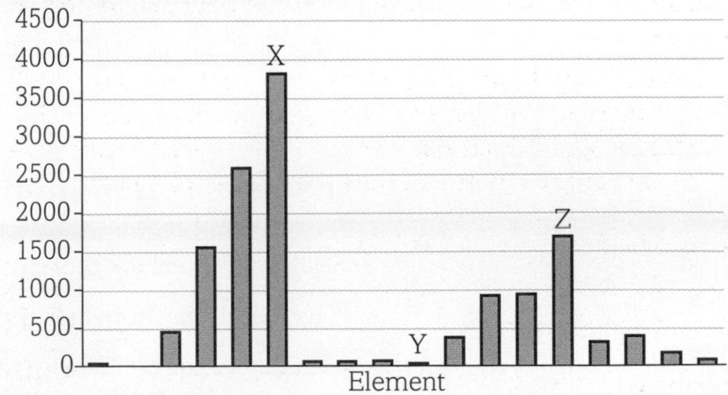

Figure 5 Melting points of the period 2 and 3 elements.

Group 2 elements: redox reactions

By the end of this topic, you should be able to demonstrate and apply your knowledge and understanding of:

* the outer shell s^2 electron configuration and the loss of these electrons in redox reactions to form 2+ ions
* the relative reactivities of the group 2 elements Mg $\rightarrow$ Ba shown by their redox reactions with: (i) oxygen, (ii) water, (iii) dilute acids
* the trend in reactivity in terms of the first and second ionisation energies of group 2 elements down the group

The group 2 elements

All five elements in group 2 have hydroxides that are alkaline, which is reflected in the common name for this group – the alkaline earth metals. They are shown in Figure 1.

Figure 1 Group 2 elements – from the left beryllium, magnesium, calcium, strontium and barium.

The atoms of the elements have a tendency to lose two electrons. The reactivity of the elements increases moving down the group – beryllium is the least reactive, barium the most.

Physical properties

The group 2 elements have similar general physical properties – they:

- have reasonably high melting points and boiling points
- are light metals with low densities
- form colourless (white) compounds.

Electronic configuration and ionisation energy

The elements in group 2 have their highest-energy electrons in an s sub-shell. Together with group 1, they form the s-block of the periodic table. Each group 2 element has:

- two electrons more than the electronic configuration of a noble gas
- an outer shell containing two electrons.

This means they need to lose two electrons to achieve noble gas configuration. The ionisation energy values for the elements (see Table 1) show that these two outer electrons are lost more easily going down the group to barium. This, in turn, means that reactivity increases going down the group. This is due to the fact that each successive element has its outer electrons in a higher energy level, has a larger atomic radius and feels more shielding from the positive pull of the nucleus.

Element	Electron configuration	First ionisation energy/kJ mol^{-1}	Second ionisation energy/kJ mol^{-1}
Be	[He] $2s^2$	900	1757
Mg	[Ne] $3s^2$	736	1450
Ca	[Ar] $4s^2$	590	1145
Sr	[Kr] $5s^2$	548	1064
Ba	[Xe] $6s^2$	502	965

Table 1 Electron configurations and ionisation energies.

Reactivity of the group 2 elements

The group 2 elements are reactive metals and strong reducing agents. Group 2 elements are oxidised in their reactions to form 2+ ions. The ionisations of the elements can be written using M to represent any group 2 metal:

$M \rightarrow M^+ + e^-$ (+1 oxidation state: first ionisation)

$M^+ \rightarrow M^{2+} + e^-$ (+2 oxidation state: second ionisation)

LEARNING TIP

The group 2 elements all lose 2 electrons from their outermost s sub-shell during reactions to form 2+ ions, by a first ionisation and a subsequent second ionisation.

Reactions between group 2 elements and oxygen

The group 2 elements react vigorously with oxygen – this is a redox reaction. The product is an ionic oxide with the general formula MO. For example, calcium reacts with oxygen to form calcium oxide:

$2Ca(g) + O_2(g) \rightarrow 2CaO(s)$

0 → +2 oxidation (oxidation number increases)

0 → −2 reduction (oxidation number decreases)

Reactions between group 2 elements and water

All the group 2 elements, except beryllium, react with water to form hydroxides with the general formula $M(OH)_2$. Hydrogen gas is also formed.

Mg reacts with water very slowly. Moving further down the group, each metal reacts more vigorously with water. For example, calcium reacts with water to produce calcium hydroxide and hydrogen gas:

$Ca(s) + 2H_2O(l) \rightarrow Ca(OH)_2(aq) + H_2(g)$

0 → +2 oxidation

+1 → +1

+1 → +1

+1 → 0 reduction

+1 → 0

The reaction between group 2 metals and water is also a redox reaction – the metal is oxidised and one hydrogen atom from each water molecule is reduced.

Reactions between group 2 elements and dilute acids

All the group 2 elements, except Be, react with dilute acids to form a salt and hydrogen gas. The reaction becomes more vigorous moving down the group. For example, calcium reacts with hydrochloric acid to form calcium chloride and hydrogen gas:

$Ca(s) + 2HCl(aq) \rightarrow CaCl_2 + H_2$

0 → +2 oxidation

+1 → 0 reduction

+1 → 0

LEARNING TIP

Half-equations can be used to show exactly what happens to electrons during redox reactions. For example, in the reaction between calcium and water:

$Ca \rightarrow Ca^{2+} + 2e^-$ oxidation (loss of electrons)

$2H_2O + 2e^- \rightarrow 2OH^- + H_2$ reduction (gain of electrons)

Questions

1 The following reaction is a redox process:

$Mg + 2HCl \rightarrow MgCl_2 + H_2$

(a) Identify the changes in oxidation number.

(b) Which species is being oxidised and which is being reduced?

2 (a) Write down equations for the following reactions:

(i) the reaction of barium with water

(ii) the reaction of strontium with oxygen

(iii) the reaction of magnesium with sulfuric acid.

(b) Use oxidation numbers to identify what has been oxidised and what has been reduced in reactions (a)(i) and (a)(ii).

3 Explain why magnesium will only react with boiling water but calcium will react with cold water.

Group 2 compounds: reactions

By the end of this topic, you should be able to demonstrate and apply your knowledge and understanding of:

* the action of water on group 2 oxides and the approximate pH of any resulting solutions, including the trend of increasing alkalinity
* uses of some group 2 compounds as bases, including equations, for example (but not limited to):
  (i) $Ca(OH)_2$ in agriculture to neutralise acid soils
  (ii) $Mg(OH)_2$ and $CaCO_3$ as 'antacids' in treating indigestion

Reactions between group 2 oxides and water

Group 2 oxides react with water to form metal hydroxides. The general reaction for this, where the group 2 metal is M, is:

$$MO(s) + H_2O(l) \rightarrow M(OH)_2(aq)$$

The metal hydroxides are soluble in water, and form alkaline solutions with water because they release OH^- ions. The typical pH of these solutions is between 10 and 12.

Solubility of group 2 metal hydroxides

However, not all of the group 2 metal hydroxides that are formed when the metal oxide reacts with water have the same solubility – some are better at dissolving than others.

The solubility of the hydroxides in water increases down the group. When a hydroxide is more soluble than another, it will release more OH^- ions, and so will make a more alkaline solution, with a higher pH.

- Beryllium is at the top of group 2; beryllium oxide is insoluble in water.
- Magnesium forms $Mg(OH)_2(s)$, which is only slightly soluble in water – the resulting solution is dilute with a comparatively low $OH^-(aq)$ concentration.
- $Ba(OH)_2(s)$ is much more soluble in water than $Mg(OH)_2(s)$, and so has a higher $OH^-(aq)$ concentration. The resulting solution is more alkaline than a solution of $Mg(OH)_2$.

This trend is summarised in Figure 1.

LEARNING TIP

More OH^- ions will make a solution more alkaline – meaning it will have a higher pH value.

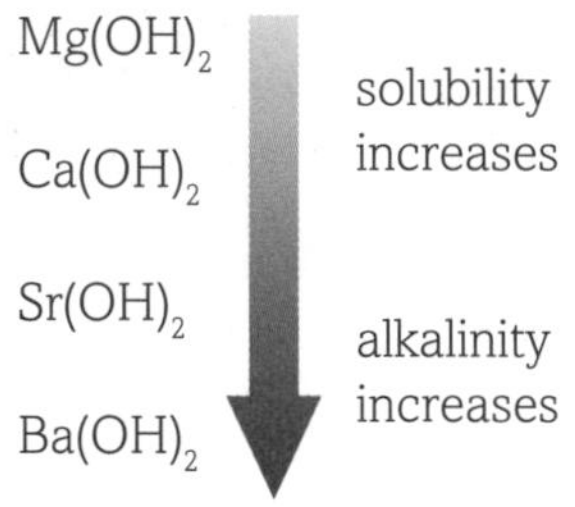

$Mg(OH)_2$

$Ca(OH)_2$

$Sr(OH)_2$

$Ba(OH)_2$

solubility increases

alkalinity increases

Figure 1 The trend in alkalinity of group 2 hydroxides.

Uses of group 2 compounds

The oxides, carbonates and hydroxides of group 2 metals are basic – this means they will react with acids to form a salt and water. For example:

$$MgO(s) + 2HCl(aq) \rightarrow MgCl_2(aq) + H_2O(l)$$

$$Ca(OH)_2(s) + 2HCl(aq) \rightarrow CaCl_2(aq) + 2H_2O(l)$$

As these reactions with hydrochloric acid take place, you will see a solid oxide or hydroxide 'dissolve'.

The alkalinity of group 2 metal compounds has resulted in them being used for a number of purposes, where neutralising acids is important. They are also very important substances in the construction industry.

Neutralising acidic soils

Calcium hydroxide, $Ca(OH)_2$, is used by farmers and gardeners as 'lime' – it is used to reduce the acidity levels of soil (Figure 2).

Figure 2 Spreading lime on crop fields.

Indigestion remedies

Human stomachs contain a small amount of hydrochloric acid. When we suffer from indigestion, there is a build-up of too much of this acid.

Many indigestion remedies are available – for example 'milk of magnesia', which contains magnesium hydroxide, $Mg(OH)_2$. The magnesium hydroxide in the milk of magnesia neutralises the excess acid producing a salt and water:

$$Mg(OH)_2 + 2HCl \rightarrow MgCl_2 + 2H_2O$$

Building and construction uses

One of the group 2 metal carbonates, calcium carbonate ($CaCO_3$), is an extremely useful building material – it is present in both limestone and marble (Figure 3). It is by far the most important calcium compound industrially – it is used in the building trade and in the manufacture of glass and steel.

Figure 3 Limestone quarry in the Derbyshire Peak District, UK.

A major drawback of the use of group 2 carbonates as building materials is they readily react with acids. For example:

$$CaCO_3(s) + 2HCl(l) \rightarrow CaCl_2(aq) + H_2O(l) + CO_2(g)$$

Most rainwater has an acidic pH, which leads to gradual erosion of objects made using limestone or marble, such as buildings or statues.

Questions

1 (a) Explain why the reaction between group 2 metal oxides and water results in an alkaline solution.
(b) Describe the trend in pH of the solutions formed moving down group 2.

2 Write balanced equations for the following reactions:
(a) barium oxide with hydrochloric acid
(b) calcium oxide with nitric acid
(c) calcium oxide with water.

3.1 10 Group 17: the halogens

By the end of this topic, you should be able to demonstrate and apply your knowledge and understanding of:

- explanation of existence of halogens as diatomic molecules and the trend in the boiling points of Cl_2, Br_2 and I_2 in terms of induced dipole–dipole interactions (London forces)
- explanation of the outer shell s^2p^5 electron configuration and the gaining of one electron in many redox reactions to form 1– ions
- explanation of the trend in reactivity of the halogens Cl_2, Br_2 and I_2 illustrated by reaction with other halide ions
- explanation of the trend in reactivity from the decreasing ease of forming 1– ions, in terms of attraction, atomic radius and electron shielding
- explanation of the term *disproportionation* as oxidation and reduction of the same element, illustrated by:
  - (i) the reaction of chlorine with water as used in water purification
  - (ii) the reaction of chlorine with cold, dilute aqueous sodium hydroxide, as used to form bleach
  - (iii) reactions analogous to these
- the benefits of chlorine use in water treatment (killing bacteria) contrasted with associated risks (e.g. hazards of toxic chlorine gas and possible risks from formation of chlorinated hydrocarbons)

Properties of the halogens

The group 17 elements, generally known as the halogens, are fluorine (F), chlorine (Cl), bromine (Br), iodine (I) and astatine (At). They:

- have low melting points and boiling points
- exist as diatomic molecules, X_2 where X represents the halogen.

Electronic configuration

The group 17 elements have 7 outer shell electrons, five being in the p sub-shell. The elements form part of the p-block in the periodic table (see Figure 1).

F	...... $2s^22p^5$
Cl	...... $3s^23p^5$
Br	.... $4s^24p^5$
I	...... $5s^25p^5$
At	.... $6s^26p^5$

Figure 1 Electronic configurations of the halogen atoms.

Trend in boiling points

Moving down the group, the boiling point increases and the physical state changes from gas, to liquid, to solid. This is because each successive element has an extra shell of electrons – this leads to a higher level of London forces (intermolecular induced dipole–dipole forces) between the molecules (Figure 2).

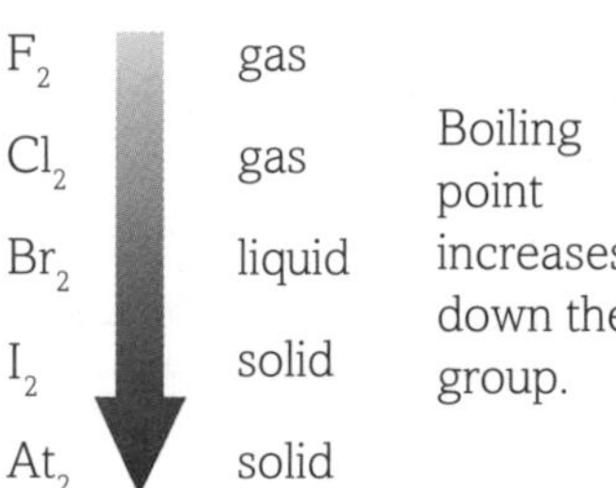

Figure 2 The changes in boiling points of the halogens, and therefore the physical state, moving down the group is due to increasing London forces.

Reactivity of the halogens

The halogens are very reactive and highly electronegative. They are very good at attracting and capturing electrons – this means that they are strong oxidising agents. During reactions each atom gains an electron to form 1– ions (known as halide ions) and obtain a noble gas configuration.

The reactivity and oxidising power of the halogens decrease moving down the group. This is because:

- the atomic radius increases (the nuclear pull is further away from the incoming electrons)
- the electron shielding increases
- the ability to gain an electron in the p sub-shell and form 1– ions decreases.

This decrease in reactivity can be shown in the reactions that halogens undergo with other halide ions.

Redox reactions

Redox reactions occur between aqueous solutions of halide ions – $Cl^-(aq)$, $Br^-(aq)$ and $I^-(aq)$ – and aqueous solutions of halogens – $Cl_2(aq)$, $Br_2(aq)$ and $I_2(aq)$.

A more reactive halogen will oxidise and displace a halide of a less reactive halogen – this is called a displacement reaction.

Halogens form different-coloured solutions, so colour changes indicate that redox reactions have occurred. The mixture is usually shaken with an organic solvent, such as cyclohexane, to help distinguish between bromine and iodine. Table 1 shows the colours of solutions of Cl_2, Br_2 and I_2 in water and in cyclohexane. The colours in a non-polar solvent are shown in Figure 3.

Halogen	Colour in water	Colour in cyclohexane
Cl_2	pale green	pale green
Br_2	orange	orange
I_2	brown	violet

Table 1 Colours of halogen solutions in different solvents.

Chlorine oxidises both Br^- and I^- ions:

- $Cl_2(aq) + 2Br^-(aq) \rightarrow 2Cl^-(aq) + Br_2(aq)$ orange in water and also in cyclohexane
- $Cl_2(aq) + 2I^-(aq) \rightarrow 2Cl^-(aq) + I_2(aq)$ brown in water; violet in cyclohexane

The changes in oxidation numbers for the reaction of chlorine with bromide ions are:

$Cl_2(aq) + 2Br^-(aq) \rightarrow 2Cl^-(aq) + Br_2(aq)$

0 → 2(−1) chlorine reduced

2(−1) → 0 bromine oxidised

Bromine can only oxidise I^- ions:

- $Br_2(aq) + 2I^-(aq) \rightarrow 2Br^-(aq) + I_2(aq)$ brown in water; violet in cyclohexane

Iodine does not oxidise either Cl^- or Br^- ions.

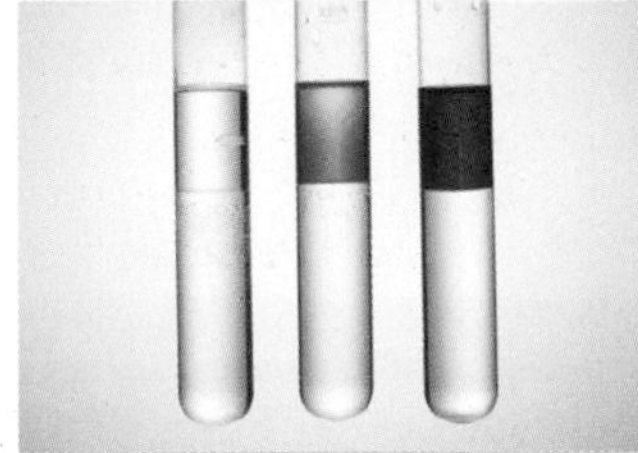

Figure 3 Three test tubes containing solutions of halogens. The lower layer is water and the upper layer is an organic solvent, in which the halogens are far more soluble. The halogens can be identified by their characteristic colours in each solvent.

Disproportionation

Disproportionation is a reaction in which the same element is both reduced and oxidised. It occurs in the following reactions.

KEY DEFINITION

Disproportionation is the oxidation and reduction of the same element in a redox reaction.

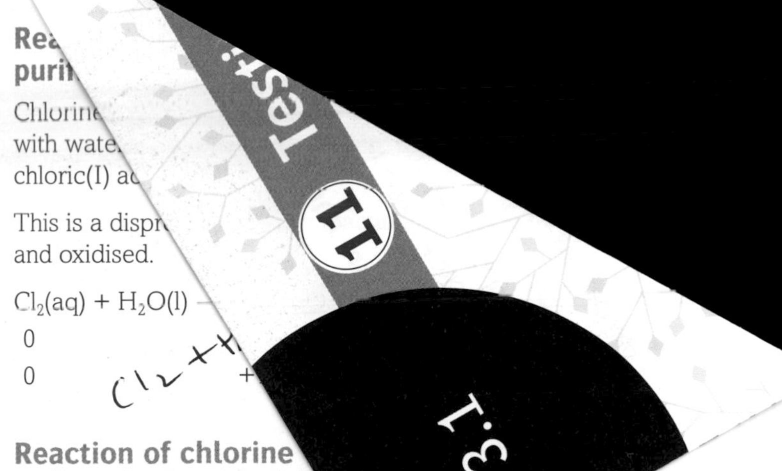

Rea… puri…

Chlorine… with wate… chloric(I) a…

This is a dispr… and oxidised.

$Cl_2(aq) + H_2O(l)$ …

0

0

Reaction of chlorine … sodium hydroxide (ble…

Chlorine is only slightly solubl… mild bleaching action. Household bleach is for… …te aqueous sodium hydroxide and chlorine react toge… room temperature. Chlorine is both oxidised and reduced in this reaction:

$Cl_2(aq) + 2NaOH(aq) \rightarrow NaCl(aq) + NaClO(aq) + H_2O(l)$

0 → −1 chlorine reduced

0 → +1 chlorine oxidised

DID YOU KNOW?

Should chlorine be added to water?

Many countries around the world choose to add chlorine to their water supplies – in the UK it has been routinely added since the 1890s. Chlorine kills bacteria and is credited with virtually eradicating some water-carried diseases, such as cholera and dysentery.

Some people believe that it should not be added, and it is not fair that individual people do not get to choose whether or not it is present in their water. Chlorine is a toxic gas and some environmentalists believe that chlorine reacts with organic matter to form traces of chlorinated hydrocarbons, suspected of causing cancer. In 1991, the government of Peru thought that this was an unacceptable risk and stopped adding chlorine to drinking water. Unfortunately, an outbreak of cholera followed that affected a million people and caused 10 000 deaths. The Peruvian government swiftly began adding chlorine to their drinking water again. Other methods are available for water purification including filtration, aeration, coagulation and treatment with ultraviolet light.

Questions

1. Explain how, and why, the reactivity and physical state of the halogens change moving down the group.
2. Write an ionic equation to show what would happen if an aqueous solution of chlorine were mixed with an aqueous solution of potassium bromide. Suggest what final colours you would observe if the reaction was carried out in a mixture of water and cyclohexane.
3. Bleach is formed in the disproportionation reaction between chlorine and cold aqueous sodium hydroxide. Use oxidation numbers to show if the reaction between chlorine and hot sodium hydroxide to form sodium chloride, sodium chlorate(V) and water is a disproportionation reaction.

   $6NaOH(aq) + 3Cl_2(aq) \rightarrow 5NaCl + NaClO_3 + 3H_2O$

...ing for ions

By the end of this topic, you should be able to demonstrate and apply your knowledge and understanding of:

* the precipitation reactions, including ionic equations, of the aqueous anions Cl^-, Br^- and I^- with aqueous silver ions, followed by aqueous ammonia, and their use as a test for different halide ions
* qualitative analysis of ions on a test-tube scale; processes and techniques needed to identify the following ions in an unknown compound:
  (i) anions – CO_3^{2-}, by reaction with H^+(aq) forming CO_2(g); SO_4^{2-}, by precipitation with Ba^{2+}(aq); Cl^-, Br^-, I^-
  (ii) cations – NH_4^+, by reaction with warm NaOH(aq) forming NH_3

Testing unknown solutions

Many chemicals look the same – in particular many are clear, colourless solutions with no smell. If you are trying to find the identity of a substance, its appearance may be of little help. Chemists use simple tests to identify parts, usually ions, in substances or to rule them out – these are called qualitative tests. Qualitative tests tell you which ions are present, but not how much.

Many qualitative tests involve a precipitation reaction – in a precipitation reaction an insoluble solid forms, usually from two aqueous solutions.

Qualitative tests are usually carried out on a small scale, in a test tube with only 1 or 2 cm^3 of the unknown substance. This is especially important if not much of the substance is available – you do not want to waste too much of it on identification tests.

Identifying anions

Anions are negative ions. The following can be used to test for carbonate ions (CO_3^{2-}), sulfate ions (SO_4^{2-}) and halide ions (Cl^-, Br^-, I^-).

Carbonate ions, CO_3^{2-}

Carbonate compounds, which contain CO_3^{2-} ions, react with acids:

$$CO_3^{2-}(aq) + 2H^+(aq) \rightarrow H_2O(aq) + CO_2(g)$$

Method	Positive test observations
• Add a dilute strong acid to the suspected carbonate. • Collect any gas formed and pass it through limewater.	• Fizzing/colourless gas is produced. • The gas turns limewater cloudy.

Table 1

Sulfate ions, SO_4^{2-}

Sulfate ions react with barium ions to form an insoluble salt, barium sulfate ($BaSO_4$):

$$Ba^{2+}(aq) + SO_4^{2-}(aq) \rightarrow BaSO_4(aq)$$

Method	Positive test observations
• Add dilute hydrochloric acid and barium chloride to the suspected sulfate.	• A white precipitate of barium sulfate is produced.

Table 2

LEARNING TIP

When testing unknown substances, it is important to carry the tests out in this order:

1. carbonate test
2. sulfate test
3. halide test.

Barium ions form an insoluble precipitate of $BaCO_3$ – if you hadn't already ruled out carbonate ions, you wouldn't know if the precipitate was this carbonate or barium sulfate.
Silver ions form an insoluble precipitate of Ag_2SO_4 – if you hadn't already ruled out sulfate ions, you wouldn't know if the precipitate was silver sulfate or a silver halide.

Halide ions, Cl^-, Br^-, I^-

Halide ions react with silver ions to form different-coloured silver halide precipitates, Ag*X* (where *X* represents the halide):

* silver and chloride – $Ag^+(aq) + Cl^-(aq) \rightarrow AgCl(s)$
* silver and bromide – $Ag^+(aq) + Br^-(aq) \rightarrow AgBr(s)$
* silver and iodide – $Ag^+(aq) + I^-(aq) \rightarrow AgI(s)$

Method	Positive test observations
• Dissolve suspected halide in water. • Add an aqueous solution of silver nitrate. • Note the colour of any precipitate formed. • If the colour is hard to distinguish, add aqueous ammonia (first dilute, then concentrated). • Note the solubility of the precipitate in aqueous ammonia.	• Silver chloride – white precipitate, soluble in dilute NH_3(aq). • Silver bromide – cream precipitate, soluble in concentrated NH_3(aq) only. • Silver iodide – yellow precipitate, insoluble in dilute and concentrated NH_3(aq).

Table 3

The coloured silver halide precipitates are shown in Figure 1.

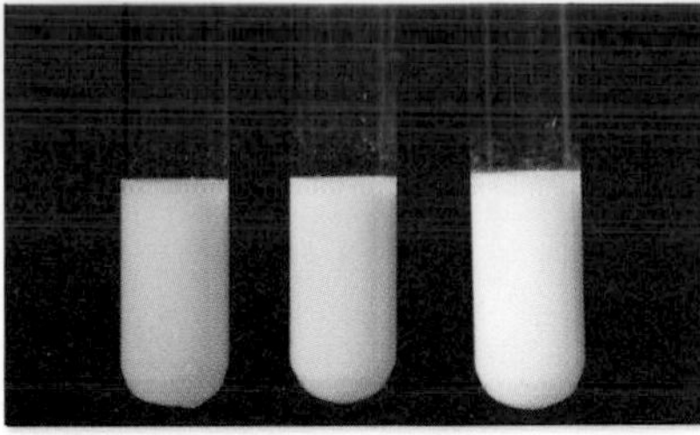

Figure 1 Halide tests using aqueous silver nitrate – silver halides are precipitated.

Identifying cations

Cations are positive ions, often containing metals ions.

Ammonium ions, NH_4^+

The ammonium ion reacts with hydroxide ions to produce ammonia and water:

$$NH_4^+(aq) + OH^-(aq) \longrightarrow NH_3(aq) + H_2O(aq)$$

Method	Positive test observations
• Add sodium hydroxide solution to the suspected ammonium compound and warm very gently. • Test any gas evolved with red litmus paper.	• Ammonia gas will turn red litmus paper blue. • Ammonia gas has a distinctive smell. (Ammonia gas is hazardous – this should be tested with care. Always follow the advice of your teacher.)

Table 4

LEARNING TIP

You need to be able to write ionic equations to describe what happens when halide ions react with silver ions – these equations are easy to write using just the ions that react together.

- $AgNO_3(aq)$ supplies $Ag^+(aq)$ ions.
- The halide compound, e.g. KBr, supplies $Cl^-(aq)$, $Br^-(aq)$ or $I^-(aq)$.
- Then $Ag^+(aq) + Cl^-(aq) \longrightarrow AgCl(s)$

You can just ignore the $NO_3^-(aq)$ ions and the ions the halide was bonded with. They don't react – they are known as spectator ions and do not need to be included in the ionic equation.

Questions

1 The following substances were tested: ammonium sulfate, ammonium chloride, potassium bromide and calcium carbonate.

Identify which of the results in Table 5, A–D, were obtained for each substance. Explain your decisions.

Substance	Method			
	• Add warm $NaOH(aq)$ solution. • Test with red litmus paper.	• Add $HCl(aq)$. • Test any gas with limewater.	• Add $HCl(aq)$ and $BaCl_2(aq)$.	• Add $AgNO_3(aq)$. • Then $NH_3(aq)$.
	Results			
A				Cream precipitate formed – this was soluble in concentrated $NH_3(aq)$ only.
B		Fizzed, gas turned limewater cloudy.		
C	Gas turned red litmus paper blue.		White precipitate formed.	
D	Gas turned red litmus paper blue.			White precipitate formed – this was soluble in dilute NH_3.

Table 5

2 Write the ionic equation for the reaction between substance D and silver nitrate.

THINKING BIGGER

NUTS ABOUT SELENIUM

Brazil nuts (see Figure 2) are a particularly rich source of the element selenium (Se). It is an essential dietary element in trace quantities, but it can be toxic in higher concentrations.

In this activity you will consider the chemistry of selenium and some of its compounds and make some predictions based on its position in the periodic table.

SOME OF OUR SELENIUM IS MISSING

Figure 1

In 1989, the multinational oil company Chevron discovered that it had cut its discharge of toxic selenium salts into the San Francisco Bay by almost three-quarters. The company should have been very pleased with itself. After all, the six oil companies in the area flush up to 3000 kilograms of selenium into the bay each year, and it seemed that by simply planting a 35-hectare wetland between its outfall and the bay, Chevron had found a practical answer to the problem.

But officials from the State Regional Water Quality Control Board were not so sanguine. They wanted to know where all the selenium was going and told Chevron to find out. The company searched in the sediment at the bottom of the wetland, it dug up the wetland plants, and it checked for a build-up of selenium compounds in the water. But despite all the effort, half of the selenium was still missing.

Alarm bells started to ring. Perhaps the local wildlife was eating the missing selenium. But there were no telltale signs – no dead or maimed animals. Then a team of biologists from the University of California at Berkeley made an educated guess. Norman Terry, Adel Zayed and their colleagues had been studying the way that some plants take toxic selenium salts from soil and water and turn them into the volatile gas dimethyl selenide. They suggested that Chevron's selenium could literally be vanishing into thin air.

- 18 November 1995 by Michelle Knott, *New Scientist*, magazine issue 2004

DID YOU KNOW?

At the start of this spread, we noted that brazil nuts are a particularly rich source of selenium, but don't go overboard! Too much selenium in the diet can cause the body to break down and excrete excess selenium in the form of the gaseous hydrogen selenide (H_2Se). This can leave the unwitting consumer of too many brazil nuts breathing out an odour not unlike garlic.

Figure 2

Where else will I encounter these themes?

1.1 2.1 2.2

Let us start by considering the nature of the writing in the article.

1. The article above was written for *New Scientist* and was based on research presented in the linked scientific paper. Consider the article and comment on the type of writing being used. Think about, for example, whether this is a scientist reporting the results of their experiments, a scientific review of a paper or a newspaper or magazine article for a specific audience. Try to answer the following questions:

a. Is there any bias present in the report? What type of words would make you think that was the case?

b. How has the use of language been adapted for the audience? Would the wording be different, for example, if this was aimed at an audience of 14–16 year olds? Or, if written by the press officer of the company Chevron?

Now we will look at the chemistry in, or connected to, this article. Do not worry if you are not ready to give answers to these questions yet. You may like to return to the questions once you have covered other topics later in the book. Use the time line at the bottom of the page to help you place this work in context with what you have already learned and what is ahead in your course.

2. Calculate the number of moles of selenium present in 3000 kg of selenium (line 6).

3. Give the electronic configuration using s, p, d notation of the Se atom and the Se^{2-} ion

4. Give the oxidation number of selenium in the following compounds:

Na_2SeO_3 Ag_2Se H_2SeO_4

5. Selenium is in group 16 of the periodic table. Give some reasons why selenium chemistry is likely to be similar to sulfur chemistry.

6. Suggest a shape for the molecule SeF_6

7. Give the formula for the molecule dimethyl selenide (paragraph 3, line 7) and use your knowledge of group 16 chemistry to suggest a shape for the molecule.

8. Dimethyl sulfide has a boiling point of 38 °C and dimethyl telluride has a boiling point of 82 °C). By drawing a suitable plot, suggest a boiling point for dimethyl selenide and explain your choice. What assumptions have you made? Consider the position of selenium relative to sulfur and tellurium in group 16 when you are drawing your plot. Make sure you use suitable axes and scale in order to obtain a reasonable suggestion for the boiling point!

Activity

In our world today there is an increasing need to communicate scientific ideas and concepts to people who consider themselves to be 'non-scientists'. Researchers may not always be the best people to communicate science (even their own!), but there are many occupations where this is an essential skill.

Imagine you need to give a 5–10 min PowerPoint presentation to the chief executive officer (CEO) of an oil company to argue the case for the use of genetically engineered plants to deal with selenium waste. How might you convince the CEO that the benefits are outweighed by any perceived concerns about genetic engineering?

Refer back to the article's final paragraph. You might also like to refer to the following document produced by the World Health Organization (WHO) and any other material you can find but you must carefully consider the nature of the source: is the material from a government website, a pressure group web site, a special interest web site, a personal blog etc?

3.1 Practice questions

1. Which of the following statements about the period 2 elements Li to Ne is true? [1]
   A. The first ionisation energy of nitrogen is higher than oxygen.
   B. Fluorine has the highest first ionisation energy in the period.
   C. Boron has a higher first ionisation energy than carbon.
   D. Lithium has the smallest atomic radius in the period.

2. Which of the following reactions are not redox reactions? [1]
   (i) $2Mg + O_2 \rightarrow 2MgO$
   (ii) $Mg + H_2O \rightarrow MgO + H_2$
   (iii) $MgCO_3 \rightarrow MgO + CO_2$
   (iv) $MgO + 2HCl \rightarrow MgCl_2 + H_2O$
   A. (i) and (iii) only
   B. (ii) only
   C. (iv) only
   D. (iii) and (iv) only

3. Look at the reaction below.
   $2NaOH(aq) + Cl_2(aq) \rightarrow NaCl(aq) + NaOCl(aq) + H_2O(l)$
   Which of the following statements about this reaction is true? [1]
   A. All of the chlorine is being oxidised.
   B. Chlorine has an oxidation number of +1 in NaCl.
   C. The reaction only takes place in the presence of ultraviolet light.
   D. Some of the chlorine is reduced.

4. Barium sulfate can be prepared in the following precipitation reaction.
   $BaCl_2(aq) + Na_2SO_4(aq) \rightarrow BaSO_4(s) + 2NaCl(aq)$
   If 25 cm^3 of 0.10 mol dm^{-3} $BaCl_2$ is added to 20 cm^3 of 0.20 mol dm^{-3} Na_2SO_4 and 80% of the dry $BaSO_4$ is recovered, what is the mass recorded (to 2 d.p.)? [1]
   A. 0.75 g
   B. 0.47 g
   C. 0.58 g
   D. 0.27 g

[Total: 4]

5. Calcium carbonate, $CaCO_3$, reacts with hydrochloric acid as shown in the equation below.
   $CaCO_3(s) + 2HCl(aq) \rightarrow CaCl_2(aq) + H_2O(l) + CO_2(g)$
   (a) 7.50×10^{-3} mol $CaCO_3$ reacts with 0.200 mol dm^{-3} HCl.
      (i) Calculate the volume, in cm^3, of 0.200 mol dm^{-3} HCl required to react with 7.50×10^{-3} mol $CaCO_3$. [2]
      (ii) Calculate the volume, in cm^3, of CO_2 formed at room temperature and pressure. [1]
   (b) When heated strongly, $CaCO_3$ decomposes.
   Write an equation, including state symbols, for the thermal decomposition of $CaCO_3$. [2]
   (c) Calcium oxide reacts with water and with nitric acid.
   State the formula of the calcium compound formed when:
      (i) calcium oxide reacts with water [1]
      (ii) calcium oxide reacts with nitric acid. [1]

[Total: 7]

[Q3, F321 June 2009]

6. Group 2 elements react with halogens.
   (a) Describe and explain the trend in reactivity of group 2 elements with chlorine as the group is descended.
   In your answer you should use appropriate technical terms, spelled correctly. [5]
   (b) A student was provided with an aqueous solution of calcium iodide.
   The student carried out a chemical test to show that the solution contained iodide ions. In this test, a precipitation reaction took place.
      (i) State the reagent that the student would need to add to the solution of calcium iodide. [1]
      (ii) What observation would show that the solution contained iodide ions? [1]
      (iii) Write an ionic equation, including state symbols, for the reaction that took place. [1]
      (iv) The student is provided with an aqueous solution of calcium bromide that is contaminated with calcium iodide.
      The student carries out the same chemical test but this time needs to add a second reagent to show that iodide ions are present.
      State the second reagent that the student would need to add. [1]

[Total: 9]

[Q4, F321 May 2013]

7. (a) Write balanced chemical equations for the following reactions:
    (i) The thermal decomposition of calcium carbonate. [2]
    (ii) The reaction of chlorine water with a solution of potassium iodide solution. [2]
    (iii) The reaction of chlorine gas with water. [2]
    (iv) The reaction of acidified silver nitrate with a solution of sodium chloride. [2]

   (b) From the above reactions in (a) identify the following:
    (i) A precipitation reaction. [1]
    (ii) A disproportionation reaction. [1]
    (iii) A redox reaction. [1]

[Total: 11]

8. Use your knowledge of bonding and atomic structure to explain the following:
   (a) The boiling points of the diatomic halogens increases in the order $F_2 < Cl_2 < Br_2 < I_2$. [3]
   (b) The first ionisation energies of the group 2 metals decreases in the order Mg > Ca > Sr > Ba. [3]
   (c) Magnesium and chlorine are both period 3 elements and yet magnesium is a solid at room temperature whereas chlorine is a gas. [3]

[Total: 9]

9. An impure sample of barium reacts with water to produce hydrogen gas and a solution of barium hydroxide. The volume of the gas collected was 350 cm^3 at S.T.P.
   (a) Write a balanced equation for the reaction of barium with water. [2]
   (b) If the mass of barium used was 2.53 g, what was the percentage barium in the sample? [4]
   (1 mole of gas occupies 22.4 dm^3 at S.T.P.)

[Total: 6]

10. Describe in detail how a pure dry sample of $MgCl_2.6H_2O$ could be prepared from $MgCO_3$ and 2.0 mol dm^{-3} hydrochloric acid. You should give details of the apparatus used in each step. [8]

[Total: 8]

MODULE 3 The periodic table and energy

CHAPTER 3.2

PHYSICAL CHEMISTRY

Introduction

Chemical reactions are accompanied by changes in energy. That is often the reason we have found out about chemical reactions – the earliest investigators were prehistoric people trying to find out how to make fire. In our modern world, we rely on energy heavily, so chemists need to be able to monitor and investigate energy changes that occur during reactions.

Industrial chemistry is the branch of chemistry responsible for making chemical substances on a huge scale, the extraction of salt, or production of ammonia for fertilisers being key examples. Industrial chemists want to be able to carry out reactions as quickly as possible, using the least amount of energy possible – as this is the recipe for the greatest profits. There is a pay-off however... fast reactions aren't always efficient reactions, and efficient reactions that give lots of products may take years, and the effect on the environment has to be considered too! In this module you will learn about the energy changes that occur during reactions, how to monitor the rate of reactions and how conditions can be controlled to make processes as economically viable, chemically feasible and sustainable as possible.

All the maths you need

To unlock the puzzles of this chapter you need the following maths:

- The units of measurement (*e.g. the mole*)
- Use of standard form and ordinary form (*e.g. quoting enthalpy values*)
- Using an appropriate amount of standard figures (*e.g. quoting bond enthalpies*)
- Changing the subject of an equation (*e.g. determining number of moles of a substance burned*)
- Substituting numerical values into algebraic equations (*e.g. calculating* ΔH)
- Solving algebraic equations (*e.g. calculating* ΔH)
- Plot and translate graphical information (*e.g. the Maxwell–Boltzmann Distribution curve*)

What have I studied before?

- That bond breaking requires energy and bond making releases energy
- Reactions can be endothermic or exothermic
- Different reactions occur at different rates
- How to monitor the rate of a reaction by collecting experimental results
- The factors that can affect the rate of a reaction and why these affect the rate of reaction

What will I study later?

- How Born–Haber cycles can be used to find out the enthalpy change associated with ionic bonding (AL)
- What entropy and free energy are and how these influence the feasibility of reactions (AL)
- The energy changes associated with fuel cells (AL)

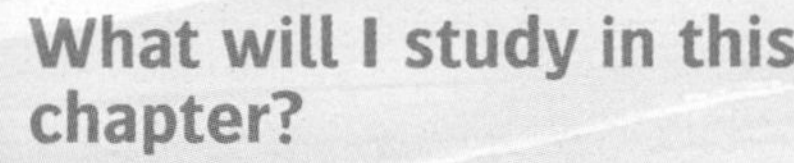

What will I study in this chapter?

- Enthalpy, both associated with bonds and specific overall reactions
- Why some reactions are endothermic and some are exothermic and how to represent these
- How enthalpy changes can be determined indirectly for theoretical reactions
- How energy is distributed amongst molecules
- How collision theory can explain why some reactions occur and others do not
- The effect of catalysts on the rate of reaction
- The equilibrium between reactants and products in some reactions and how this can be measured and altered

Enthalpy and reactions

By the end of this topic, you should be able to demonstrate and apply your knowledge and understanding of:

- **explanation that some chemical reactions are accompanied by enthalpy changes that are exothermic (ΔH, negative) or endothermic (ΔH, positive)**

What is chemical energy?

Chemical energy is a special form of potential energy that lies within chemical bonds. Chemical bonds are the forces of attraction that bind together atoms in compounds. When chemicals react to form new substances, bonds break in the reactants and new bonds are formed as products are made. This process changes the chemical energy of the substances.

During a chemical reaction, you may see changes to different forms of energy. For example, you may see chemical energy change into light energy (such as when magnesium burns in oxygen with a bright white light). However, chemists usually consider reactions in terms of changes in thermal (heat) energy. Chemists refer to thermal energy involved in reactions as enthalpy.

LEARNING TIP

Enthalpy changes are often quoted as $\Delta H^{\ominus}$, where $^{\ominus}$ means standard conditions of 100 kPa and 298 K.

Enthalpy

Enthalpy, *H*, is the thermal energy that is stored in a chemical system. It is impossible to measure directly the enthalpy of the reactants or products. However, we can measure the energy absorbed or released to the surroundings during a chemical change. The form of this energy can vary but chemists usually measure energy changes in reactions by monitoring thermal energy. Thermal energy can be monitored using temperature changes.

Systems and surroundings

You will often hear the system and the surroundings being discussed when enthalpy changes are considered:

- The system is the actual chemical reaction; that is, the atoms and bonds involved.
- The surroundings are everything else.

Consider the following reaction:

$$HCl(aq) + NaOH(aq) \rightarrow NaCl(aq) + H_2O(l)$$

The system would be only those atoms and bonds shown in the equation above. The surroundings are everything else, including the aqueous solution that all the substances are dissolved in. When you take the temperature of a reaction, including the one above, you are usually taking the temperature of the surroundings.

DID YOU KNOW?

The chemical system is the reactants and products. The surroundings are what is outside the chemical system. We determine the thermal energy (heat) exchange between the chemical system and the surroundings.

Conservation of energy

If thermal energy is released, the amount of energy that leaves a chemical system is exactly the *same* as the amount that goes into the surroundings. No energy is lost. It just transfers from one place to another, and some energy might change from one form to another. This is called the law of *conservation of energy.*

In this topic, energy changes will be treated as heat transfers. If heat is released or absorbed, we can easily measure the effect by recording the temperature change with a thermometer.

This means that:

- heat loss in a chemical system = heat gain to the surroundings (accompanied by a temperature increase)
- heat gain in a chemical system = heat loss from the surroundings (accompanied by a temperature decrease).

LEARNING TIP

The **law of conservation of energy** states that energy cannot be created or destroyed, only moved from one place to another.

Enthalpy change

In some reactions, the products of a reaction have *more* chemical energy than the reactants. In others, the products have *less* chemical energy than the reactants.

An enthalpy change, ΔH, is:

- the heat exchange with the surroundings during a chemical reaction, at constant pressure;
- the difference between the enthalpy of the products and the enthalpy of the reactants:

$$\Delta H = H_{\text{products}} - H_{\text{reactants}}$$

In general, all chemical reactions either release heat (exothermic reactions) or absorb heat (endothermic reactions).

Exothermic reactions

In an exothermic reaction:

- the enthalpy of the products is *smaller* than the enthalpy of the reactants
- there is a heat *loss* from the chemical system to the surroundings
- ΔH has a *negative* sign because heat has been lost by the chemical system.

An exothermic enthalpy change is shown in Figure 1.

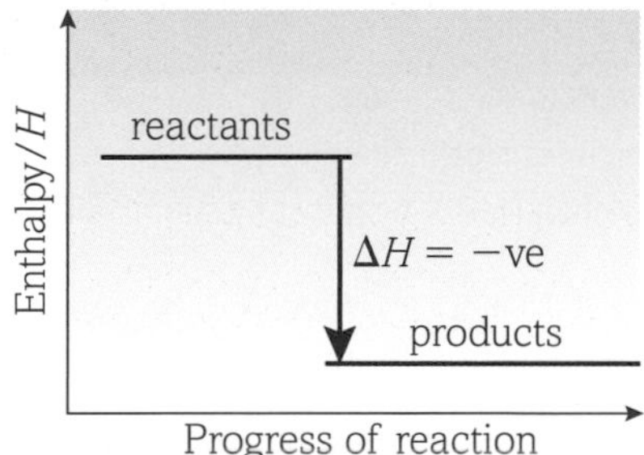

Figure 1 An exothermic enthalpy change.

Endothermic reactions

In an endothermic reaction:

- the enthalpy of the products is *greater* than the enthalpy of the reactants
- there is a heat *gain* to the chemical system from the surroundings
- ΔH has a *positive* sign because heat has been gained by the chemical system.

An endothermic enthalpy change is shown in Figure 2.

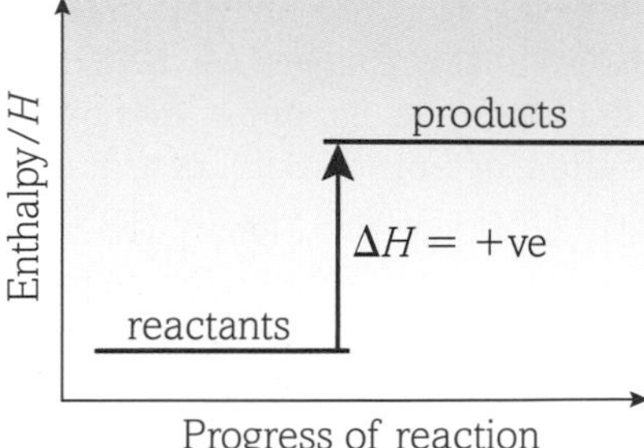

Figure 2 An endothermic enthalpy change.

LEARNING TIP

Exothermic reactions release energy and have a negative ΔH.
Endothermic reactions absorb energy and have a positive ΔH.

DID YOU KNOW?

Energy is measured in kilojoules (kJ).
Enthalpy changes are usually measured in kJ mol^{-1} (kilojoules per mole).
If an enthalpy change has a negative sign in front of it, the enthalpy change is exothermic, as the system is losing heat to the surroundings.
If an enthalpy change has a positive sign in front of it, the enthalpy change is endothermic, as the system is gaining heat from the surroundings.

Questions

1. Why are chemical reactions described in terms of enthalpy changes rather than energy changes?
2. What allows us to classify a reaction as exothermic or endothermic?
3. Photosynthesis occurs as follows:
   $6CO_2(g) + 6H_2O(l) \longrightarrow C_6H_{12}O_6(aq) + 6O_2(g) \quad \Delta H = +2801 \text{ kJ mol}^{-1}$
   Is photosynthesis an exothermic or an endothermic process? Use your answer to suggest why plants stop photosynthesising during the night.

2 Enthalpy profile diagrams

By the end of this topic, you should be able to demonstrate and apply your knowledge and understanding of:

* construction of enthalpy profile diagrams to show the difference in the enthalpy of reactants compared with products
* qualitative explanation of the term *activation energy*, including use of enthalpy profile diagrams

Simple enthalpy profile diagrams

Reactions and their enthalpy changes can be illustrated in an enthalpy profile diagram. The diagram shows what happens to the enthalpies of the reactants and products during the course of the reaction. It also emphasises the exothermic or endothermic nature of the reaction.

Exothermic reactions

The enthalpy profile diagram for the combustion of methane is shown in Figure 1.

- $H_{\text{products}} < H_{\text{reactants}}$
- The enthalpy change, ΔH, is negative.
- Excess energy is released into the surroundings as heat, so the temperature will rise.

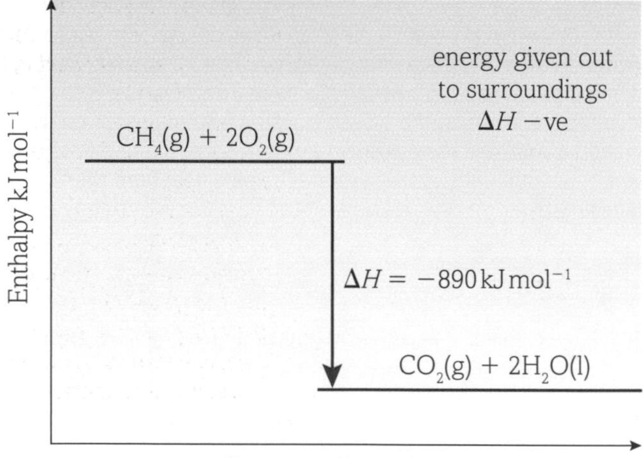

Figure 1 Enthalpy profile diagram for an exothermic reaction.

DID YOU KNOW?

Remember that H is the symbol for enthalpy.
In an exothermic reaction:

- ΔH is negative
- heat is given out to the surroundings
- the reacting chemicals lose energy
- heat lost by chemicals = heat gained by surroundings.

Endothermic reactions

The enthalpy profile diagram for the decomposition of calcium carbonate is shown in Figure 2.

- H products > H reactants.
- The enthalpy change, ΔH, is positive.
- Energy is required in order for the reaction to happen. This energy has to be removed from the surroundings, so the temperature will decrease.

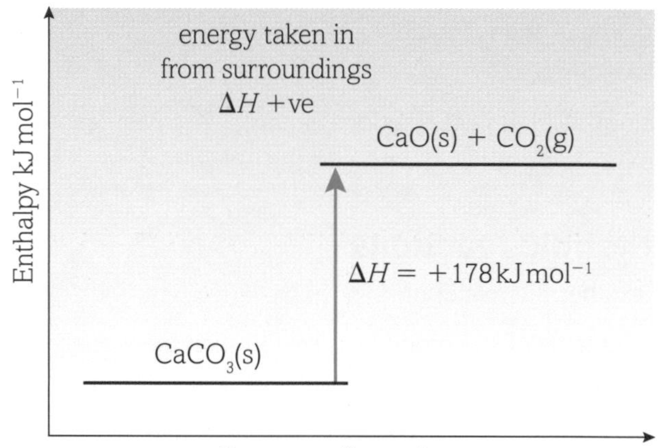

Figure 2 Enthalpy profile diagram for an endothermic reaction.

DID YOU KNOW?

In an endothermic reaction:

- ΔH is positive
- heat is taken in from the surroundings
- the reacting chemicals gain energy
- heat gained by chemicals = heat lost by surroundings.

Activation energy

Chemical reactions have an energy barrier that prevents many reactions from taking place spontaneously. This is called the **activation energy**, E_a, and it is required in order to break bonds in the reactants.

We can show activation energies in enthalpy profile diagrams.

KEY DEFINITION

Activation energy is the minimum energy required to start a reaction by breaking bonds in the reactants.

Exothermic reactions

Even though the products have a *lower* energy than the reactants, there still has to be an input of energy to break the first bond and start the reaction. The activation energy is often supplied by a spark or by heating the chemicals. For example, when natural gas combusts it needs a spark in order to overcome the activation energy of the combustion reaction. Once this has been overcome, the net output of energy provides the activation energy, so the reaction can continue. This is true of all *exothermic* reactions – once they begin, the activation energy is regenerated and the reaction becomes self-sustaining.

If activation energy did not exist, exothermic reactions would take place spontaneously. Fuels would not exist because they would immediately spontaneously combust!

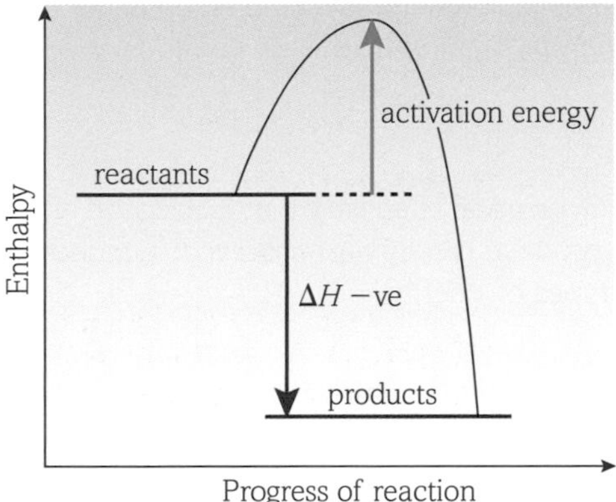

Figure 3 Activation energy for an exothermic reaction.

Endothermic reactions

Figure 4 shows the activation energy for an *endothermic* reaction.

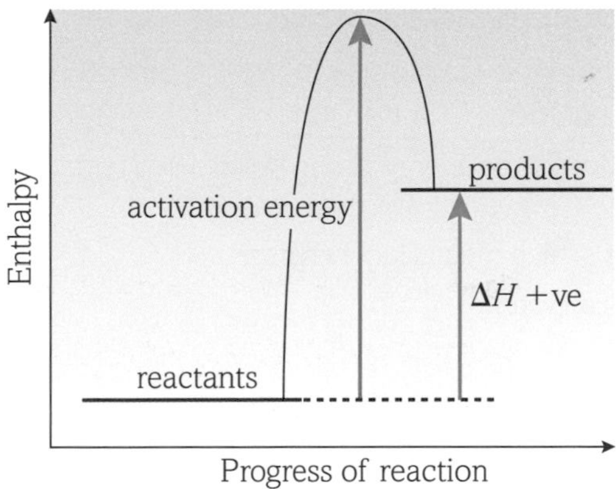

Figure 4 Activation energy for an endothermic reaction.

LEARNING TIP

It is important to show the activation energy and enthalpy change using arrows pointing in the correct direction:

- the activation energy is positive and the arrow always points up.
- the activation energy is always shown leading from the reactants to the top of the energy barrier. This is easier to show for exothermic reactions than for endothermic reactions, so take care!

Questions

1. You are given the data for the following reaction:

   $CO(g) + NO_2(g) \longrightarrow CO_2(g) + NO(g)$

   $\Delta H = -226\ kJ\ mol^{-1}$ $\quad E_a = +134\ kJ\ mol^{-1}$

   Draw an enthalpy profile diagram for this reaction, showing both of these energy changes. Identify and explain whether this reaction is endothermic or exothermic.

2. You are given the data for the following reaction:

   $H_2(g) + I_2(g) \longrightarrow 2HI(g)$

   $\Delta H = +53\ kJ\ mol^{-1}$ $\quad E_a = +183\ kJ\ mol^{-1}$

   Draw an enthalpy profile diagram for this reaction, showing both of these energy changes. Identify and explain whether this reaction is endothermic or exothermic.

3.2

3 Enthalpy terms

By the end of this topic, you should be able to demonstrate and apply your knowledge and understanding of:

* explanation and use of the terms:
  (i) *standard conditions* and *standard states* (physical states under standard conditions)
  (ii) *enthalpy change of reaction* (enthalpy change associated with a stated equation, $\Delta_r H$)
  (iii) *enthalpy change of formation* (formation of 1 mol of a compound from its elements, $\Delta_f H$)
  (iv) *enthalpy change of combustion* (complete combustion of 1 mol of a substance, $\Delta_c H$)
  (v) *enthalpy change of neutralisation* (formation of 1 mol of water from neutralisation, $\Delta_{neut} H$)

Thermodynamics

Thermodynamics is a branch of physical chemistry that focuses on energy in a chemical system, which is the reactants and the products. Energy is conserved; it cannot be created or destroyed. But it can be changed into different types of energy.

Standard conditions

For chemists to carry out meaningful calculations, all values must be measured under an agreed set of conditions. These are known as standard conditions. Standard conditions are:

- 100 kPa – this is the same as 100 000 Pa (0.986 atm).
- 273 K – remember, add 273 to convert from °C to K.

All substances are in their standard states, i.e. the most stable form. For example, carbon would be graphite rather than diamond.

Standard conditions are given the plimsoll line symbol, $^{\ominus}$.

LEARNING TIP

Some data books and textbooks quote a standard pressure of one atmosphere, i.e. 1 atm (1 atm = 101.325 kPa). This is not in agreement with IUPAC recommendations. For this reason, we will use 100 kPa as the standard pressure for thermochemical measurements.

Enthalpy change of reaction

Enthalpy change of reaction is the energy change associated with a given reaction. It has the symbol $\Delta_r H^{\ominus}$, where:

- Δ is the Greek letter delta, which means large change
- H is enthalpy or energy in a system
- $_r$ stands for reaction
- $^{\ominus}$ means under standard conditions.

A thermodynamic equation includes both the balanced chemical equation and enthalpy data.

Consider the endothermic thermal decomposition of copper carbonate, which can be represented by the following thermodynamic equation:

$$CuCO_3(s) \rightarrow CuO(s) + CO_2(g) \qquad \Delta_r H^{\ominus} = +1146\ kJ\ mol^{-1}$$

Figure 1 The thermal decomposition of copper carbonate is an endothermic reaction.

Enthalpy change of formation

Enthalpy change of formation is the energy change that takes place when 1 mole of a compound is formed from its constituent elements in their standard state under standard conditions. It has the symbol $\Delta_f H^{\ominus}$, where $_f$ stands for formation.

When writing a thermodynamic equation you may need to use fractions to balance the equation, as you must ensure there is only 1 mole of product.

Consider the exothermic reaction for the formation of water:

$$H_2(g) + \tfrac{1}{2}O_2(g) \rightarrow H_2O(l) \qquad \Delta_f H^{\ominus} = -286\ kJ\ mol^{-1}$$

Figure 2 When the activation energy is provided to a mixture of hydrogen and oxygen, an explosion produces water.

The standard enthalpy change of formation for an element in its standard state is 0 kJ mol^{-1}. This is because according to the definition there is no change (as no compound is formed), so no energy is released or taken in.

Enthalpy change of combustion

Enthalpy change of combustion is the energy change that takes place when 1 mole of a substance is completely combusted. It has the symbol $\Delta_c H^\ominus$, where $_c$ stands for combustion.

The standard enthalpy of combustion for ethane can be represented by the following thermodynamic equation:

$$C_2H_6(g) + 3\tfrac{1}{2}O_2(g) \rightarrow 2CO_2(g) + 3H_2O(l) \qquad \Delta_c H^\ominus = -1560\ \text{kJ mol}^{-1}$$

Enthalpy change of neutralisation

Enthalpy change of neutralisation is the energy change associated with the formation of 1 mole of water from a neutralisation reaction. It has the symbol $\Delta_{neut}H^\ominus$, where $_{neut}$ stands for neutralisation.

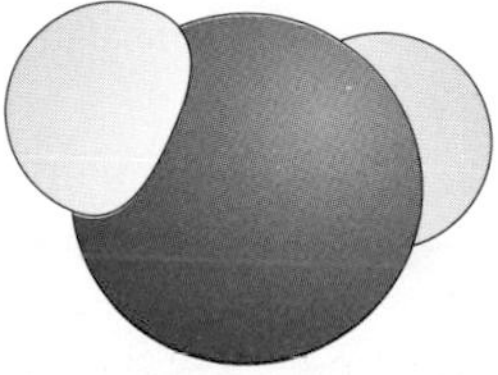

Figure 3 Enthalpy change of neutralisation is measured in terms of 1 mole of water produced.

KEY DEFINITIONS

The **enthalpy change of reaction**, $\Delta_r H^\ominus$, is the energy change associated with a given reaction.
The **enthalpy change of formation**, $\Delta_f H^\ominus$, is the energy change that takes place when 1 mole of a compound is formed from its constituent elements in their standard state under standard conditions.
The **enthalpy change of combustion**, $\Delta_c H^\ominus$, is the energy change that takes place when 1 mole of a substance is completely combusted.
The **enthalpy change of neutralisation**, $\Delta_{neut} H^\ominus$, is the energy change associated with the formation of 1 mole of water from a neutralisation reaction.

DID YOU KNOW?

The enthalpy of combustion of hydrogen is the same thermodynamic equation as the enthalpy of formation of water.

$$H_2(g) + \tfrac{1}{2}O_2(g) \rightarrow H_2O(l) \qquad -286\ \text{kJ mol}^{-1}$$

LEARNING TIP

Thermodynamic equations contain a balanced symbol equation and the ΔH value.

Questions

1. Write the equations for the enthalpy of combustion for:
   (a) ethanol
   (b) propane.

2. Write the equations for the enthalpy of formation for:
   (a) methane
   (b) methanol.

3. Write the equation for the enthalpy of neutralisation for the reaction between sulfuric acid and sodium hydroxide.

4 Calorimetry

By the end of this topic, you should be able to demonstrate and apply your knowledge and understanding of:

* determination of enthalpy changes directly from appropriate experimental results, including use of the relationship: $q = mc\Delta T$
* the techniques and procedures used to determine enthalpy changes directly and indirectly

Determination of enthalpy changes

You cannot measure enthalpy directly but you can measure enthalpy change. The temperature of a chemical system is monitored. If there is a temperature rise, an exothermic reaction has caused heat to be gained by the surroundings. If a temperature drop is recorded, an endothermic reaction has taken in heat from the surroundings.

Calorimetry uses a mathematical relationship to calculate enthalpy change from experimental quantitative data. The expression used is:

$$q = mc\Delta T$$

where:

- q is the heat exchanged with the surroundings, usually expressed in joules (J)
- m is the mass of the substance heated or cooled, usually expressed in grams (g)
- c is the **specific heat capacity** of the substance that is heated or cooled; it is the energy required to raise the temperature of 1 g of a substance by 1 K, usually expressed as $J\,g^{-1}\,K^{-1}$
- ΔT is the change in temperature, measured in kelvin (K).

WORKED EXAMPLE

0.25 g of ethane was used to heat 100 cm³ of water from 19.0 °C to 42.2 °C. What was the molar enthalpy change of this reaction? (Water has a specific heat capacity of $4.18\,J\,g^{-1}\,K^{-1}$.)

Use $q = mc\Delta T$, where:

m = mass (as 1 cm^3 of water has a mass of 1 g, the mass is 100 g)

$c = 4.18\,J\,g^{-1}\,K^{-1}$

ΔT = change in temperature measured in K

Convert into K by adding 273, then find the temperature change:
$(273 + 42.2) - (273 + 19) = 23.2$ K

So, $q = 100 \times 4.18 \times 23.2 = 9697.6$ J, released by 0.25 g of ethane.
This data can be used to generate the molar enthalpy change.

Ethane has a relative molecular mass of 30, so using $n = \frac{m}{M_r}$ there is 0.008 moles of ethane used in this reaction.

9697.6 J was released by 0.008 moles of ethane, so $\frac{9697.6}{0.008} = 1\,212\,200$ J $= 1212.2\,kJ\,mol^{-1}$ of energy released.

As this is an exothermic reaction, enthalpy change is negative and $\Delta H_c = -1212.2\,kJ\,mol^{-1}$.

However, the reported value is $-1560\,kJ\,mol^{-1}$. The experimental value is not accurate as the experiment is unlikely to be completed under standard conditions and incomplete combustion could have occurred. But the main error is due to heat lost to the surroundings.

KEY DEFINITIONS

Calorimetry is the quantitative study of energy in a chemical reaction.
Specific heat capacity is the energy required to raise the temperature of 1 g of a substance by 1 K.

Direct measurement of enthalpy of a reaction

A calorimeter can be used to study reactions that involve two chemicals in solution simply being mixed together.

INVESTIGATION

A coffee cup calorimeter

The simplest calorimeter is a plastic coffee cup, like the one shown in Figure 1.

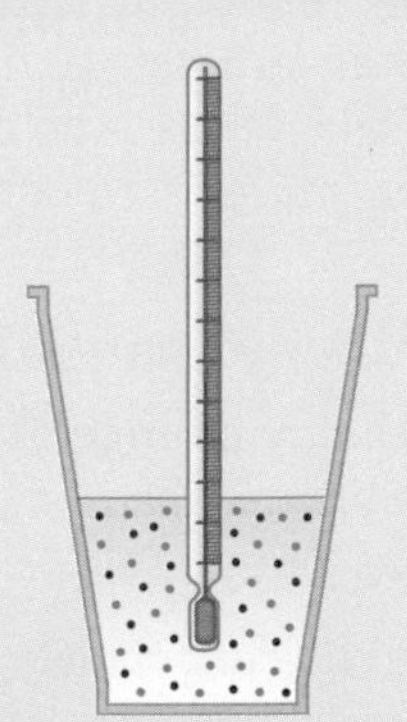

Figure 1 A coffee cup calorimeter.

The reaction vessel can be the insulated coffee cup held in a beaker. You could use a polystyrene lid with a hole for the thermometer to reduce heat loss to the surroundings.

- Add a measured mass of the first liquid reactant. Take the temperature every minute until it is stable. This usually takes around 4 minutes.
- At 5 minutes, add the second reactant. Do not take or record the temperature for the fifth minute.
- Monitor the temperature of the reaction mixture every minute for a further 5 minutes.
- Plot a graph to infer the maximum temperature change generated by the reaction.

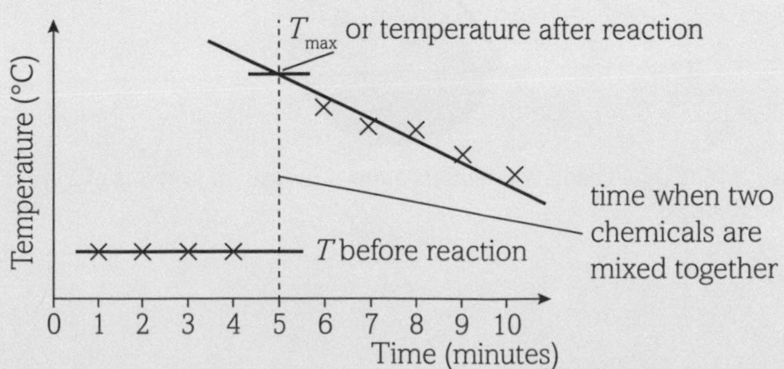

Figure 2 Graph to show how to infer the maximum temperature change of a reaction using coffee cup calorimetry.

DID YOU KNOW?

Remember, $q = mc\Delta T$ uses the mass of *all* the chemicals that are heated. So, when the two solutions are mixed, use the mass of both liquids, not just of the first reactant.
Often the specific heat capacity of water is used in the calculation, as solutions contain a lot of water and therefore have a value similar to this.

Direct measurement of enthalpy of combustion

INVESTIGATION

Copper calorimetry

When a fuel is combusted, the heat energy can be used to increase the temperature of a known mass of water.

- Measure the starting mass of the fuel.
- Add a known mass of water to a copper calorimeter.
- Mount the copper calorimeter over the fuel and take the starting temperature of the water.
- Combust the fuel for a few minutes and take the final temperature of the water.
- Take the mass of the unused fuel and calculate the mass of the fuel burnt.

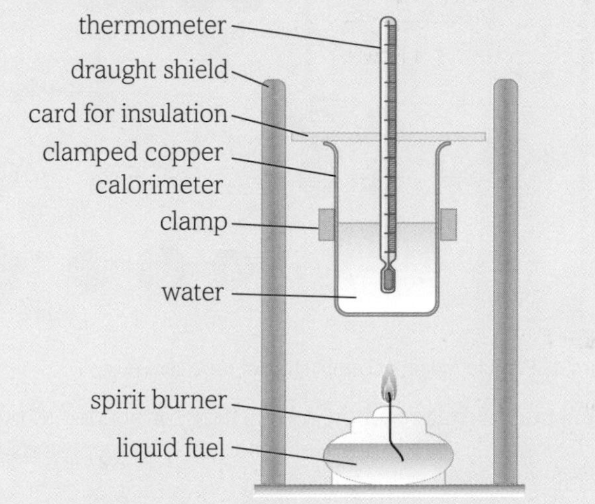

Figure 3 Copper calorimetry.

DID YOU KNOW?

Remember, $q = mc\Delta T$ uses the mass of the water in the calorimeter.

LEARNING TIP

Heat loss to the surroundings is the biggest experimental error. You can reduce it by insulating the equipment with draught shields.

A bomb calorimeter

A bomb calorimeter is a sophisticated piece of equipment that minimises heat loss as much as possible. It uses pure oxygen, to ensure complete combustion is achieved.

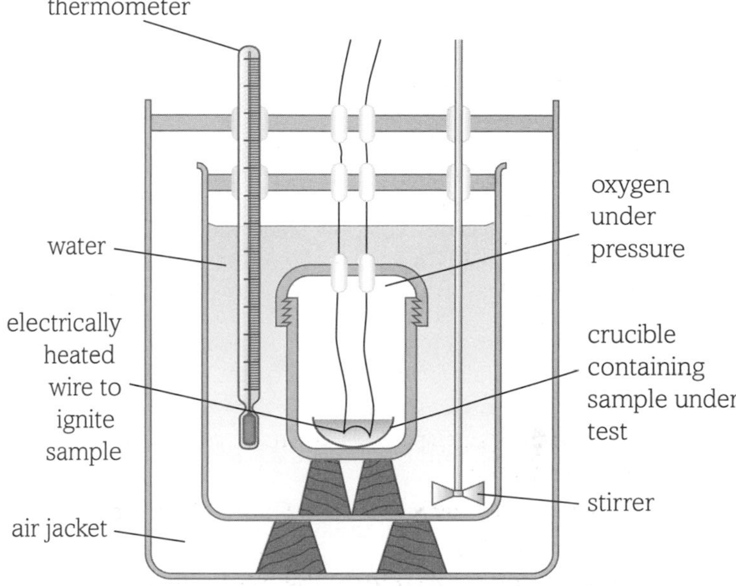

Figure 4 A bomb calorimeter.

Questions

1. 0.327 g of zinc powder was added to 55.0 cm^3 of aqueous copper(II) sulfate at 22.8 °C. The temperature rose to 32.3 °C. The copper sulfate was in excess. Find the enthalpy change of the reaction, shown by the following equation:

   $Zn(s) + CuSO_4(aq) \longrightarrow ZnSO_4(aq) + Cu(s)$

2. Calculate the enthalpy change of combustion when the combustion of 1.51 g of butan-1-ol heats 300 cm^3 of water by 42.0 °C.

Bond enthalpies

By the end of this topic, you should be able to demonstrate and apply your knowledge and understanding of:

- explanation of the term *average bond enthalpy* (breaking of 1 mol of bonds in gaseous molecules)
- explanation of exothermic and endothermic reactions in terms of enthalpy changes associated with the breaking and making of chemical bonds
- use of average bond enthalpies to calculate enthalpy changes and related quantities

Bond enthalpy

Energy is needed to break a chemical bond and the same amount of energy is released when the same bond is made. You get information about the strength of a chemical bond from its bond enthalpy. Bond enthalpies tell you how *much* energy is needed to break each different bond. You can then compare the strengths of different bonds.

Note that energy is *needed* to *break* bonds – the change is *endothermic*. When bonds *form*, the same quantity of energy is *released* – the change is *exothermic*.

The equations below show the bond enthalpies for H–H and H–Cl bonds:

$$H\text{–}H(g) \longrightarrow 2H(g) \qquad \Delta H = +436\ \text{kJ mol}^{-1}$$

$$H\text{–}Cl(g) \longrightarrow H(g) + Cl(g) \qquad \Delta H = +432\ \text{kJ mol}^{-1}$$

The H–H bond enthalpy value is always the same because a H–H bond can only ever exist in a H_2 molecule. Similarly, the H–Cl bond enthalpy applies only to a HCl molecule.

Average bond enthalpies

Unlike H–H and H–Cl bonds, some bonds can occur in different molecules. For example, almost every organic molecule contains C–H bonds. The C–H bond strength will vary across the different environments in which it is found.

Table 1 shows some **average bond enthalpies**. These are averaged over a number of typical chemical species containing that type of bond.

Bond	Average bond enthalpy/kJ mol^{-1}
C–H	+413
C=O	+805
H–O	+463
O=O	+497

Table 1 Values of some average bond enthalpies.

KEY DEFINITION

Average bond enthalpy is the mean energy needed for 1 mole of a given type of gaseous bonds to undergo homolytic fission.

Breaking and making bonds

A chemical reaction is often modelled as a three-step process:

1. Reactant bonds are broken. This process takes in energy and so is endothermic.
2. Atoms rearrange to form products.
3. Product bonds are formed. This releases energy and is an exothermic change.

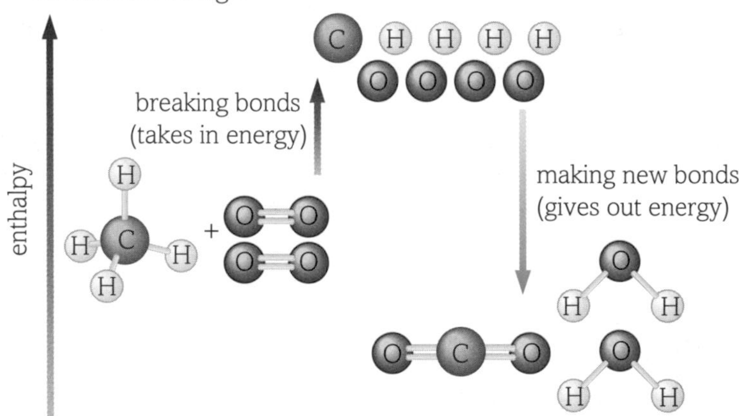

Figure 1 Particle model of combustion of methane.

In an endothermic reaction, more energy is needed to break the reactant bonds than is released when the product bonds are made. Overall the reaction takes in energy.

In an exothermic reaction, more energy is released when product bonds are formed than is needed to break reactant bonds. Overall the reaction releases energy.

Using average bond enthalpy data to predict enthalpy change

You can predict the enthalpy change for a reaction by using the average bond enthalpy data for the reactants and products. The average bond enthalpy is the mean energy needed for 1 mole of a given type of gaseous bonds to undergo homolytic fission. The actual bond enthalpy is individual to each molecule. However, an average value is taken across a variety of molecules.

So, the enthalpy change for a reaction can be calculated from average bond enthalpy data using the following expression:

$$\Delta H = \sum(\text{bond enthalpies of reactants}) - \sum(\text{bond enthalpies of products})$$

where $\sum$ is the Greek letter sigma and means 'sum of'.

WORKED EXAMPLE

Use bond enthalpy data to calculate the enthalpy of combustion for methane.

- Step 1: Write a balanced chemical equation using displayed formula: $CH_4 + 2O_2 \rightarrow CO_2 + 2H_2O$
- Step 2: Tally up the number and type of bond (see Table 1 for reference).

Bonds broken in the reactants = 4(C–H) + 2(O=O)
Bonds made in the reactants = 2(C=O) + 4(O–H)

- Step 3: Substitute into equation $\Delta H = \sum$(bond enthalpies of reactants) – $\sum$(bond enthalpies of products)

$\Delta H = [(4 \times 413) + (2 \times 497)] - [(2 \times 805) + (4 \times 463)] = -816\ \text{kJ mol}^{-1}$

The negative value of the enthalpy change shows that the reaction is exothermic. This is the expected outcome as combustion reactions always release heat and so are exothermic.

DID YOU KNOW?

In an endothermic reaction, the reactants have less energy than the products and so there is a positive enthalpy change.
In an exothermic reaction, the reactants have more energy than the products and so there is a negative enthalpy change.

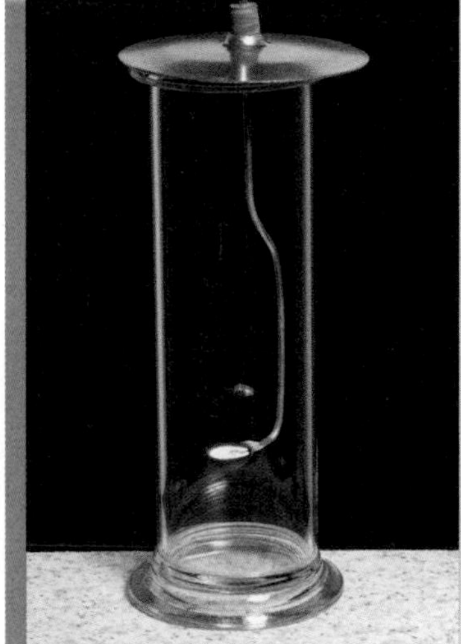

Figure 2 It is not possible to measure the standard enthalpy of formation for carbon monoxide.

Calculating enthalpy change for incomplete combustion

The enthalpy change for an incomplete combustion reaction is difficult or impossible to measure directly. This is because there is often complete and incomplete combustion happening simultaneously.

However, using average bond enthalpy data, this enthalpy change can be calculated. Remember, though, that the values are averages for the bond in a variety of environments in different molecules, so it is not as accurate as direct measurement using precise and sensitive equipment like a bomb calorimeter.

LEARNING TIP

When you have calculated ΔH using bond enthalpy data just check the sign. Combustion should always be a negative ΔH as it is an exothermic reaction.

Questions

1 Use the average bond enthalpy data in the table below to calculate:

(a) the enthalpy of combustion of hydrogen
(b) the enthalpy change associated with the direct hydration of ethene
(c) the enthalpy of formation of ethane
(d) the enthalpy of formation of ammonia.

Bond	Average bond enthalpy/kJ mol^{-1}
C–H	+413
C–C	+347
C=C	+612
C=O	+805
O–H	+463
H–H	+436
N≡N	+945
N–H	+391
C–O	+336
O=O	+497

6 Hess' law and enthalpy cycles

By the end of this topic, you should be able to demonstrate and apply your knowledge and understanding of:

* Hess' law for construction of enthalpy cycles and calculations to determine indirectly:
  (i) an enthalpy change of reaction from enthalpy changes of combustion
  (ii) an enthalpy change of reaction from enthalpy changes of formation
  (iii) enthalpy changes from unfamiliar enthalpy cycles
* the techniques and procedures used to determine enthalpy changes directly and indirectly

Measuring enthalpy changes indirectly

In a chemical reaction, new substances are made and an energy or enthalpy change occurs. Some enthalpy changes are impossible to measure directly (see topic 3.2.4). For these reactions, enthalpy cycles can be used to indirectly calculate the enthalpy change.

Hess' law and enthalpy cycles

Hess' law states that the enthalpy change in a chemical reaction is independent of the route it takes.

Consider a chemical reaction where reactant chemicals A and B form product C. This can be written as a balanced chemical equation:

$$A + B \rightarrow C$$

Imagine that sometimes A and B can react to make D, and then D can react to become C. This can be shown by two balanced chemical equations:

$$A + B \rightarrow D, \text{ then } D \rightarrow C$$

These chemical reactions can be joined in an **enthalpy cycle**. An enthalpy cycle shows alternative routes between reactants and products.

A simple enthalpy cycle should look like a triangle. Each corner of the shape should have chemicals. All the corners should have the same amount of each atom, so you may need to add numbers to balance them.

The sides of the triangle are made of arrows. These show the direction of a chemical reaction, from the reactants to the products, and the enthalpy change can be written on the arrow.

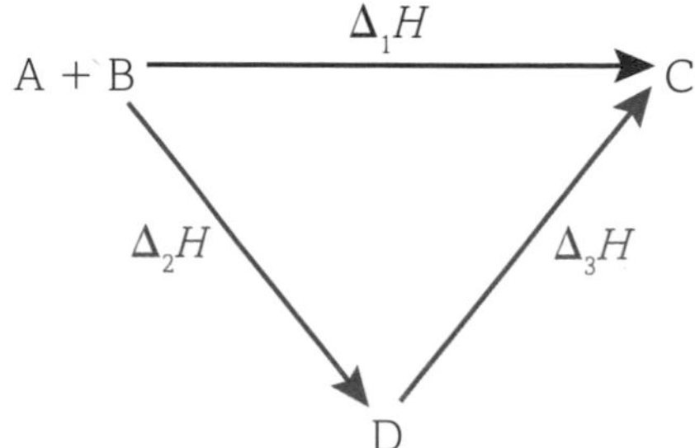

Figure 1 An enthalpy cycle.

The enthalpy change to go directly from A and B to C is the same as if you go through the intermediate D. So, $\Delta_1H = \Delta_2H + \Delta_3H$.

KEY DEFINITIONS

Hess' law states that the enthalpy change in a chemical reaction is independent of the route it takes.
An **enthalpy cycle** is a pictorial representation showing alternative routes between reactants and products.

DID YOU KNOW?

Germain Henri Hess (1802–1850) was a chemist and medical doctor. He wrote a paper on thermochemistry that included the idea of Hess' law. (A paper is a method that researchers use to communicate their work and findings.)
Hess' ideas were later developed into the first law of thermodynamics. This is an application of the conservation of energy principle (energy cannot be created or destroyed) to heat and chemical reactions.

WORKED EXAMPLE 1

Enthalpy change of reaction from enthalpy changes of combustion
When ethene and hydrogen are heated to about 150 °C with a nickel catalyst, ethane is made. Using enthalpy of combustion data and an enthalpy cycle, we can calculate the enthalpy change for this reaction.

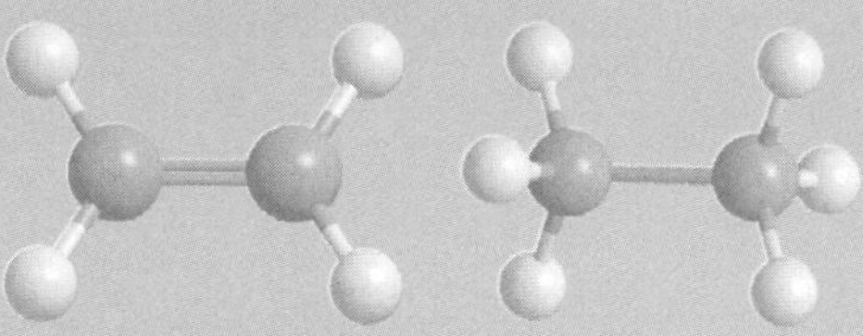

Figure 2 Ball and stick diagrams of ethene and ethane.

Ethene can be completely combusted to form carbon dioxide and water only:

$$C_2H_4(g) + 3O_2(g) \rightarrow 2CO_2(g) + 2H_2O(g)\ \Delta_cH^{\ominus} = -1393\ \text{kJ mol}^{-1}$$

Ethane can be completely combusted, forming the same products, but in different ratios:

$$C_2H_6(g) + 3\tfrac{1}{2}O_2(g) \rightarrow 2CO_2(g) + 3H_2O(l)\Delta_cH^{\ominus} = -1561\ \text{kJ mol}^{-1}$$

This means that an enthalpy cycle can be written. One corner is combustion products and the other two are the chemicals that the enthalpy of combustion is for. Remember, each corner of the triangle needs to have the same number and type of each atom. So, oxygen needs to be added to balance all the equations.

$$\Delta H = (-1393) + (-286) - 1561 = -118\ \text{kJ mol}^{-1}$$

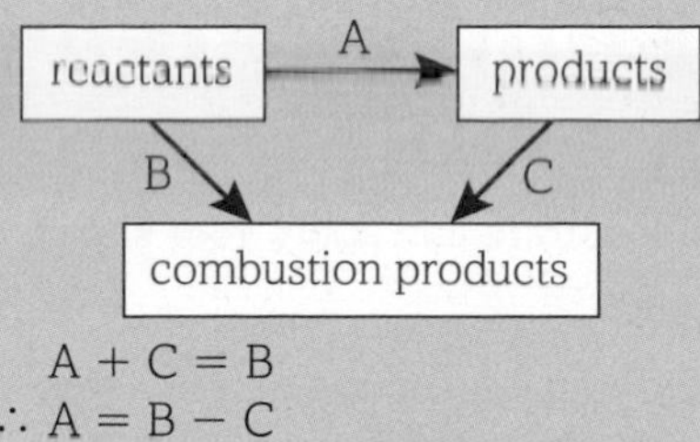

Figure 3 Using $\Delta_c H^{\ominus}$ values, the general enthalpy cycle can be drawn.

For hydrogen, $H_2(g) + \frac{1}{2}O_2(g) \rightarrow H_2O(l)$ $\Delta_c H^{\ominus} = -286\ \text{kJ mol}^{-1}$.
These data and the enthalpy cycle can be used to calculate the enthalpy change for the hydrogenation of ethene:

$$C_2H_4(g) + 3\tfrac{1}{2}O_2(g) + H_2(g) \longrightarrow C_2H_6(g) + 3\tfrac{1}{2}O_2(g)$$

$$\searrow \qquad \swarrow$$

$$2CO_2(g) + 3H_2O(g)$$

Figure 4 Enthalpy cycle from combustion data.

LEARNING TIP

If you are a good mathematician, you may not need to draw the cycle. With enthalpy of combustion data, the unknown enthalpy change can be calculated using the following formula:

$\Delta H = \sum \Delta_c H^{\ominus}(\text{reactants}) - \sum \Delta_c H^{\ominus}(\text{products})$,

where $\sum$ is the Greek letter sigma and means 'sum of'.

WORKED EXAMPLE 2

Enthalpy change of reaction from enthalpy changes of formation

In the Contact process, sulfur trioxide, SO_3, is formed from burning sulfur dioxide, SO_2, and oxygen at high temperatures, with V_2O_5 as the catalyst. Enthalpy of formation data can be used to calculate the enthalpy change of this reaction.

Substance	$\Delta_f H^{\ominus}/\text{kJ mol}^{-1}$
$SO_2(g)$	−297
$SO_3(l)$	−441

Table 1 Enthalpy of formation data.

Step 1: Write the thermodynamic equations for all the data:

$SO_2(g) + \frac{1}{2}O_2(g) \rightarrow SO_3(l)$ $\Delta_r H^{\ominus} = ?$

$S(s) + O_2(g) \rightarrow SO_2(g)$ $\Delta_f H^{\ominus} = -297\ \text{kJ mol}^{-1}$

$S(s) + 1\frac{1}{2}O_2(g) \rightarrow SO_3(l)$ $\Delta_f H^{\ominus} = -441\ \text{kJ mol}^{-1}$

Step 2: Construct the enthalpy cycle.
Remember that a simple enthalpy cycle should look like a triangle. This time, one corner should be the elements and the other corners should be the compounds that you have formation data for. It is important to balance every side of the cycle, even if you need to add fractions of an element. Then identify the two routes for the enthalpy change.

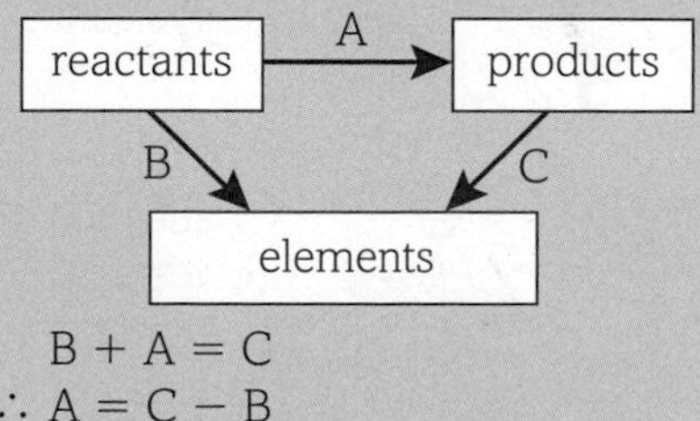

Figure 5 Using $\Delta_f H^{\ominus}$ values, the general enthalpy cycle can be drawn.

Step 3: Write a mathematical expression for the enthalpy cycle.

$\Delta_f H^{\ominus}(SO_2) + \Delta_r H^{\ominus} = \Delta_f H^{\ominus}(SO_3)$

Step 4: Rearrange to make the desired enthalpy change the subject of the equation. Then, substitute in the data and calculate a value.

$\Delta_r H^{\ominus} = \Delta_f H^{\ominus}(SO_3) - \Delta_f H^{\ominus}(SO_2) = -441 - (-297) = -144\ \text{kJ mol}^{-1}$

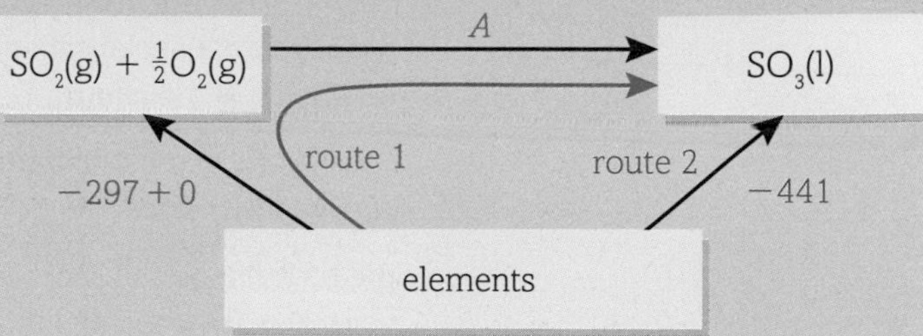

Figure 6 Use the enthalpy cycle to generate an expression to calculate the enthalpy change.

LEARNING TIP

With enthalpy of formation data, the unknown enthalpy change can be calculated using the following formula:

$\Delta H = \sum \Delta_f H^{\ominus}(\text{products}) - \sum \Delta_f H^{\ominus}(\text{reactants})$

Questions

1. Use the following enthalpy change of combustion data to calculate the enthalpy change of formation of:
   (a) butane
   (b) ethanol.

Substance	$\Delta_c H^{\ominus}/\text{kJ mol}^{-1}$
C(s)	−394
$H_2(g)$	−286
$C_4H_{10}(g)$	−2877
$C_2H_5OH(l)$	−1367

2. Use the following enthalpy change of formation data to calculate the enthalpy change for:
   (a) $2NO(g) + O_2(g) \rightarrow 2NO_2(g)$
   (b) $4NH_3(g) + 5O_2(g) \rightarrow 4NO(g) + 6H_2O(l)$

Substance	$\Delta_f H^{\ominus}/\text{kJ mol}^{-1}$
$NO_2(g)$	+33
$NH_3(g)$	−46
NO(g)	+90
$H_2O(l)$	−286

Collision theory and rates of reaction

By the end of this topic, you should be able to demonstrate and apply your knowledge and understanding of:

* **the effect of concentration, including the pressure of gases, on the rate of a reaction, in terms of frequency of collisions**
* **calculation of reaction rate from the gradients of graphs measuring how a physical quantity changes with time**
* **the techniques and procedures used to investigate reaction rates including the measurement of mass, gas volumes and time**

Rate of reaction

Reactions can be very rapid, taking only a fraction of a second (for example an explosion) or very slow, taking millions of years (for example graphite turning into diamond). The rate of reaction is defined as the change in concentration of a reactant or a product in a given time:

$$\text{rate} = \frac{\text{change in concentration}}{\text{time}} \quad \text{Units: mol dm}^{-3}/\text{s} = \text{mol dm}^{-3}\text{s}^{-1}$$

Usually:

- concentrations of reactants are highest at the start of a reaction when the rate is at its *fastest* (often referred to as the *initial rate*)
- the rate slows down as reactant concentrations *decrease* (reactants get used up)
- when any reactant has a concentration of zero (it is used up), the reaction *stops* and the rate is zero.

LEARNING TIP

You should read the units of rate, mol dm^{-3} s^{-1}, as 'moles per decimetre cubed per second'.

Factors affecting the rate of a chemical reaction

Reaction rates are affected by a number of factors. These include:

- temperature (increased temperature → increased rate)
- pressure, for gaseous reactants only (increased pressure → increased rate)
- concentration (increased concentration → increased rate)
- surface area (increased surface area → increased rate)
- adding a catalyst (catalysts increase the rate).

Collision theory is used to explain the effect of these changes. For two molecules to react, they must first *collide*. This collision must have sufficient energy to overcome the activation energy of the reaction, and the collision must have the correct orientation (direction) (see Figure 1).

The effect of concentration on reaction rate

- Increased concentration gives more molecules in the same volume.
- The molecules will be closer together and so there is a greater chance of them colliding with sufficient energy to overcome the activation energy.
- Collisions will be more frequent – more collisions will occur in a certain length of time.

Therefore, rate increases with increased concentration.

The effect of pressure on reaction rate

- When the pressure of a gas is increased, the molecules are pushed closer together.
- The same number of molecules occupies a smaller volume.
- For a gaseous reaction, increasing the pressure is the same as increasing the concentration.
- More collisions are likely to occur with sufficient energy to overcome the activation energy.

Therefore, rate increases with increased pressure (for gaseous reactions only).

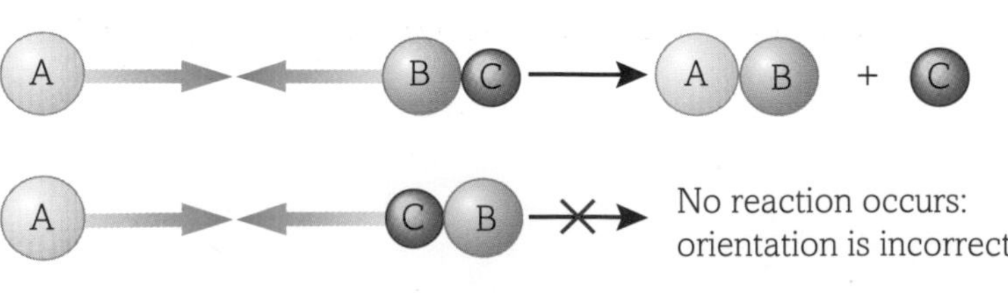

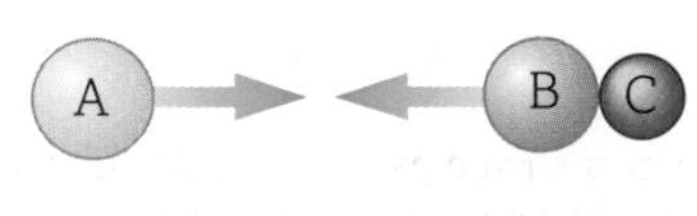

A head-on collision will provide sufficient energy to overcome the activation energy.

A glancing blow will not provide sufficient energy to overcome the activation energy and no reaction will occur.

Figure 1 The effect of orientation and energy of collisions on rate of reaction for the reaction A + BC → AB + C.

LEARNING TIP

Make sure you can use collision theory to explain how concentration and pressure affect the rate of a reaction. Describe what is happening to the particles and how the number of collisions, the energy of the collisions and their orientation is affected.

Calculating the rate of a reaction

The rate of a reaction can be calculated by monitoring changes in physical quantities during a reaction. These could include:

- concentration of a reactant or product (e.g. using a titration)
- gas volume of products (e.g. using a gas syringe to collect gases formed)
- mass of substances formed, or decreasing mass of reactants (e.g. monitoring a reaction on a balance).

These values are plotted against time on a graph and the rate is equal to the gradient.

Figure 2 shows a graph of a reactant's concentration over time. The curve is a downward curve, as the reactant is gradually being used up.

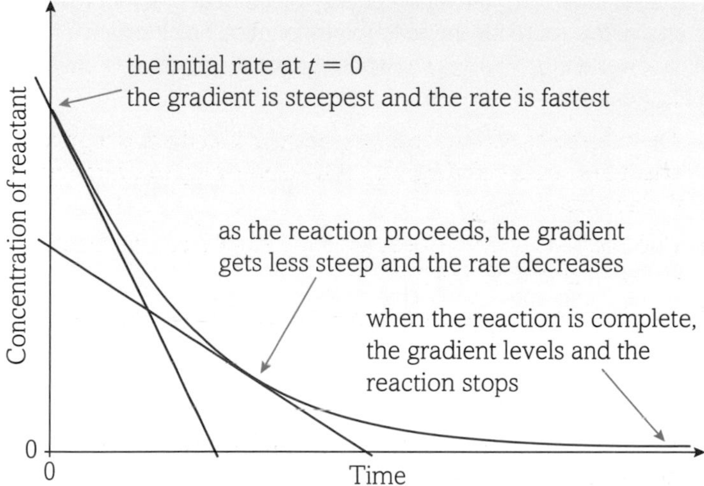

Figure 2 A graph showing the decrease in concentration of a reactant over time. The gradient has been taken in two places: at the very start of the reaction (known as $t = 0$) and during the reaction. The gradient of the curve is steepest at $t = 0$, showing the rate of reaction was fastest at this point.

LEARNING TIP

Remember that the gradient of the line is equivalent to the rate of the reaction.

$$\text{gradient} = \frac{\text{change on } y\text{-axis}}{\text{change on } x\text{-axis}}$$

You can remember this as $\frac{\text{'rise'}}{\text{'run'}}$

If you need to calculate the gradient of a curved part of a graph, you will first need to draw a flat line against it, called a tangent line. You then work out the 'rise' and 'run' of this tangent line.

Figure 3 shows a graph of a product's concentration over time. Notice that this has an upward curve, as the product's concentration rises as it is produced. Any physical quantity can be used to plot a graph to calculate gradient and therefore rate, as long as it changes during the reaction and can be monitored and recorded.

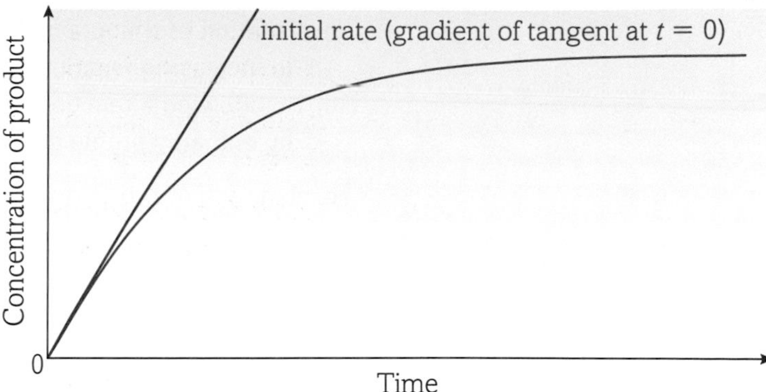

Figure 3 A graph showing the increase in concentration of a product over time. The gradient of the curve is equivalent to the rate of reaction.

LEARNING TIP

Reaction rates

Chemists must be able to monitor the changes in concentrations of substances over time in order to determine reaction rates. Some of the techniques they could use are:

- collecting gas in a gas syringe during a reaction
- monitoring a reaction's mass during a reaction
- recording the time it takes for a particular change to occur, such as a precipitate forming
- using a colorimeter to monitor changes in concentrations of coloured reactants or products.

Questions

1 Using collision theory, explain how the following changes affect the rate of a reaction:

(a) an increase in concentration of a reactant

(b) an increase in pressure.

2 A student recorded the volume of gas produced during a reaction. Her results were:

Time (minutes)	0	1	2	3	4	5	6	7	8	9	10
Volume of gas (cm^3)	0	2.5	6.0	7.0	7.6	7.8	7.9	8.0	8.0	8.0	8.0

(a) Plot a graph of her results and use it to calculate the gradient at $t = 2$ minutes and $t = 8$ minutes (t = time). Express the times as seconds.

(b) Express this gradient as a rate, using the units $mol\,dm^{-3}\,s^{-1}$.

8 Catalysts

By the end of this topic, you should be able to demonstrate and apply your knowledge and understanding of:

* explanation of the role of a catalyst:
  (i) in increasing reaction rate without being used up by the overall reaction
  (ii) in allowing a reaction to proceed via a different route with lower activation energy, as shown by enthalpy profile diagrams
* explanation of the terms *homogeneous* and *heterogeneous* catalysts
* explanation that catalysts have great economic importance and benefits for increased sustainability by lowering temperatures and reducing energy demand from combustion of fossil fuels with resulting reduction in CO_2 emissions

What is a catalyst?

We have known about catalysts for many centuries. Bread-making and the fermentation of wine both involve catalysts, although the people who developed these techniques probably knew little about the Chemistry involved!

LEARNING TIP

A **catalyst** is a substance that increases the rate of a reaction without being used up during the process.

The Swedish scientist Jöns Jakob Berzelius (1779–1848) was the first person to use the term catalyst. He was carrying out experiments on hydrogen peroxide, H_2O_2, which breaks down slowly into water and oxygen as follows:

$$2H_2O_2(l) \rightarrow 2H_2O(l) + O_2(g)$$

Berzelius discovered that the addition of certain compounds made the hydrogen peroxide break down faster than it did when left alone. He thought that these *catalysts* broke down hydrogen peroxide by loosening the bonds within it.

Figure 1 Catalase, an enzyme found in raw liver, accelerates the decomposition of hydrogen peroxide, H_2O_2. An enzyme is a biological catalyst, increasing reaction rates without itself reacting or changing.

A catalyst works by increasing the rate of a chemical reaction without being used up in the process:

- A catalyst may react to form an intermediate.
- The catalyst is later regenerated, so it does not undergo any permanent change.

A catalyst *lowers* the activation energy of the reaction by providing an alternative route for the reaction to follow. The alternative route has a lower energy. This is shown in the reaction pathway diagram in Figure 2.

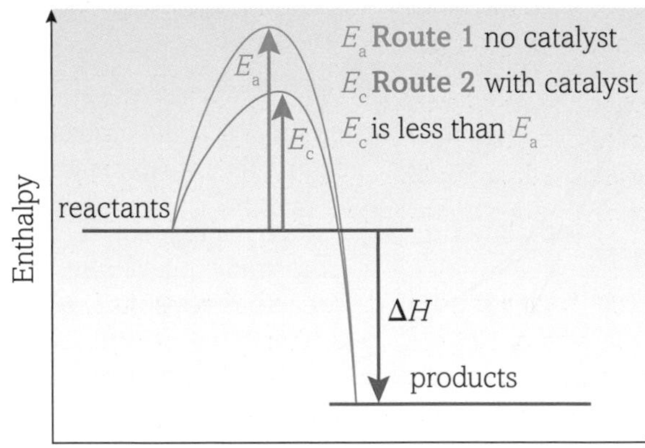

Figure 2 Reaction pathway diagram showing a catalyst lowering the activation energy.

Types of catalysts

Homogeneous catalysts

If a catalyst for a reaction is in the same phase as the reactants, it is called a *homogeneous* catalyst. This could include, for example, a liquid catalyst that is mixed with liquid reactants. It could also include a gaseous catalyst with gaseous reactants.

Enzymes within bodily fluids such as blood or saliva are examples of homogeneous catalysts.

Heterogeneous catalysts

If a catalyst for a reaction is in a different phase from the reactants, it is called a *heterogeneous* catalyst. A typical example for this would be a solid catalyst used in liquid reactants, or gaseous reactants passed over a solid catalyst.

Phases do not just include the states of solid, liquid and gas. If all the chemicals are immiscible liquids (that is, they do not mix), the catalyst could still be heterogeneous, as it could be in a different liquid layer from the reactants.

A catalytic converter in a car is an example of a heterogeneous catalyst. It includes a solid metal structure that has been coated in platinum, rhodium and palladium, which act as the catalysts. Gaseous exhaust fumes pass over this and are converted into less harmful substances.

The economic importance of catalysts

Catalysts help to lower the energy demands of processes. By doing so, they reduce costs and also help the environment. Less fossil fuel needs to be burned to generate the required energy, and this also means lower carbon dioxide emissions.

For example, the Haber process for the production of ammonia is of great economic importance. The ammonia produced is used as the basis for fertiliser manufacture, improving crop yields to feed the ever-increasing world population.

Ammonia is made by reacting together nitrogen and hydrogen:

$$N_2(g) + 3H_2(g) \rightleftharpoons 2NH_3(g)$$

A lot of energy is required to break the triple bond in nitrogen, N≡N, contributing to a high activation energy. Iron is used to catalyse this reaction, weakening the N≡N bond and lowering the activation energy and therefore the costs – both financial and environmental.

Another important industrial application of catalysts is within catalytic convertors. Catalytic converters play an important role in improving our air quality by reducing toxic emissions from vehicles and preventing photochemical smog.

Whole branches of chemistry are based on the development of successful catalytic processes. It is a profitable business!

DID YOU KNOW?

Catalysts are found in nature – we call them enzymes. Enzymes are of huge importance; for example, they help to catalyse reactions in the body and are important commercially too. They can be found in household products such as washing powders and detergents, helping to make washing clothes a less energy-intensive process. Some enzymes, called cellulose enzymes, even help to make clothes feel softer.

DID YOU KNOW?

Catalysts often contain toxic substances or substances that are hard to obtain. Chemists have to consider this against the potential benefits they have, such as environmental sustainability and cost. An example of how choices over catalysts can change is the production of ethanoic acid. Ethanoic acid is a very important chemical, widely used in the production of goods such as cleaning products and foods. It is important that it can be made easily, at the lowest possible cost and with as little impact on the natural environment as possible.

Most ethanoic acid, CH_3COOH, used to be made by oxidising butane or light naphtha or via the hydration of ethene. Butane was heated in air, using a catalyst of manganese, cobalt or chromium ions, to produce ethanoic acid:

$$2C_4H_{10}(g) + 5O_2(g) \rightarrow 4CH_3COOH(l) + 2H_2O(l)$$

This process was inefficient, with a low percentage yield of ethanoic acid. Today, the Monsanto process is one of the commercial processes used to produce ethanoic acid. In this process, methanol is reacted with carbon monoxide:

$$CH_3OH(l) + CO(g) \rightarrow CH_3COOH(l)$$

Developed by BASF in 1960, the Monsanto process was originally carried out at a temperature of 300 °C and a pressure of 700 atmospheres, with a catalyst of cobalt and iodide ions. However, at 700 atmospheres, safety is a serious concern and there were high energy costs to generating the high temperature and pressure.

Catalyst research led to the use of rhodium, rather than cobalt, as the catalyst. This new catalyst needs a lower temperature of 150 °C to 200 °C and a much lower pressure of about 30 atmospheres. The process has an atom economy of 100%.

The Cativa process has since superseded the Monsanto process. In this process an iridium catalyst is used, which is cheaper than rhodium, thereby cutting costs. The Cativa process also releases less CO_2 into the atmosphere, with obvious environmental benefits.

Questions

1. Explain how a catalyst increases the rate of a chemical reaction.

2. In the stratosphere, gaseous chlorine radicals from substances such as CFCs catalyse the breakdown of the gaseous ozone layer. Explain whether chlorine radicals would be classed as a heterogeneous or homogeneous catalyst.

3. Draw an enthalpy profile diagram for an exothermic reaction. Clearly label the axes and indicate both the enthalpy change and the activation energy. Show how the activation energy changes when a catalyst is added.

9 The Boltzmann distribution

By the end of this topic, you should be able to demonstrate and apply your knowledge and understanding of:

* qualitative explanation of the Boltzmann distribution and its relationship with activation energy
* explanation, using Boltzmann distributions, of the qualitative effect on the proportion of molecules exceeding the activation energy and hence the reaction rate, for:
  (i) temperature changes
  (ii) catalytic behaviour

What is the Boltzmann distribution?

For a gas or liquid, the molecules move around inside a container, colliding with each other and with the walls of the container. The collisions are assumed to be elastic and so, when the molecules do collide, no energy is lost from the system. As the molecules are moving, they have kinetic energy.

In a sample of a gas or liquid:

- some molecules move *fast* and have *high* energy
- some molecules move *slowly* and have *low* energy
- the majority of the molecules have an *average* energy.

The Boltzmann distribution shows the distribution of molecular energies in a gas at constant temperature, as illustrated in Figure 1.

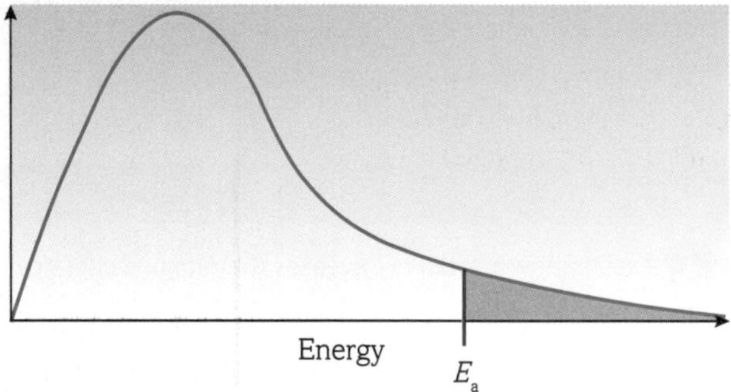

Figure 1 Distribution of molecular energies, known as a Boltzmann distribution. If a certain reaction has the activation energy E_a, then only those particles in the green shaded area would be able to go on and react.

KEY DEFINITION
The **Boltzmann distribution** is the distribution of energies of molecules at a particular temperature, often shown as a graph.

Important features of the Boltzmann distribution

- The area under the curve is equal to the total number of molecules in the sample. The area does not change with conditions.
- There are no molecules in the system with zero energy – the curve starts at the origin.
- There is no maximum energy for a molecule – the curve gets close to, but does not touch or cross, the energy axis.
- Only the molecules with an energy greater than the activation energy, E_a, are able to react.

This last point is extremely important and forms the basis of our understanding of how chemical reactions take place following collision. It also helps us to understand why not *all* collisions lead to a reaction.

The effect of temperature on reaction rate

At higher temperatures, the kinetic energy of all the molecules increases. The Boltzmann distribution flattens and shifts to the right. The number of molecules in the system does *not* change, so the area under the curve remains the same.

In Figure 2, temperature T_2 is greater than temperature T_1.

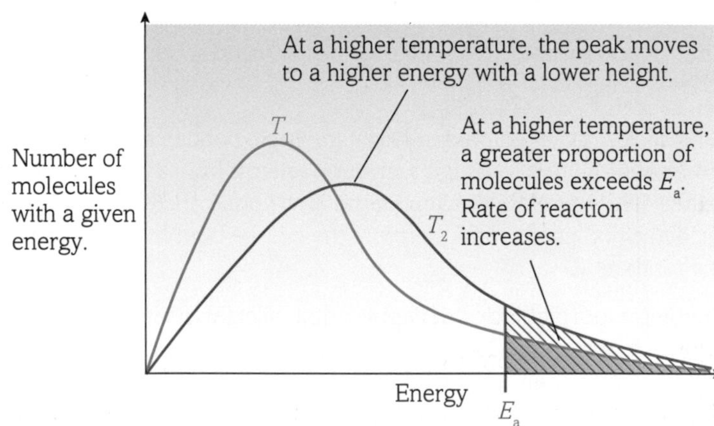

Figure 2 The effect of temperature on reaction rate. Notice how, at higher temperatures, the peak moves to a higher energy with a lower height.

LEARNING TIP
Whenever you draw a Boltzmann curve, make sure you start it at the origin and that the curve never touches the *x*-axis.

As the temperature of the reaction is increased, the rate of the reaction *increases*. This is because:

- more collisions take place in a certain length of time – the molecules are moving *faster* and have *more* kinetic energy
- a higher proportion of molecules have an energy that is *greater* than the activation energy – *more* collisions will lead to a chemical reaction.

This means that, on collision, *more* molecules in the system will overcome the activation energy of the reaction. There will be *more* successful collisions in a certain length of time and the rate of reaction will *increase*.

The effect of a catalyst on reaction rate

Catalysts lower the activation energy of reactions. They do not change the distribution of energy within molecules. However, by lowering the activation energy, more particles will automatically be above the activation energy barrier.

On collision, *more* molecules in the system will overcome the new *lower* activation energy of the reaction. There will be *more* successful collisions in a certain length of time and the rate of reaction will *increase*.

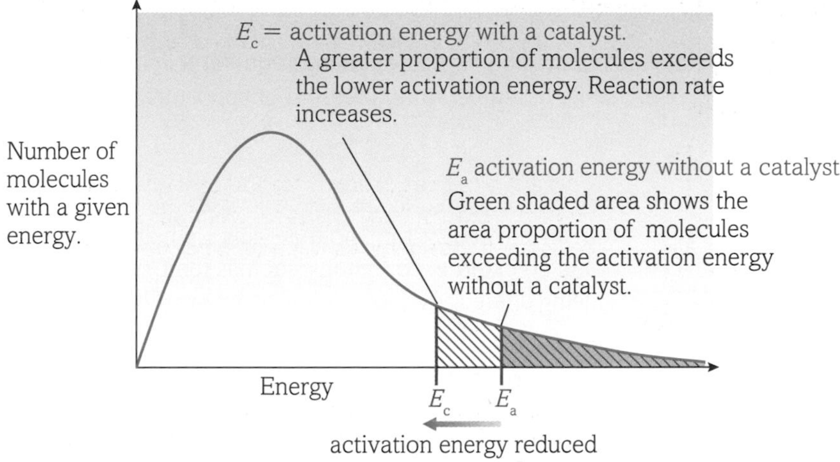

Figure 3 Effect of a catalyst on reaction rate. Notice that when a catalyst is present, the energy of the particles does not change but the activation barrier is lowered.

LEARNING TIP

Make sure you use the concept of collisions when describing how rates are affected by temperature changes and the use of a catalyst. You must discuss both the energy of collisions and the number of collisions. Both these factors influence rates of reactions.

You can use Boltzmann curves to illustrate these changes, as the shape of the curve will look different for each set of conditions. The Boltzmann curve 'models' what is happening to particles, and therefore how this affects rates of reactions. Make sure you can draw a curve for the energy distribution at different temperatures and when a catalyst is in use.

Questions

1. What do you understand by the term activation energy, E_a?
2. Explain how and why the rate of a chemical reaction is affected by an increase in temperature.
3. (a) Sketch a Boltzmann distribution curve for a sample of a gas at constant temperature. Label the axes.
   (b) Explain what would happen to the Boltzmann distribution curve and the rate if a catalyst was added.
   (c) Add to your diagram a second distribution curve for a lower temperature.

10 Chemical equilibrium

By the end of this topic, you should be able to demonstrate and apply your knowledge and understanding of:

* **explanation that a dynamic equilibrium exists in a closed system when the rate of the forward reaction is equal to the rate of the reverse reaction and the concentrations of reactants and products do not change**
* **le Chatelier's principle and its application for homogeneous equilibria in order to deduce qualitatively the effect of a change in temperature, pressure or concentration on the position of equilibrium**

Reversible reactions

You will have seen many reactions during your Chemistry studies. One example you may have seen is when a reactive metal reacts with an excess of acid to produce a salt and hydrogen. For example:

$Mg(s) + H_2SO_4(aq) \rightarrow MgSO_4(aq) + H_2(g)$

- The reaction stops when all of the metal has reacted.
- The reaction is said to have gone to completion, indicated by the $\rightarrow$ sign.

However, some reactions are reversible. They can take place in either the forward or the reverse direction. For example:

$2SO_2(g) + O_2(g) \rightleftharpoons 2SO_3(g)$

This reaction is reversible. The $\rightleftharpoons$ sign is used to show when a reaction is at *equilibrium*.

Dynamic equilibrium

What does equilibrium mean?

When a system is in a state of equilibrium, there is no observable change – nothing appears to be happening. However, the system is *dynamic*; that is, it is in constant motion. As fast as the reactants are converted into products, the products are being converted back into reactants. Although you cannot see anything happening, reactions are still taking place in both directions.

Here are a couple of everyday scenarios to help you visualise what equilibrium means:

- Filling a paddling pool with a hole in it – if you are filling the pool at the same rate as the water is running out of the hole, the situation would be at equilibrium. The pool would never fill but would also not empty.
- Drying washing in the rain – if the washing is drying at the same rate as the rain is making it wet again, the situation would be at equilibrium. The washing would not get any wetter but would also not get any drier.

A chemical system is in dynamic equilibrium when:

- the *concentrations* of the reactants and the products remain *constant*
- the rate of the *forward* reaction is the *same* as the rate of the *reverse* reaction.

An equilibrium only applies as long as the system remains *isolated*. In an isolated system, no materials are being added or taken away and no external conditions, such as temperature or pressure, are being altered.

The position of equilibrium

At equilibrium, the extent of a reaction is called the *position of equilibrium*. Equilibrium is not established until the reaction has been occurring for a period of time (the actual time taken can vary from being almost immediate to days). In the example at the start of the topic, sufficient $SO_3(g)$ would need to be formed before it could start breaking down to make the reverse reaction occur. It is once the concentrations of the reactants and products have become stable and the forward and reverse rates are the same that equilibrium is said to have been established.

Factors affecting the position of equilibrium

A reversible reaction only remains in dynamic equilibrium when it is *isolated* in a closed system. This means that *nothing* must be allowed in or out. The position of equilibrium can be altered by changing the following in the system:

- concentrations of the reactants or products
- pressure in reactions involving gases
- temperature.

Le Chatelier's principle

The effect of a change can be predicted using a principle known as le Chatelier's principle. This principle states that when a system in dynamic equilibrium is subjected to a change, the position of equilibrium will shift to minimise the change. Changing conditions of the reaction will cause the reaction to adopt the pathway that minimises the change; that is, carry out the forward or reverse reaction more, whichever helps minimise the changes.

Figure 1 NO_2 and N_2O_4 exist in equilibrium. NO_2 is brown and N_2O_4 is colourless. Changing the conditions changes the colour observed as the equilibrium shifts to minimise the changes.

LEARNING TIP

Sometimes changes in equilibrium will be accompanied by colour changes. For example, if a reaction has colourless reactants and a brown gaseous product, an increase in the forward reaction would produce more brown gas and the reaction mixture would appear to get darker.

DID YOU KNOW?

Le Chatelier's principle is used every day by chemists to help them work out what the optimum conditions will be for reactions they use. There is a pay-off to consider, however. The optimum conditions for a reaction according to le Chatelier's principle will not always be the most suitable in practice. For example, a reaction may be favoured by dangerously high pressures, or by extremely low temperatures that would make it incredibly slow.
This is illustrated by the Haber process, used to produce ammonia. Ideally, a very low temperature and very high pressure are required to give the optimum yield, but these conditions cannot be used as they would be expensive, unsafe and very slow to produce ammonia. This will be covered in more detail in topic 3.2.11.

The effect of concentration on equilibrium

For a system at equilibrium, a change in the concentration of reactants or products will result in a shift in the position of equilibrium. Consider this equilibrium:

$$\underset{\text{ethanol}}{CH_3CH_2OH(l)} + \underset{\text{propanoic acid}}{CH_3CH_2COOH(l)} \rightleftharpoons \underset{\text{ethyl propanoate}}{CH_3CH_2COOCH_2CH_3(l)} + \underset{\text{water}}{H_2O(l)}$$

Increasing the concentration of a reactant (either ethanol or propanoic acid) causes the equilibrium to shift in the direction that decreases this reactant's concentration:

- The system opposes the change by decreasing the concentration of the reactant by removing it.
- The position of equilibrium moves to the *right-hand* side, forming *more* products.

Increasing the concentration of a *product* (either ethyl propanoate or water) causes the equilibrium to shift in the direction that *decreases* this product's concentration:

- The system opposes the change by decreasing the concentration of the product by removing it.
- The position of equilibrium will move to the *left-hand* side, forming *more* reactants.

The effect of pressure on equilibrium

Changing the total pressure of a system will only change the position of equilibrium if there are *gases* present. Consider this equilibrium:

$$\underbrace{N_2(g) + 3H_2(g)}_{\substack{\text{4 mol of gas}\\ \text{higher pressure}}} \rightleftharpoons \underset{\substack{\text{2 mol of gas}\\ \text{lower pressure}}}{2NH_3(g)}$$

In total, there are 4 moles of gas on the left-hand side and 2 moles of gas on the right-hand side.

- The side with more moles of gas is the side at the higher pressure.

Increasing the total pressure of the system causes the position of equilibrium to move to the side with *fewer* gas molecules, as this will *decrease* the pressure. In our equation, the position of equilibrium will move to the *right*.

Decreasing the total pressure of the system causes the position of equilibrium to move to the side with the *greater* number of gas molecules, as this will *increase* the pressure. In our equation, the position of equilibrium will move to the *left*.

Increasing the pressure of one of the gases in the system is the same as increasing its concentration.

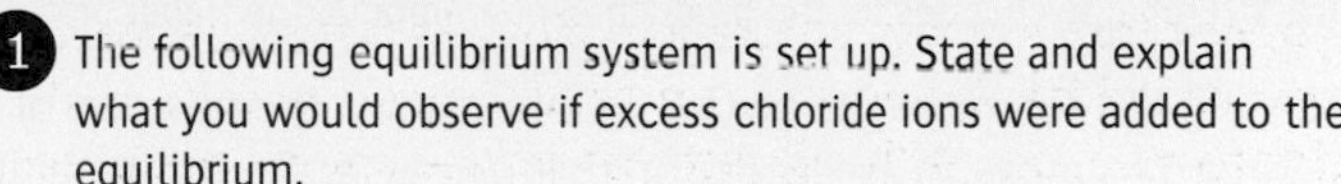

Questions

1. The following equilibrium system is set up. State and explain what you would observe if excess chloride ions were added to the equilibrium.

$$\underset{\text{pale blue solution}}{[Cu(H_2O)_6]^{2+}(aq) + 4Cl^-(aq)} \rightleftharpoons \underset{\text{yellow solution}}{[CuCl_4]^{2-}(aq) + 6H_2O(l)}$$

2. The production of nitric acid from ammonia involves the oxidation process shown below as one of the essential steps:

$$4NH_3(g) + 5O_2(g) \rightleftharpoons 4NO(g) + 6H_2O(g)$$

(a) State and explain how an increase in the total pressure of the system would affect the position of equilibrium.

(b) Nitrogen monoxide, NO, is removed from the system as it is made. How would this change affect the position of equilibrium?

3.2

11 Equilibrium and industry

By the end of this topic, you should be able to demonstrate and apply your knowledge and understanding of:

- le Chatelier's principle and its application for homogeneous equilibria to deduce qualitatively the effect of a change in temperature, pressure or concentration on the position of equilibrium
- explanation that a catalyst increases the rate of both forward and reverse reactions in an equilibrium by the same amount resulting in an unchanged position of equilibrium
- explanation of the importance to the chemical industry of a compromise between chemical equilibrium and reaction rate in deciding the operational conditions

The effect of temperature on equilibrium

The effect of changing the temperature on the position of equilibrium depends on the enthalpy sign. For the equilibrium shown in Figure 1:

- the forward reaction is exothermic (gives out heat)
- the reverse reaction is endothermic (takes in heat).

In this equilibrium, the forward reaction is exothermic with a ΔH value of $-92\,\text{kJ mol}^{-1}$.

The reverse reaction will be endothermic, with an enthalpy change of $+92\,\text{kJ mol}^{-1}$.

The forward and reverse reactions have the same magnitude of ΔH, but opposite signs.

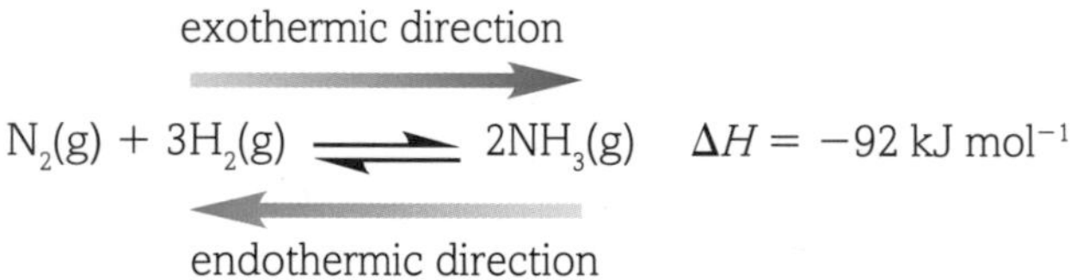

Figure 1 Enthalpy changes for the production of ammonia (the Haber process).

Increasing the temperature of the system causes the position of equilibrium to move in the direction that *decreases* the temperature:

- The system opposes the change by *taking in* heat and the position of equilibrium moves to the left.
- The position of equilibrium moves in the endothermic (ΔH is positive) direction.

Decreasing the temperature of the system causes the position of equilibrium to move in such a way as to *increase* the temperature:

- The system opposes the change by *releasing* heat and the position of equilibrium moves to the right.
- The position of equilibrium moves in the exothermic (ΔH is negative) direction.

LEARNING TIP

When discussing changes in temperature, it is important to state whether the reaction is exothermic or endothermic in the *forward* direction. Remember, the forward and reverse reactions will have the same magnitude of ΔH, but opposite signs.

The effect of a catalyst on equilibrium

A catalyst does *not* alter the position of equilibrium or the composition of an equilibrium system:

- A catalyst speeds up the rate of the forward and reverse reactions *equally*.
- A catalyst increases the rate at which equilibrium is *established* but does not affect the position of the equilibrium.

Equilibrium vs yield

Many important chemical processes exist as equilibrium systems. Examples include:

- the preparation of ammonia from nitrogen and hydrogen in the Haber process
- the conversion of sulfur dioxide into sulfur trioxide in the Contact process.

In industry, chemists strive to achieve the highest possible yield of a desired product. However, this has to be balanced against the optimum position of equilibrium, which allows industrial processes to be as cheap and energy-efficient as possible. After all, industrial chemistry exists to generate money.

The production of ammonia through the Haber process is an excellent illustration of this point.

The Haber process

Chemical equation: $N_2(g) + 3H_2(g) \rightleftharpoons 2NH_3(g) \qquad \Delta H = -92\,\text{kJ mol}^{-1}$

The raw materials, nitrogen and hydrogen, must be readily available:

- Nitrogen is obtained from the air by fractional distillation.
- Hydrogen is prepared by reacting together methane (from natural gas) and water.

Ammonia is produced by the *forward* reaction in this equilibrium. The optimum conditions are high pressure and low temperature. This is because:

- the forward reaction produces *fewer* gas molecules (4 molecules $\rightarrow$ 2 molecules), so is favoured by using *high* pressure.
- the forward reaction is exothermic (ΔH is negative), so is favoured by using *low* temperature.

However, there are drawbacks to using these theoretical conditions:

- Although a low temperature should produce a high equilibrium yield, the reaction would take place at a very low rate. At low temperatures, comparatively few N_2 and H_2 molecules have enough energy to overcome the required activation energy.
- A high pressure increases the concentration of the gases, increasing the reaction rate. So, a high pressure should produce both a high equilibrium yield and a high rate. However, large quantities of energy are required to compress gases, adding significantly to the running costs. There are also safety implications – any failure in the systems could potentially allow chemicals to leak into the environment, endangering those working on site.

The modern ammonia plant

A modern ammonia plant needs to produce a sufficient yield of ammonia at a reasonable cost and in as short a time as possible. In practice, a *compromise* is made between yield and rate:

- Temperature – this must be high enough to allow the reaction to proceed at a realistic rate, whilst still producing an acceptable equilibrium yield. A temperature of 400–500 °C is typically used.
- Pressure – a high pressure must be used, but it must not be so high that the workforce is put in danger or the environment threatened. A pressure of 200 atmospheres is typically used.
- Catalyst – an iron catalyst is added to speed up the rate of reaction, allowing the equilibrium to be established faster and lower temperatures to be used. Less energy is needed to generate heat, reducing costs.

The actual compromise conditions used convert only 15 per cent of nitrogen and hydrogen into ammonia. The ammonia produced is liquefied and removed. Unreacted nitrogen and hydrogen gases are then passed through the reactor again. Eventually, virtually all the nitrogen and hydrogen will have been converted into ammonia.

LEARNING TIP

Remember to refer to le Chatelier's principle when predicting optimum conditions for any equilibrium process.

Questions

1. Synthesis gas, a mixture of CO and H_2 gases, is made by passing steam over hot coke. The equation for the process is:

   $C(s) + H_2O(g) \rightleftharpoons CO(g) + H_2(g) \quad \Delta H = +132\ \text{kJ mol}^{-1}$

   State and explain the effect of a temperature rise on the composition of the equilibrium mixture.

2. The Contact process is used in the manufacture of sulfuric acid, H_2SO_4. The equilibrium in the Contact process is shown below:

   $2SO_2(g) + O_2(g) \rightleftharpoons 2SO_3(g) \quad \Delta H = -196\ \text{kJ mol}^{-1}$

   (a) State and explain the effect of a decrease in pressure on the position of equilibrium.

   (b) What conditions would be used to obtain the maximum theoretical yield of $SO_3(g)$?

   (c) In the industrial process, a catalyst of vanadium pentoxide is used. What is the effect of the catalyst on the position of equilibrium and on the time taken to reach equilibrium?

12 The equilibrium constant, K_c

By the end of this topic, you should be able to demonstrate and apply your knowledge and understanding of:

- the techniques and procedures used to investigate changes to the position of equilibrium for changes in concentration and temperature
- expressions for the equilibrium constant, K_c for homogeneous reactions and calculations of the equilibrium constant, K_c from provided equilibrium concentrations
- estimation of the position of equilibrium from the magnitude of K_c

Equilibrium concentrations

When a reaction has reached equilibrium, the concentrations of each species involved remain constant. The reaction is still occurring, but the overall concentrations do not change. The rates of the forward and reverse reactions are equal.

This does not mean that the concentrations of all the species are the same. The reactants may be present in very high concentrations and the products in very low concentrations, or vice versa, or they may have similar concentrations.

If the concentration of the products is much higher than that of the reactants, we say the equilibrium *lies to the right* or *favours the products*. If the concentration of the reactants is much higher than that of the products, we say the equilibrium *lies to the left* or *favours the reactants*.

The equilibrium constant, K_c, and the equilibrium law

Chemists consider the position of an equilibrium using the equilibrium constant, K_c. K_c gives a measure of where the equilibrium lies, essentially by giving the ratio between products and reactants.

$$K_c = \frac{\text{products}}{\text{reactants}}$$

Figure 1 K_c gives an indication of the position of equilibrium.

The *equilibrium law* tells us that if reactants A and B and products C and D are in the equilibrium:

$$aA + bB \rightleftharpoons cC + dD$$

then they are related by the following equation:

$$K_c = \frac{[C]^c[D]^d}{[A]^a[B]^b}$$

The capital letters in the equilibrium $aA + bB \rightleftharpoons cC + dD$ represent each species. The lowercase letters represent the number of moles of each species.

LEARNING TIP

You will only be expected to work out K_c values for homogenous reactions; that is, reactions where all the species are in the same physical state.

WORKED EXAMPLE 1

Writing an expression for K_c

N_2O_4 and NO_2 can exist in the following equilibrium:

$$N_2O_4(g) \rightleftharpoons 2NO_2(g)$$

The equilibrium expression would be written as:

$$K_c = \frac{[NO_2(g)]^2}{[N_2O_4(g)]}$$

LEARNING TIP

Notice how the original equilibrium equation showed 1 mole of N_2O_4 in equilibrium with 2 moles of NO_2. This means in the expression for K_c, N_2O_4 is raised to the power of 1, so nothing is written outside the brackets. NO_2 must be raised to the power of 2, so a 2 is written outside the brackets.

Square brackets are always used in expressions for K_c, as shown in the equilibrium law expression. Square brackets mean 'concentration'. So, $[N_2O_4]$ means concentration of N_2O_4.

The size of K_c will give an indication of the position of equilibrium. Once an expression for K_c has been determined, concentrations of each species involved in the equilibrium must be established. This is done experimentally.

INVESTIGATION

Practical techniques used to study equilibrium

There are many practical techniques available to monitor reactions that exist at equilibrium and to determine concentrations of each species involved.

Many equilibrium reactions will involve coloured compounds that will be present in different quantities (and therefore intensities of colour) as equilibria shift. These can be monitored using colorimeters. A colorimeter measures how much light is absorbed by coloured substances. Absorption is dependent on the concentration of the coloured substances. Concentrations of substances can then be put into the expression for K_c.

Another technique that is often used is titration. Samples of an equilibrium mixture are removed at set intervals and titrated against known standard solutions to find out how much of the substance reacts. This can help determine the concentration of each chemical involved in the equilibrium and these values can then be put into the expression for K_c.

These and other techniques can be used to find out how equilibria are affected by changes in concentration, temperature and the presence of a catalyst.

The significance of the K_c value

The magnitude of K_c indicates the extent of a chemical reaction.

A K_c value of 1 would indicate that the position of equilibrium is halfway between reactants and products. This would mean the ratio of the product concentration to reactant concentration was 1:1 (but remember, this does not mean the actual concentrations have the value 1).

When K_c is greater than 1:

- the reaction favours the *products* (lies to the right)
- the products on the *right-hand side* predominate at equilibrium.

When K_c is smaller than 1:

- the reaction favours the *reactants* (lies to the left)
- the reactants on the *left-hand side* predominate at equilibrium.

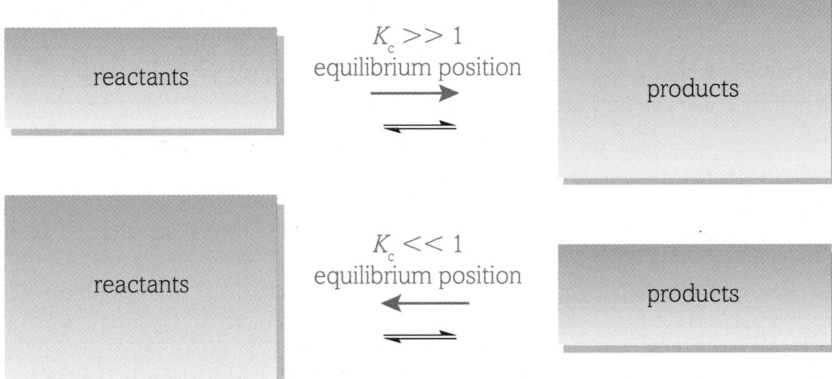

Figure 2 A visual illustration of what K_c values tell us about the position of equilibrium.

WORKED EXAMPLE 2

Calculating the value of K_c

$H_2(g)$, $I_2(g)$ and $HI(g)$ exist in equilibrium in a closed system:

$$H_2(g) + I_2(g) \rightleftharpoons 2HI(g)$$

The equilibrium concentrations are:

$H_2(g)$ 0.140 mol dm^{-3}; $I_2(g)$ 0.0400 mol dm^{-3}; HI(g) 0.320 mol dm^{-3}

Calculate K_c under these conditions.

Write the expression for K_c and comment on the position of the equilibrium:

$$K_c = \frac{[HI(g)]^2}{[H_2(g)]\,[I_2(g)]}$$

Use the equilibrium concentrations to calculate a value for K_c:

$$K_c = \frac{0.320^2}{(0.140 \times 0.0400)}$$
$$= 18.3$$

The value for K_c is greater than 1, so the reaction favours the products; HI will predominate in the equilibrium.

Questions

1 For each of the following, write down the expression for K_c.

(a) $2NO(g) + O_2(g) \rightleftharpoons 2NO_2(g)$

(b) $H_2(g) + Br_2(g) \rightleftharpoons 2HBr(g)$

(c) $N_2(g) + 3H_2(g) \rightleftharpoons 2NH_3(g)$

2 In a closed system, $N_2(g)$, $H_2(g)$ and $NH_3(g)$ exist in equilibrium: $N_2(g) + 3H_2(g) \rightleftharpoons 2NH_3(g)$
The equilibrium concentrations are $N_2(g)$ 1.20 mol dm^{-3}, $H_2(g)$ 2.00 mol dm^{-3} and $NH_3(g)$ 0.876 mol dm^{-3}

(a) Calculate the value for K_c under these conditions. The units for the answer will be dm^6 mol^{-2}.

(b) Explain whether the equilibrium favours the products or the reactants.

WHICH FUEL AND WHY?

Our dependence on organic fuels has necessitated the exploitation of renewable biofuels. One of the key aims behind the use of biofuels is to reduce carbon emissions, but with limited land space an increasing area of research is the use of waste biomass. Read the following extract and answer the questions.

ORGANIC CHEMISTS CONTRIBUTE TO RENEWABLE ENERGY

Background – Why is this important?

> Biofuels play an essential role in reducing the carbon emissions from transportation. The development of 'drop in' fuels produced from lignocellulosic raw materials will increase both the availability of biofuels and the sustainability of the biofuel industry.
>
> Adrian Higson – Energy Consultant

Biofuels can be either liquid or gaseous fuel. They can be produced from any source that can be replenished rapidly, e.g. plants, agricultural crops and municipal waste. Current biofuels are produced from sugar and starch crops such as wheat and sugar cane, which are also part of the food chain.

One of the key targets for energy researchers is a sustainable route to biofuels from non-edible lignocellulosic (plant) biomass, such as agricultural wastes, forestry residues or purpose grown energy grasses. These are examples of so-called advanced biofuels.

Current biofuels, such as ethanol, have a lower energy content (volumetric energy density) compared with conventional hydrocarbon fuels, petroleum and natural gas. The aim is to produce fuels that have a high carbon content and therefore have a higher volumetric energy density. This can be achieved by chemical reactions that remove oxygen atoms from biofuel chemical compounds. This process produces a so called 'drop-in biofuel', i.e. a fuel that can be blended directly with existing hydrocarbon fuels that have similar combustion properties.

What did the organic chemists do?

Efficient synthesis of renewable fuels remains a challenging and important line of research. Levulinic acid and furfural are examples of potential 'platform molecules', i.e. molecules that can be produced from biomass and converted into biofuels. Levulinic acid can be produced in high yield (>70%) from inedible hexose bio-polymers such as cellulose, which is a polymer of glucose and the most common organic compound on Earth. Furfural has been produced industrially for many years from pentose-rich agricultural wastes and can also act as a platform molecule.

Recent reports have highlighted the use of organic chemistry to convert platform molecules like levulinic acid and furfural into potential advanced biofuels. Specifically, changing parts of the molecules that are responsible for their structure and function. This process is called 'functional group interconversion' and is part of the basic toolkit of organic chemistry. For example, researchers have described a process for converting levulinic acid into so-called 'valeric biofuels'. One of these biofuels, ethyl valerate, is claimed to be a possible advanced bio-gasoline molecule with several advantages over bio-ethanol.

A second method to create hydrocarbons involves Dumesic's approach via a decarboxylation of gamma-valerolactone, which can be produced in one step from levulinic acid by hydrogenation.

You will need to be familiar with skeletal formula for this section.

Corma's synthesis [2] is one of the more recent and ingenious examples of C5 or C6 decomposition products being used as precursors to biofuel molecules. This method generates a C15 hydrocarbon molecule from furfural via a molecule called, 2-methylfuran (a C5 molecule). The key step in this process is an acid catalysed, water mediated trimerisation of 2-methylfuran to give a trimer. A catalysed hydrogenation was then used to deoxygenate this trimer resulting in the C15 hydrocarbon. This molecule is a potential biodiesel molecule with advantages over first generation biodiesels both in terms of fuel quality and sustainability.

What is the impact?

Biodiesel is likely to be the second most important biofuel after ethanol in the short to medium term. Global production of biodiesel is expected to increase from 11 billion litres to reach 24 billion litres by 2017. For organic chemists, there are significant opportunities associated with further developing energy crops and producing advanced biofuels from new sources such as algae, industrial or post-consumer waste.

http://www.rsc.org/Membership/Networking/InterestGroups/OrganicDivision/organic-chemistry-case-studies/organic-chemistry-biofuels.asp

Where else will I encounter these themes?

1.1 2.1 2.2

Let us start by considering the nature of the writing in the article.

1. This article is written for members of an international chemistry association and thus relies on the audience having a high degree of scientific literacy. Having read the article a few times attempt one of the following questions.

   a. Have a go at re-writing the article for a less scientifically literate article. Can you get the main ideas across without using chemical structures and terminology to the same degree?

or

   b. Re-write the article in such a way as to present a strongly positive argument in favour of biofuels. You will probably notice that the article itself shies away from using polemical language, but can you present such a positive argument without altering the facts as presented?

Questions 2–4 ask questions about the chemistry in, or connected to, this article. Some of the questions will build on ideas that you covered at GCSE and others will be covered later in the course. The timeline below tells you about other chapters in this book that are relevant to this question: do *not worry* if you are not ready to give answers to these questions at this stage. You may like to read through the activity at this initial stage and return to the questions once you have covered other topics later in the book.

2. The extract mentions the compounds levulinic acid and furfural.
   a. Calculate the molar mass of each compound
   b. Calculate the percentage mass of oxygen in each molecule
   c. Give a chemical test, and its result, that would enable you to distinguish between the two molecules

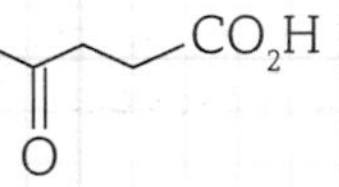

Figure 1 Levulinic acid.

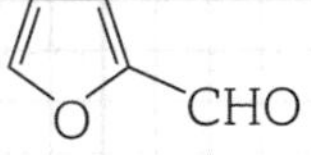

Figure 2 Furfural.

3. In this question you will compare four different fuels (shown in the table below.)

Molecule	Molecular mass in g mol^{-1}	Standard molar enthalpy of combustion	Number of kJ of energy per mole of CO_2 produced	Number of kJ of energy per g of fuel	State at 25°C and 1 atm pressure.
CH_4	16	−890			gas
C_8H_{18}	114	−5470			liquid
C_2H_5OH	46	−1367			liquid
$C_{15}H_{32}$	212	−10 490			liquid

   a. Write balanced equations for the complete combustion of C_8H_{18} and C_2H_5OH.
   b. Complete the table.

4. Why can't the C-15 hydrocarbon synthesised by the process detailed above be described as a completely carbon neutral solution to the energy shortage problem. Refer to the text and any other web resources to help support your answer.

Activity

The metropolitan borough of East Grimshire is about to buy a fleet of 60 buses but must choose one of the four fuels in Table 1 on which they are to run. Select one of the fuels and deliver a presentation on why your chosen fuel is the best option. Think about pros and cons of your chosen fuel. You should attempt to reference a range of source materials in support of your choice. Your presentation you should consider the following: sustainability; energy efficiency; carbon footprint; the pros and cons of your chosen fuel.

Your presentation should be between 4 and 8 slides and last no more than 10 min but be prepared to defend your presentation in a 5 min Q&A session!

There will be plenty of source material on the internet but be critical about who is presenting the material. For example, an oil company will have vested interests!

3.2 Practice questions

1. What is the correct equation representing the enthalpy of formation of methanol? [1]
   A. $C(s) + O(g) + 4H(g) \rightarrow CH_3OH(l)$
   B. $C(g) + O(g) + 2H_2(g) \rightarrow CH_3OH(l)$
   C. $C(s) + \frac{1}{2}O_2(g) + 2H_2(g) \rightarrow CH_3OH(g)$
   D. $C(s) + \frac{1}{2}O_2(g) + 2H_2(g) \rightarrow CH_3OH(l)$

2. The table below shows enthalpies of combustion.

Substance	Standard enthalpy of combustion in $kJ\,mol^{-1}$
C_2H_6	−1560
C_2H_4	−1411
H_2	−286

   Look at the reaction below.
   $C_2H_4(g) + H_2(g) \rightarrow C_2H_6(g)$
   What is the enthalpy change for this reaction? Use the data in the table to help you. [1]
   A. $-137\,kJ\,mol^{-1}$
   B. $+137\,kJ\,mol^{-1}$
   C. $-149\,kJ\,mol^{-1}$
   D. $+149\,kJ\,mol^{-1}$

3. Hydrogen gas can be produced on an industrial scale by steam reformation, according to the following equation.
   $CH_4(g) + H_2O(g) \rightleftharpoons CO(g) + 3H_2(g)$
   $\Delta H_c = +210\,kJ\,mol^{-1}$
   Given the information above, which of the following statements is true? [1]
   A. The forward reaction gives out heat energy.
   B. The forward reaction is favoured by low pressure.
   C. A catalyst will speed up the forward reaction.
   D. The reverse reaction is favoured by low temperature.

4. Consider the following reaction at equilibrium.
   $2NOCl_2(g) \rightleftharpoons 2NO(g) + Cl_2(g)$
   Which of the following statements is true for this reaction? [1]
   A. It is an example of a heterogeneous equilibrium.
   B. The K_c for this reaction will increase as pressure increases.
   C. A change in temperature will not affect the K_c value for this reaction.
   D. Increasing the pressure will push the equilibrium in the reverse direction.

[Total: 4]

5. Nitrogen monoxide is an atmospheric pollutant, formed inside car engines by the reaction between nitrogen and oxygen.
   $N_2(g) + O_2(g) \rightarrow 2NO(g)$
   $\Delta H = +66\,kJ\,mol^{-1}$
   (a) This reaction is endothermic.
   (i) Explain the meaning of the term *endothermic*. [1]
   (ii) What is the value for the enthalpy change of formation of nitrogen monoxide? [1]
   (b) (i) Complete the enthalpy profile diagram for the reaction between nitrogen and oxygen.
   On your diagram:
   - add the product
   - label the activation energy as E_a
   - label the enthalpy change as ΔH. [3]

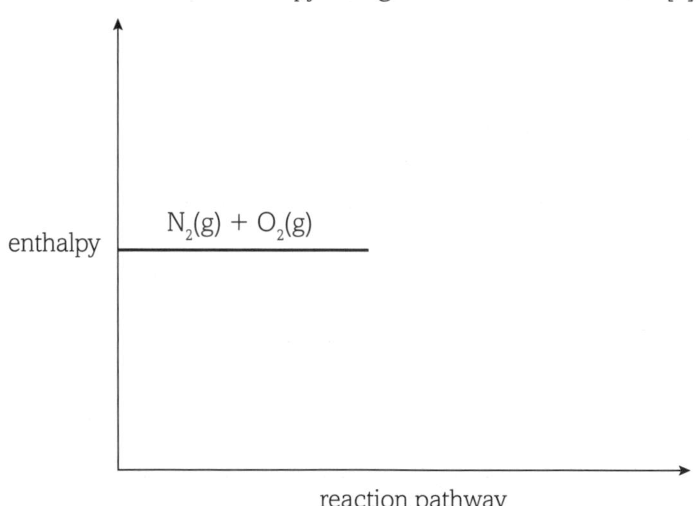

   (ii) Explain the meaning of the term *activation energy*. [1]

(c) A research chemist investigates the reaction between nitrogen and oxygen.
She mixes nitrogen and oxygen gases in a sealed container. She then heats the container at a constant temperature for one day until the gases reach a dynamic equilibrium.
   (i) Explain, in terms of the rate of the forward reaction and the rate of the backward reaction, how the mixture of N_2(g) and O_2(g) reaches a dynamic equilibrium containing N_2(g), O_2(g) **and** NO(g). [2]
   (ii) The research chemist repeats the experiment at the same temperature using the same initial amounts of N_2(g) and O_2(g). This time she carries out the experiment at a much **higher pressure**.
   Suggest why:
   - much less time is needed to reach dynamic equilibrium
   - the composition of the equilibrium mixture is the same as in the first experiment. [5]

   (iii) The reaction between nitrogen and oxygen in a car engine does not reach a dynamic equilibrium.
   Suggest why not. [1]

[Total: 14]

[From Q3, F322 May 2011]

6. For some chemical reactions, an increase in 10°C can result in the rate of reaction doubling. Using appropriate sketches of Maxwell–Boltzmann distributions of molecular energies *and* your knowledge of collision theory, explain why this can be the case. [5]

[Total: 5]

7. (a) Define the term *standard enthalpy of combustion.* [3]
(b)

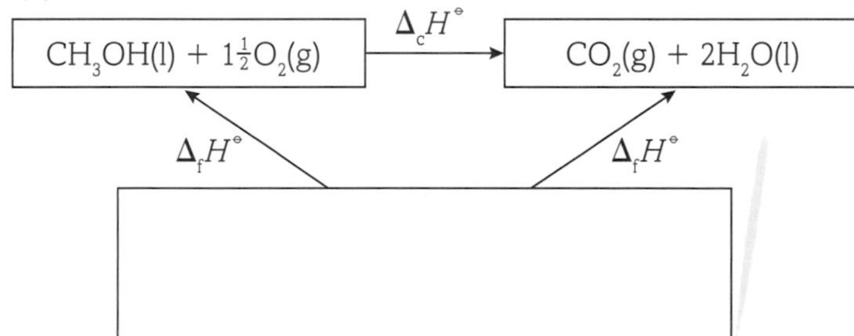

Complete the Hess's cycle above using standard enthalpies of formation using correct state symbols. [2]

(c) Use the data in the table below to calculate a value for the standard enthalpy of combustion of methanol.

Compound	$\Delta_f H^\circ$/kJ mol^{-1}
CO_2(g)	−394
H_2O(l)	−286
CH_3OH(l)	−239

[3]

(d) A calculation using standard bond enthalpies gives an enthalpy of combustion value that is 12% lower than the values calculated by Hess's law. Suggest why this is the case. [1]

[Total: 9]

8. Syngas is a mixture of gases largely made from carbon monoxide and hydrogen. It is made from the reaction between methane and steam according to the following equation:

$$CH_4(g) + H_2O(g) \rightleftharpoons CO(g) + 3H_2(g) \quad \Delta_r H = +206\,\text{kJ mol}^{-1}$$

The industrial conditions used for this process are
(a) A nickel-based catalyst, (b) 750 °C reaction temperature and (c) 5–10 atm pressure.

Discuss each of these reaction conditions relating them to the information shown above. Why do you think different industrial process for the manufacture of syngas might use different conditions? [10]

[Total: 10]

MODULE 4

Core organic chemistry

CHAPTER 4.1

BASIC CONCEPTS AND HYDROCARBONS

Introduction

Life on Earth is based on carbon. We are made of thousands of carbon-based chemicals like amino acids, fatty acids and keratin; and all of these are made with many carbon atoms. These substances are called organic chemicals.

In the past, it was believed that carbon-based compounds could only be made by living organisms like plants and animals. But now scientists can use a variety of techniques and chemical reactions to synthesise organic chemicals in a laboratory.

Not only are we are made of organic chemicals but many everyday products like our clothes, food, cosmetics and medicines are too. There are more than ten million known carbon compounds and an estimated 300 000 new carbon compounds are discovered each year. Organic chemistry is the study of how carbon atoms bond to make compounds. This important branch of chemistry affects all areas of our lives.

All the maths you need

To unlock the puzzles of this chapter you need the following maths:

- Be able to use general formula to generate a molecular formula (*e.g.* C_nH_{2n+2} *becomes* C_2H_6)
- The units of measurement (*e.g. the mole*)
- Use of standard form and ordinary form (*e.g. representing the amount of an atom*)
- Using an appropriate amount of standard figures (*e.g. in multi-step calculations*)
- Changing the subject of an equation (*e.g. finding a molecular mass*)
- Substituting numerical values into algebraic equations (*e.g. using* $n = \frac{m}{M_r}$)

What have I studied before?

- Understood what crude oil is and how it can be separated
- Know the uses of crude oil fractions
- Know how and why cracking is used on heavy fractions
- Define alkanes and alkenes
- Draw the structural formula for the first five straight chain alkanes, and the first three straight-chain alkenes
- Draw structural isomers of butane, pentane and butene

What will I study later?

- How analytical techniques can be used to differentiate between organic molecules
- More complex reaction mechanisms

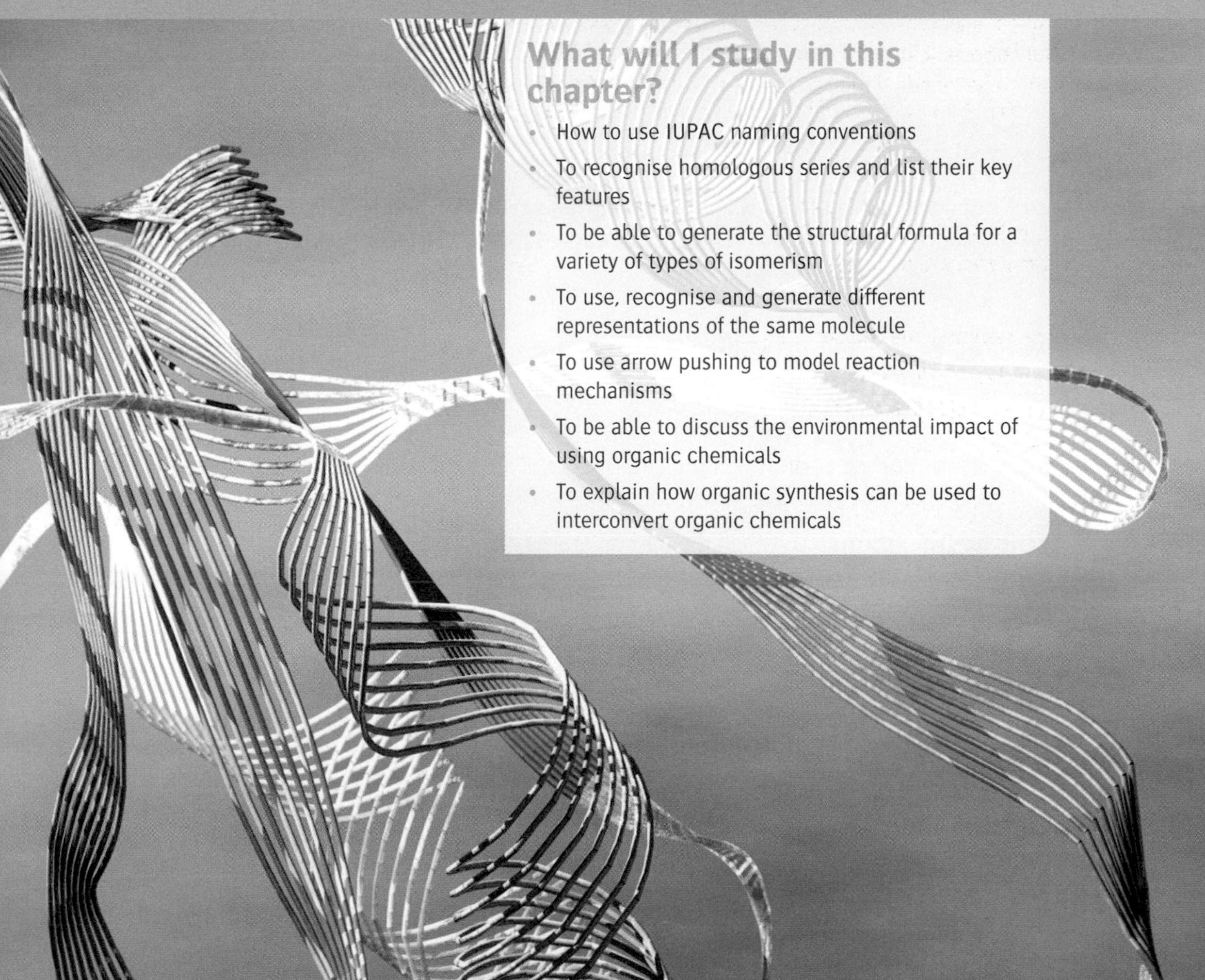

What will I study in this chapter?

- How to use IUPAC naming conventions
- To recognise homologous series and list their key features
- To be able to generate the structural formula for a variety of types of isomerism
- To use, recognise and generate different representations of the same molecule
- To use arrow pushing to model reaction mechanisms
- To be able to discuss the environmental impact of using organic chemicals
- To explain how organic synthesis can be used to interconvert organic chemicals

1 Naming organic chemicals

By the end of this topic, you should be able to demonstrate and apply your knowledge and understanding of:

- interpretation and use of the terms:
  - (i) *homologous series* (a series of organic compounds having the same functional group but with each successive member differing by CH_2)
  - (ii) *functional group* (a group of atoms responsible for the characteristic reactions of a compound)
  - (iii) *alkyl group* (of formula C_nH_{2n+1})
  - (iv) *aliphatic* (a compound containing carbon and hydrogen joined together in straight chains, branched chains or non-aromatic rings)
  - (v) *alicyclic* (an aliphatic compound arranged in non-aromatic rings with or without side chains)
  - (vi) *aromatic* (a compound containing a benzene ring)
- use of the general formula of a homologous series to predict the formula of any member of the series
- application of the IUPAC rules of nomenclature for systematically naming organic compounds
- use systematic nomenclature to avoid ambiguity
- appreciate the role of IUPAC in developing a systematic framework for chemical nomenclature

DID YOU KNOW?

IUPAC

The International Union of Pure and Applied Chemistry has derived a set of systematic rules for naming organic compounds which they have published in the 'Blue Book'. This is a systematic naming system to ensure that every organic chemical has a unique name which allows it to be recognised anywhere in the world. This organisation is the governing body for naming newly discovered elements and also publishes the 'Red Book' containing information about how to name inorganic compounds.

There are many millions of organic compounds. Some of these share the same molecular formula but have different arrangements of atoms, and so different properties. Scientists communicate in many different languages, so it is important that each substance has its own unique name.

The International Union of Pure and Applied Chemistry, IUPAC, has developed a naming system for organic chemicals – the system of naming compounds is called **nomenclature**.

Homologous series

Families of organic chemicals are called **homologous series**. To be members of a homologous series the substances must have the same functional group and successive members differ only by a CH_2 unit.

Aliphatic, alicyclic and aromatic hydrocarbons

Hydrocarbons are made up of molecules that contain only hydrogen atoms and carbon atoms. When you look at their structures (Figure 1) you can classify them as:

- **aliphatic hydrocarbons** – in which the carbon atoms are joined together in either straight (unbranched) chains or branched chains
- **alicyclic hydrocarbons** – in which the carbon atoms are joined together in a ring structure but are not aromatic
- **aromatic hydrocarbons** – in which there is at least one benzene ring in the structure.

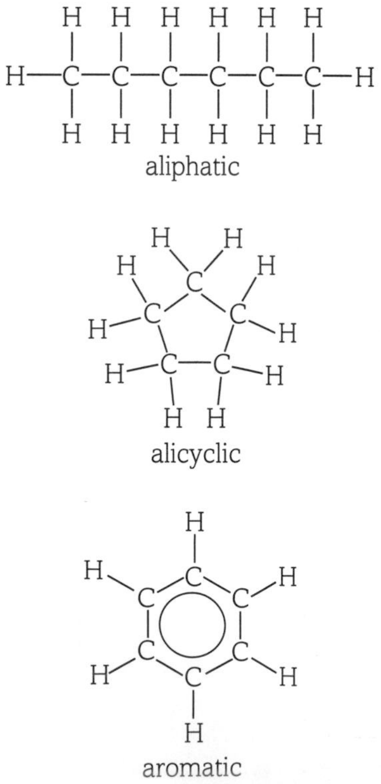

aliphatic

alicyclic

aromatic

Figure 1 Aliphatic, alicyclic and aromatic hydrocarbons.

KEY DEFINITIONS

Nomenclature is the naming system for compounds.

A **homologous series** is a series of organic compounds that have the same functional group with successive members differing by CH_2.

An **aliphatic hydrocarbon** is a hydrocarbon with carbon atoms joined together in straight or branched chains.

An **alicyclic hydrocarbon** is a hydrocarbon with carbon atoms joined together in a ring structure.

Aromatic hydrocarbons contain at least one benzene ring.

Naming straight-chain alkanes

The simplest homologous series is the alkanes. They are aliphatic **saturated** hydrocarbons, so they contain only carbon and hydrogen joined by single covalent bonds.

Table 1 lists the first 10 members of the alkanes homologous series. You will need to learn these names.

Number of carbons	Name	Formula	Displayed formula
1	methane	CH_4	$\begin{array}{ccccc} & & H & & \\ & & \vert & & \\ H & - & C & - & H \\ & & \vert & & \\ & & H & & \end{array}$
2	ethane	C_2H_6	$\begin{array}{ccccccc} & & H & & H & & \\ & & \vert & & \vert & & \\ H & - & C & - & C & - & H \\ & & \vert & & \vert & & \\ & & H & & H & & \end{array}$
3	propane	C_3H_8	$\begin{array}{ccccccccc} & & H & & H & & H & & \\ & & \vert & & \vert & & \vert & & \\ H & - & C & - & C & - & C & - & H \\ & & \vert & & \vert & & \vert & & \\ & & H & & H & & H & & \end{array}$
4	butane	C_4H_{10}	$\begin{array}{ccccccccccc} & & H & & H & & H & & H & & \\ & & \vert & & \vert & & \vert & & \vert & & \\ H & - & C & - & C & - & C & - & C & - & H \\ & & \vert & & \vert & & \vert & & \vert & & \\ & & H & & H & & H & & H & & \end{array}$
5	pentane	C_5H_{12}	$\begin{array}{ccccccccccccc} & & H & & H & & H & & H & & H & & \\ & & \vert & & \vert & & \vert & & \vert & & \vert & & \\ H & - & C & - & C & - & C & - & C & - & C & - & H \\ & & \vert & & \vert & & \vert & & \vert & & \vert & & \\ & & H & & H & & H & & H & & H & & \end{array}$
6	hexane	C_6H_{14}	$\begin{array}{ccccccccccccccc} & & H & & H & & H & & H & & H & & H & & \\ & & \vert & & \vert & & \vert & & \vert & & \vert & & \vert & & \\ H & - & C & - & C & - & C & - & C & - & C & - & C & - & H \\ & & \vert & & \vert & & \vert & & \vert & & \vert & & \vert & & \\ & & H & & H & & H & & H & & H & & H & & \end{array}$
7	heptane	C_7H_{16}	$\begin{array}{ccccccccccccccccc} & & H & & H & & H & & H & & H & & H & & H & & \\ & & \vert & & \vert & & \vert & & \vert & & \vert & & \vert & & \vert & & \\ H & - & C & - & C & - & C & - & C & - & C & - & C & - & C & - & H \\ & & \vert & & \vert & & \vert & & \vert & & \vert & & \vert & & \vert & & \\ & & H & & H & & H & & H & & H & & H & & H & & \end{array}$
8	octane	C_8H_{18}	$\begin{array}{ccccccccccccccccccc} & & H & & H & & H & & H & & H & & H & & H & & H & & \\ & & \vert & & \vert & & \vert & & \vert & & \vert & & \vert & & \vert & & \vert & & \\ H & - & C & - & C & - & C & - & C & - & C & - & C & - & C & - & C & - & H \\ & & \vert & & \vert & & \vert & & \vert & & \vert & & \vert & & \vert & & \vert & & \\ & & H & & H & & H & & H & & H & & H & & H & & H & & \end{array}$
9	nonane	C_9H_{20}	$\begin{array}{ccccccccccccccccccccc} & & H & & H & & H & & H & & H & & H & & H & & H & & H & & \\ & & \vert & & \vert & & \vert & & \vert & & \vert & & \vert & & \vert & & \vert & & \vert & & \\ H & - & C & - & C & - & C & - & C & - & C & - & C & - & C & - & C & - & C & - & H \\ & & \vert & & \vert & & \vert & & \vert & & \vert & & \vert & & \vert & & \vert & & \vert & & \\ & & H & & H & & H & & H & & H & & H & & H & & H & & H & & \end{array}$
10	decane	$C_{10}H_{22}$	$\begin{array}{ccccccccccccccccccccccc} & & H & & H & & H & & H & & H & & H & & H & & H & & H & & H & & \\ & & \vert & & \vert & & \vert & & \vert & & \vert & & \vert & & \vert & & \vert & & \vert & & \vert & & \\ H & - & C & - & C & - & C & - & C & - & C & - & C & - & C & - & C & - & C & - & C & - & H \\ & & \vert & & \vert & & \vert & & \vert & & \vert & & \vert & & \vert & & \vert & & \vert & & \vert & & \\ & & H & & H & & H & & H & & H & & H & & H & & H & & H & & H & & \end{array}$

Table 1 The first ten members of the alkanes.

Naming branched alkanes

To name an organic compound you need to consider three rules:

- stem – the main part of the name; generated by identifying the longest carbon chain or parent chain
- suffix – the end of the name; identifies the most important functional group
- prefix – the front part of the name; identifies the other functional groups and the carbon atoms they are attached to.

When you remove a hydrogen atom from an alkane, an alkyl group is formed – see Table 2.

Number of carbons	Alkyl group (stem)	Formula
1	methyl	CH_3
2	ethyl	C_2H_5
3	propyl	C_3H_7
4	butyl	C_4H_9
5	pentyl	C_5H_{11}
6	hexyl	C_6H_{13}
7	heptyl	C_7H_{15}
8	octyl	C_8H_{17}
9	nonyl	C_9H_{19}
10	decyl	$C_{10}H_{21}$

Table 2 Nomenclature terms for the side-chains of organic compounds.

LEARNING TIP

Remember, when naming compounds:

- numbers are used to indicate which carbon atom an alkyl group is joined to
- dashes are used to separate numbers from words
- commas are used to separate numbers
- prefixes are written in alphabetical order using the smallest numbers possible to indicate which carbon atoms they are joined to.

WORKED EXAMPLE 1

Look carefully at the displayed formulae below and follow the three rules above to name each compound.

1.

```
    H   CH3 H   H
    |   |   |   |
H — C — C — C — C — H
    |   |   |   |
    H   H   H   H
```

Answer

- Stem – the longest chain of carbon atoms is four, so the stem is 'but-'
- Suffix – this is a saturated hydrocarbon, so the suffix is '-ane'
- Prefix – there is just one $-CH_3$ group on the second carbon atom, so the prefix is '2-methyl'.

The name of the compound is 2-methylbutane.

2.

```
              CH3
              |
CH3 — CH2 — CH — CH — CH2 — CH2 — CH3
                   |
                   CH2 — CH3
```

Answer

When there is more than one type of side-chain, they are written in alphabetical order.

- Stem – the longest chain of carbon atoms is seven, so the stem is 'hept-'
- Suffix – this is a saturated hydrocarbon, so the suffix is '-ane'
- Prefix – there is one $-CH_3$ group on the third carbon atom and a $-C_2H_5$ group on the fourth carbon atom, so the prefix is '4-ethyl-3-methyl'.

The name of the compound is 4-ethyl-3-methylheptane.

3.

```
    H   H   CH3 CH3 H
    |   |   |   |   |
H — C — C — C — C — C — H
    |   |   |   |   |
    H   H   CH3 H   H
```

Answer

If there is more than one atom with the same side-chain, they are written with commas between the numbers.

- Stem – the longest chain of carbon atoms is five, so the stem is 'pent-'
- Suffix – this is a saturated hydrocarbon, so the suffix is '-ane'
- Prefix – there are three methyl groups, one on the second carbon atom and two on the third carbon atom, so the prefix is '2,3,3-trimethyl'.

The name of the compound is 2,3,3-trimethylpentane.

Functional groups

Functional groups are detailed in either the suffix if it is the most important group, or the prefix, if it is an additional functional group.

To help with pronunciation, if the suffix starts with a vowel then the final '-e' is removed from the alkane part of the name. Numbers are added to indicate which carbon atom the functional group is attached to.

LEARNING TIP

No number is needed when naming aldehydes because this functional group is always on a terminal (end) carbon atom.

For organic compounds with more than one identical group, add:

- 'di-' to show two of the same group
- 'tri-' to indicate three of the same group
- 'tetra-' to describe four identical groups.

Table 3 lists functional groups and their prefix or suffix.

Type of compound	Formula	Prefix	Suffix
alcohol	–OH	hydroxy-	-ol
aldehyde	–CHO –C(=O)–H		-al
alkane	C–C		-ane
alkene	C=C		-ene
carboxylic acid	–COOH –C(=O)–OH		-oic acid
haloalkane	–F –Cl –Br –I	fluoro- chloro- bromo- iodo-	
ketone	C–CO–C C–C(=O)–C		-one

Table 3 Nomenclature terms for the prefix and suffix of organic compounds.

KEY DEFINITIONS

Saturated compounds have only single bonds.
A **functional group** is a group of atoms that is responsible for the characteristic chemical reactions of a compound.

WORKED EXAMPLE 2

Look carefully at the displayed formulae below and follow the three rules (stem, suffix, prefix) to name each compound.

1.

```
    H   H   H   H   H
    |   |   |   |   |
H — C — C — C — C — C — H
    |   |   |   |   |
    H   Cl  H   H   H
```

Answer

- Stem – the longest chain of carbon atoms is five, so the stem is 'pent-'
- Suffix – this is a saturated organic compound, so the suffix is '-ane'
- Prefix – there is one -Cl on the second carbon atom, so the prefix is '2-chloro'.

The name of the compound is 2-chloropentane.

2.

```
    H   H    H
    |   |    |
H — C — C  — C — OH
    |   |    |
    H   CH3  H
```

Answer

Remember that hyphens separate numbers and words.

- Stem – the longest chain of carbon atoms is three, so the stem is 'prop-'
- Suffix – this has an alcohol functional group on the first carbon, so the suffix is '-1-ol'
- Prefix – there is a one CH_3 group on the second carbon atom, so the prefix is '2-methyl'.

The name of the compound is 2-methylpropan-1-ol.

3.

```
    H   H   H   Cl  Cl
    |   |   |   |   |
H — C — C — C — C — C — Cl
    |   |   |   |   |
    H   H   Br  H   H
```

Answer

Remember that if there is more than one group, they are written in alphabetical order and commas are used to separate numbers.

- Stem – the longest chain of carbon atoms is five, so the stem is 'pent-'
- Suffix – this is a saturated organic compound, so the suffix is '-ane-'
- Prefix – there are three chlorine atoms: two on the first carbon atom, one on the second carbon and a bromine atom on the third carbon atom, so the prefix is '3-bromo-1,1,2-trichloro'.

The name of the compound is 3-bromo-1,1,2-trichloropentane.

4.

```
    H   H   CH3 H        O
    |   |   |   |      //
H — C — C — C — C — C
    |   |   |   |      \
    H   H   H   H       H
```

Answer

- Stem – the longest chain of carbon atoms, which includes the functional group, is five, so the stem is 'pent-'
- Suffix – this is an organic compound with an aldehyde group, so the suffix is '-al' (there is no need for a number because this functional group is on a terminal carbon atom)
- Prefix – the carbon atom with the main functional group is taken to be carbon1, so there is a methyl group on the third carbon atom and the prefix is '3-methyl'.

The name of the compound is 3-methylpentanal.

Questions

1 State the differences and similarities between ethane and ethene.

2 Give the molecular formula and displayed formula for these organic compounds:
(a) trichloromethane; (b) butanal;
(c) pentan-3-ol; (d) 2-bromo-1-chlorobutane.

3 Use the IUPAC rules to name the following substances:

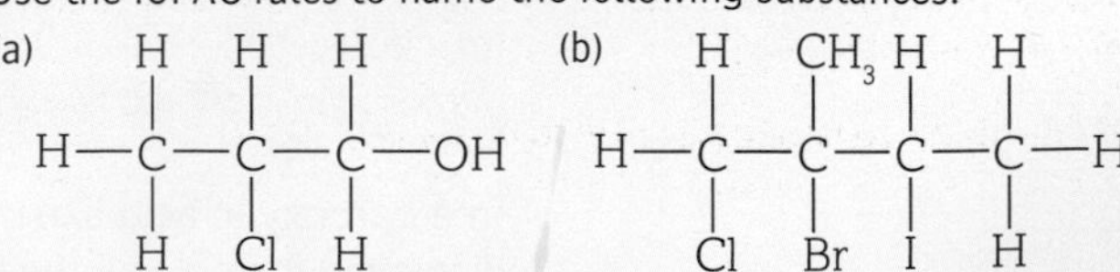

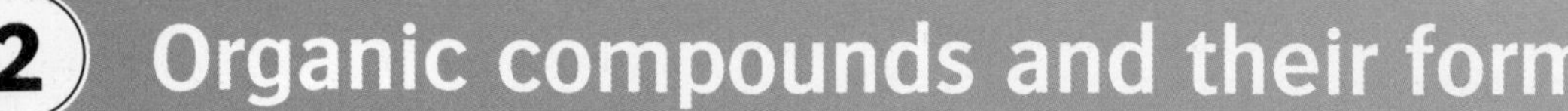

2 Organic compounds and their formulae

By the end of this topic, you should be able to demonstrate and apply your knowledge and understanding of:

- interpretation and use of the terms *general formula, structural formula* and *displayed formula*
- use of the terms *empirical formula* and *molecular formula*

General formula

All the compounds in an homologous series share the same **general formula**. This is the simplest algebraic formula for all the organic compounds in a homologous series and can be used to generate a molecular formula.

Example 1

Alkanes have the general formula C_nH_{2n+2}, where *n* is the number of carbon atoms in a molecule.

For an alkane with 4 carbon atoms (butane), $n = 4$. So the formula is $C_4H_{(2\times4)+2} = C_4H_{10}$.

```
    H   H   H   H
    |   |   |   |
H — C — C — C — C — H
    |   |   |   |
    H   H   H   H
```

Example 2

Alkenes have the general formula C_nH_{2n}, where *n* is the number of carbon atoms in a molecule.

For an alkene with 3 carbon atoms (propene), $n = 3$. So the formula is $C_3H_{(2\times3)} = C_3H_6$.

```
H       H
|       |
C = C — C — H
|   |   |
H   H   H
```

Example 3

Alcohols have the general formula $C_nH_{2n+1}OH$, where *n* is the number of carbon atoms in a molecule.

For an alcohol with 2 carbon atoms (ethanol), $n = 2$. So the formula is $C_2H_{(2\times2)+1}OH = C_2H_5OH$.

```
    H   H
    |   |
H — C — C — O — H
    |   |
    H   H
```

Displayed formula

A **displayed formula** shows the relative positions of all the atoms in a molecule and the bonds between them.

LEARNING TIP

In a displayed formula, it is important to show all the bonds, even those in functional groups. For example, an alcohol group should be shown as —O—H, not —OH.

In a carboxylic acid like ethanoic acid, the functional group is often shown as –COOH. However, in a displayed formula all the bonds and all the atoms must be shown.

The first of the formulae below is not a displayed formula because not all the bonds are shown.

```
   H  O
   |  ||
H—C—C—OH
   |
   H

   H
   |    O
H—C—C//
   |    \
   H     O—H
```

Structural formula

A **structural formula** gives the minimum detail for the arrangement of atoms in a molecule. So, butane would have the structural formula $CH_3(CH_2)_2CH_3$.

DID YOU KNOW?

Butane's full structural formula is $CH_3CH_2CH_2CH_3$.
Carboxylic acid functional groups are shortened to —COOH and esters to —COOR.

KEY DEFINITIONS

A **general formula** is the simplest algebraic formula for a homologous series.
A **displayed formula** shows the relative positions of atoms and the bonds between them.
A **structural formula** provides the minimum detail for the arrangement of atoms in a molecule.

Empirical formula

An **empirical formula** shows the smallest whole-number ratio of atoms of the elements in a compound.

Example 1

Consider hexane. It has 6 carbon atoms and 14 hydrogen atoms. The simplest ratio is 3 carbon atoms to 7 hydrogen atoms, giving the empirical formula C_3H_7.

```
   H  H  H  H  H  H
   |  |  |  |  |  |
H—C—C—C—C—C—C—H
   |  |  |  |  |  |
   H  H  H  H  H  H
```

If you know the mass of each element in a compound, you can work out the empirical formula using five steps.

- Step 1: Change the percentages to masses.
- Step 2: Calculate the number of moles of each element using
  $$\text{amount (moles)} = \frac{\text{mass (g)}}{\text{relative atomic mass}}$$
- Step 3: Divide by the smallest value.
- Step 4: Change the ratio to whole numbers.
- Step 5: Write the formula.

WORKED EXAMPLE 1

A hydrocarbon contains 82.76% carbon and 17.24% hydrogen. Calculate the hydrocarbon's empirical formula.

Answer

Step 1: Imagine you have 100 g of the substance – it would contain 82.76 g of carbon and 17.24 g of hydrogen.

Step 2: C : H

$$\frac{82.76}{12} : \frac{17.24}{1}$$

6.90 : 17.24

Step 3: $\frac{6.90}{6.90} : \frac{17.24}{6.90}$

1 : 2.5 (For this example, multiply both by 2)

Step 4: 2 : 5

Step 5: The empirical formula is C_2H_5

Molecular formula

A **molecular formula** shows the numbers and types of atoms in a compound.

You can use a three-step method to generate the molecular formula from the empirical formula and molecular mass.

- Step 1: Calculate the relative empirical mass using the empirical formula.
- Step 2: Divide the relative molecular mass by the empirical formula mass.
- Step 3: Multiply each atom in the empirical formula by the number calculated in step 2 and write the molecular formula.

WORKED EXAMPLE 2

A compound has the empirical formula CH_2O and relative molecular mass 60.0. Calculate its molecular formula.

Answer

Step 1: $12 + (2 \times 1) + 16 = 30.0$

Step 2: $\frac{60.0}{30.0} = 2$

Step 3: The empirical formula is $C_2H_4O_2$

Questions

1 A hydrocarbon contains 80.0% carbon. The compound has a relative molecular mass of 30.
(a) Calculate the empirical formula of this compound.
(b) What is the molecular formula of this compound?
(c) Draw the possible displayed formula for this compound.
(d) State the homologous series this compound belongs to and its general formula.

2 A compound contains 12.79% C, 2.15% H and 85.06% Br. Its relative molecular mass is 187.9.
(a) Calculate the empirical formula of this compound.
(b) What is the molecular formula of this compound?
(c) Draw a possible displayed formula for this compound.

3 Skeletal formulae

By the end of this topic, you should be able to demonstrate and apply your knowledge and understanding of:

- interpretation and use of the term *skeletal formula*

Selecting the best type of formula for the context

There are many different ways of representing organic molecules.

Drawing displayed formulae is useful when you are considering the shape and size of a molecule, but this is time-consuming and it is easy to draw complex molecules incorrectly.

Structural formulae give a feel for the arrangement of the atoms in a molecule and they can be drawn quickly. The physical properties of a compound often relate to chain length and branching, making this type of representation useful.

As most of the chemistry for organic compounds is related to their functional groups, skeletal formulae focus the attention on these reactive parts of molecules. They are often used in constructing mechanisms, which are like flow charts that model electron movement in chemical reactions.

Skeletal formulae

A **skeletal formula** is the briefest way of representing organic molecules. Lines are used to indicate alkyl chains, and every corner represents a carbon atom.

KEY DEFINITIONS

A **skeletal formula** is a simplified structural formula drawn by removing hydrogen atoms from alkyl chains.
Unsaturated organic chemicals contain at least one carbon–carbon double covalent bond.
Hydrocarbons are compounds that contain only hydrogen atoms and carbon atoms.
Saturated organic compounds contain only single covalent bonds.

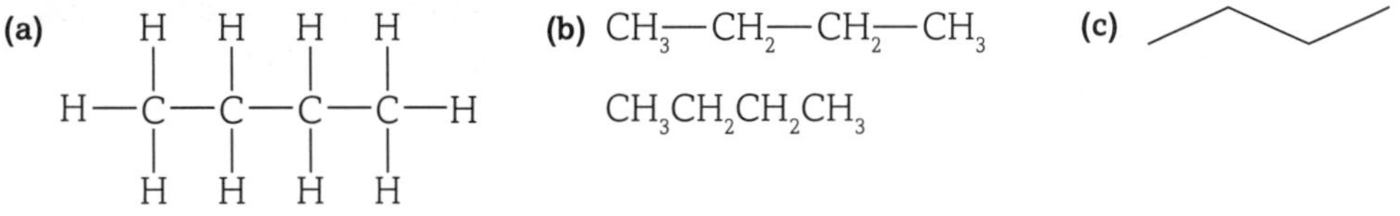

(a) (b) $CH_3—CH_2—CH_2—CH_3$ $CH_3CH_2CH_2CH_3$ (c)

Figure 1 Butane: (a) displayed formula; (b) structural formula; (c) skeletal formula.

(a) (b) $CH_2═CH—CH_3$ CH_2CHCH_3 (c)

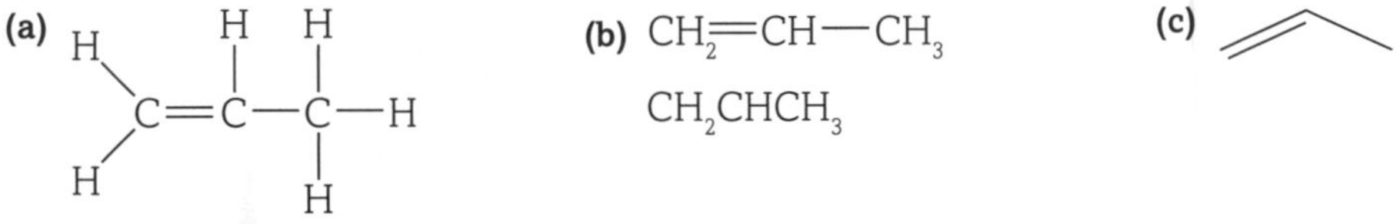

Figure 2 Propene: (a) displayed formula; (b) structural formula; (c) skeletal formula.

(a) (b) C_6H_{12} (c)

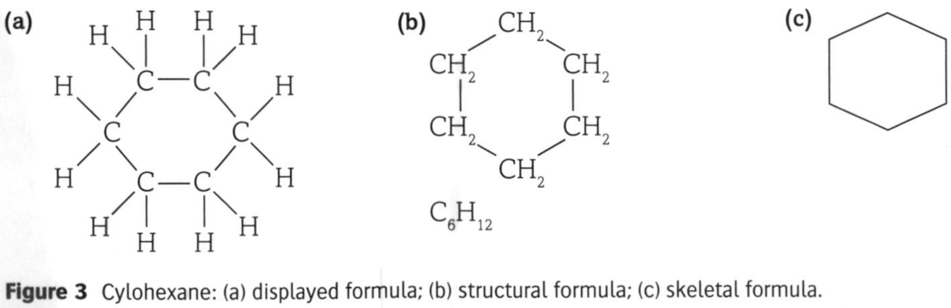

Figure 3 Cylohexane: (a) displayed formula; (b) structural formula; (c) skeletal formula.

Unsaturated hydrocarbons

Unsaturated hydrocarbons are organic compounds that contain only carbon and hydrogen, and usually have at least one C=C unit representing a double bond.

(a)

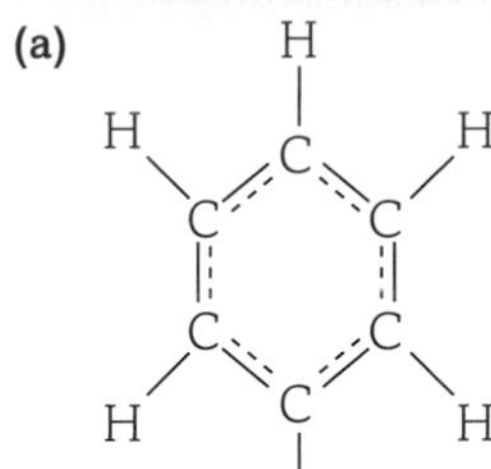

(b)

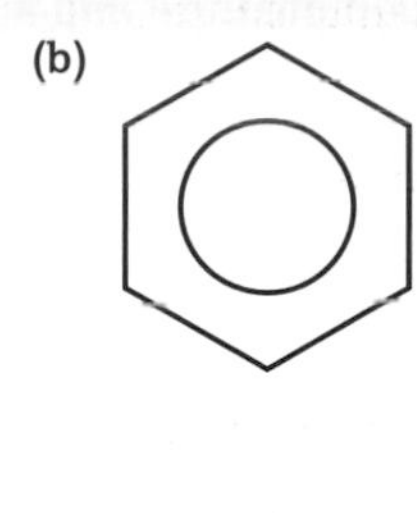

(c)

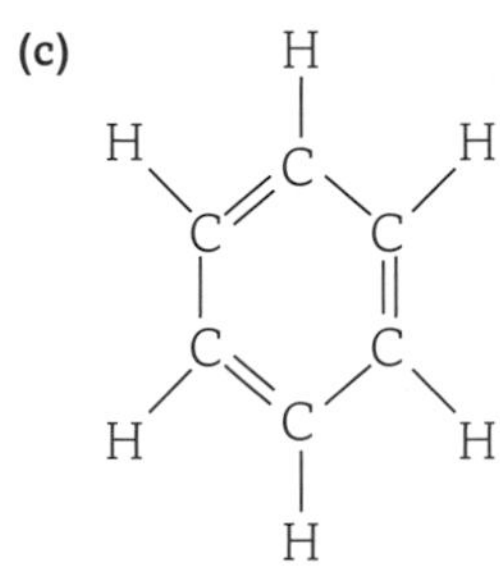

(d)

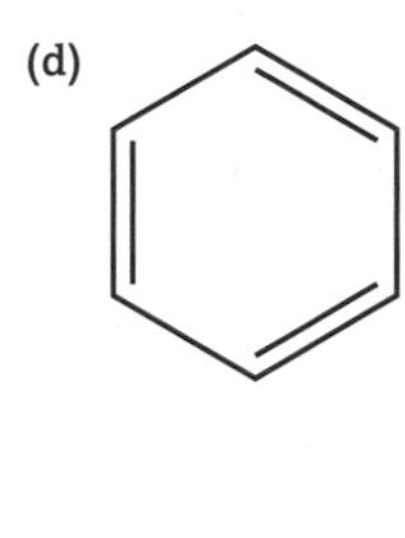

Figure 4 Benzene: (a) displayed formula showing delocalised electrons; (b) skeletal formula; (c) displayed Kekulé model; (d) skeletal formula of Kekulé model.

(a)

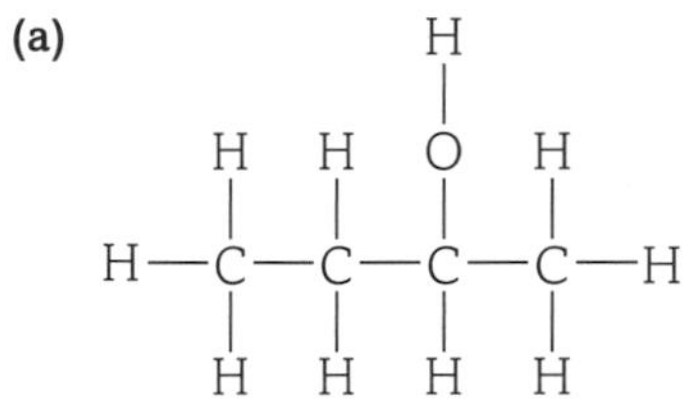

(b) $CH_3-CH(OH)-CH_2-CH_3$

(c)

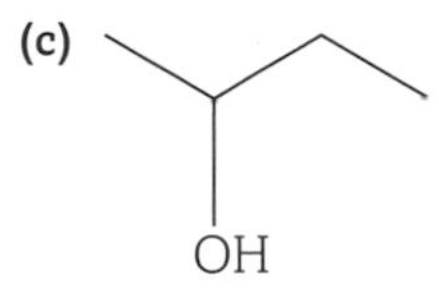

Figure 5 Butan-2-ol: (a) displayed formula; (b) structural formula; (c) skeletal formula.

(a)

(b) $CH_3-CH_2-CH_2-CH_2-CH_2-COOH$

(c)

Figure 6 Hexanoic acid: (a) displayed formula; (b) structural formula; (c) skeletal formula.

LEARNING TIP

A skeletal formula doesn't show atoms – there are carbon atoms where lines meet and also at the ends of a chain.

Cyclic compounds

Cyclic compounds are commonly shown in skeletal form. Cyclohexane has the molecular formula C_6H_{12} and is an alicyclic **saturated** hydrocarbon.

Aromatic hydrocarbons

Aromatic hydrocarbons always have a 6-carbon ring with molecular formula C_6H_6. This structure is very stable because the electrons are shared across all the carbon atoms. This part of a molecule is rarely involved in chemical reactions and is often represented using a skeletal formula.

Alcohols

The alcohol group is represented by –OH in a skeletal formula, as in structural formula. Butan-2-ol is used in some brake fluids. It is a saturated secondary alcohol containing four carbon atoms, with the functional group on the second carbon atom.

Carboxylic acids

The functional group in carboxylic acids is –COOH, and the double bond between a carbon atom and one of the oxygen atoms is drawn as double parallel lines.

Hexanoic acid is one of the components of vanilla. It contains six carbon atoms and has a carboxylic acid functional group at the end of the carbon chain.

Question

1 Draw displayed, structural and skeletal formulae for:

(a) 2-methylpent-1-ene

(b) C_5H_{10}

(c) pentanoic acid

(d) 2-methlypropan-1-ol.

Isomerism

By the end of this topic, you should be able to demonstrate and apply your knowledge and understanding of:

- explanation of the terms *structural isomers, stereoisomers, E/Z isomerism* and *cis–trans isomerism*
- interpretation and use of the terms *saturated* and *unsaturated*
- use of Cahn–Ingold–Prelog (CIP) priority rules to identify the *E* and *Z* stereoisomers
- determination of possible structural formulae of an organic molecule, given its molecular formula
- determination of possible *E/Z* or *cis–trans* stereoisomers of an organic molecule, given its structural formula

What are isomers?

Some organic compounds have the same **molecular formula** but different arrangements of their atoms. These substances are known as isomers.

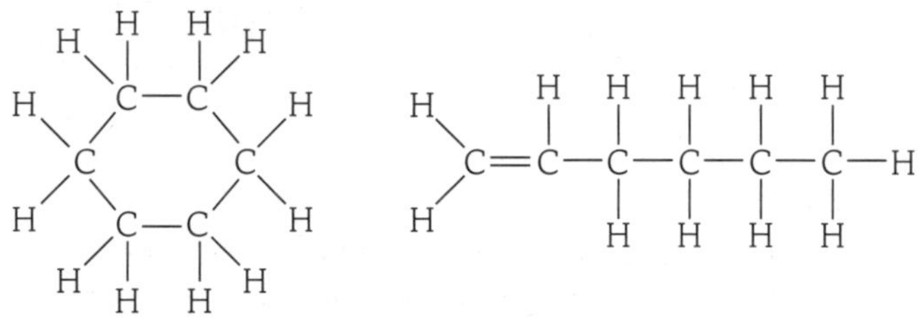

Figure 1 Cyclohexane and hexene are isomers because they have the same molecular formula (C_6H_{12}) but their atoms are arranged differently.

Structural isomers

Structural isomers are compounds with the same molecular formula but different structural formulae. There are three ways that this can happen.

- The alkyl groups are in different places. Consider a hydrocarbon with the molecular formula C_5H_{12}. There are three arrangements of atoms, so there are three structural isomers.

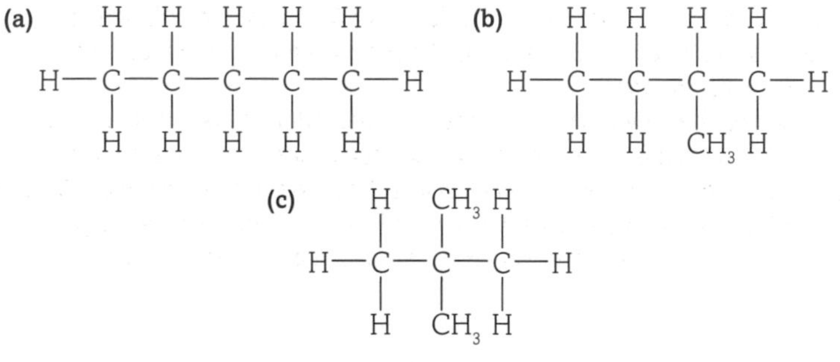

Figure 2 (a) Displayed formula of pentane. (b) Displayed formula of 2-methylbutane. (c) Displayed formula of 2,2-dimethylpropane.

- The **functional group** can be bonded to different parts of the parent chain. Consider an alcohol with the molecular formula C_3H_8O. The –OH functional group can be on an end carbon atom (making a primary alcohol) or on the second carbon atom (making a secondary alcohol). Both compounds have the same molecular formula but they will have different chemical properties.

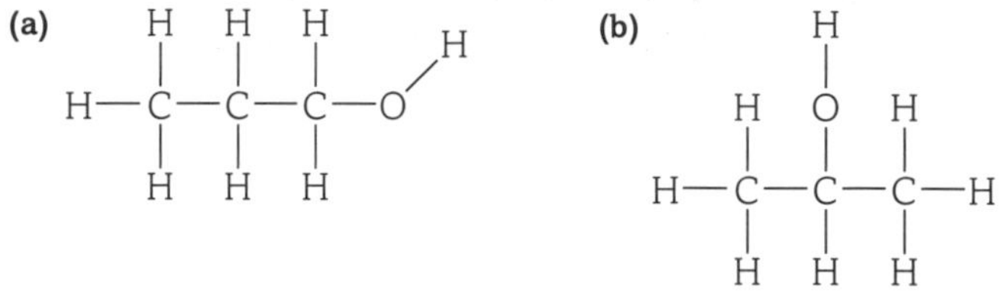

Figure 3 (a) Displayed formula of propan-1-ol. (b) Displayed formula of propan-2-ol.

- The functional group could be different. For example, when a carbonyl –C=O functional group is bonded to the end of a carbon chain, the organic compound is an aldehyde –CHO. But when the carbonyl group is bonded to an internal carbon, it is a ketone. These are isomers with different functional groups.

Figure 4 (a) Displayed formula of propanal. (b) Displayed formula of propanone.

KEY DEFINITIONS

A **molecular formula** shows the numbers and type of the atoms of each element in a compound.
Structural isomers are compounds with the same molecular formula but different structural formulae.
A **functional group** is a group of atoms responsible for the characteristic reactions of a compound.
Stereoisomers are organic compounds with the same molecular formula and structural formula but having different arrangements of atoms in space.
Cis–trans isomerism is a type of *E/Z* isomerism in which the two substituent groups attached to both carbon atoms of the C=C bond are the same.
***E/Z* isomerism** is a type of stereoisomerism caused by the restricted rotation around the double bond – two different groups are attached to both carbon atoms of the C=C double bond.

Stereoisomerism

When organic compounds have the same molecular formula and structural formula but a different arrangement of atoms in space, they are called **stereoisomers**.

Consider 1-chloroprop-1-ene – this has the molecular formula C_3H_5Cl and the structural formula $CHClCHCH_3$. The double bond will not allow the atoms to rotate freely. So there are two arrangements of the atoms in space, and therefore two stereoisomers.

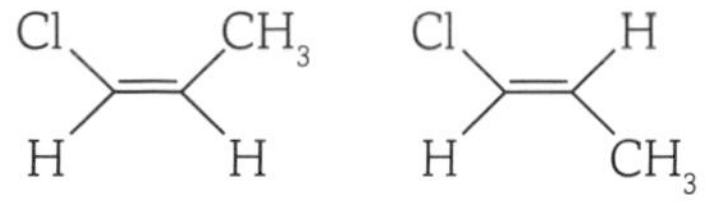

Figure 5 1-chloroprop-1-ene has two stereoisomers.

E/*Z* isomerism

***E*/*Z* isomerism** is a type of stereoisomerism. An organic molecule must have a C=C double bond and both carbon atoms must be attached to different groups. So the chloropropene stereoisomers shown in Figure 5 are examples of *E*/*Z* isomers.

Cahn–Ingold–Prelog rules

LEARNING TIP

The groups on both sides of the double bond are called substituents.

To name these compounds, chemists use the Cahn–Ingold–Prelog (CIP) rules. Use these steps to help you identify *E* and *Z* isomers.

1. Locate the C=C double bond in the molecule and redraw it to show the substituents.
2. Focus on one carbon atom and assign the priority of each substituent based on its relative atomic mass – the highest is given the highest priority. Then assign the priority of the second carbon atom.
3. If the highest priority groups are on the same side of the C=C double bond, then the isomer is *Z*. If the highest priority groups are on different sides of the C=C double bond, then the isomer is *E*.

DID YOU KNOW?

The *E*/*Z* notation has its roots in the German language. '*E*' stands for 'entgegen', which means opposite side, and '*Z*' stands for 'zusammen', which means together (or on the same side).

Cis–trans isomerism

***Cis–trans* isomerism** is a special type of *E*/*Z* isomerism where two substituents on each carbon atom are the same. In this case, the *E* isomer is also a *trans* isomer and the *Z* isomer is also a *cis* isomer.

WORKED EXAMPLE 1

Name the isomer shown in Figure 6.

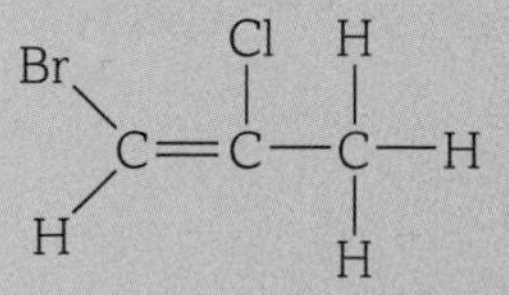

Figure 6

Answer

1. Draw the structure of the isomer, focusing on the double bond. 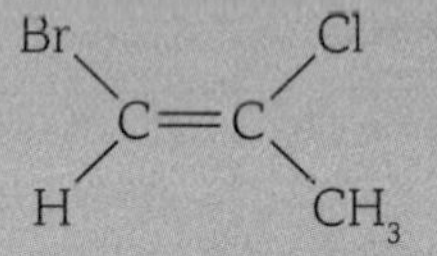
2. Assign the priorities. In the first carbon atom, bromine has higher relative atomic mass than hydrogen, so bromine has the highest priority. In the second carbon atom, chlorine has a higher relative atomic mass than the carbon of the methyl group, so chlorine has the highest priority.
3. As the two highest priority groups are on the same side of the C=C double bond, this is a *Z* isomer. The full name of this chemical is *Z*-1-bromo-2-chloropropene.

WORKED EXAMPLE 2

Name the isomers of but-2-ene shown in Figure 7.

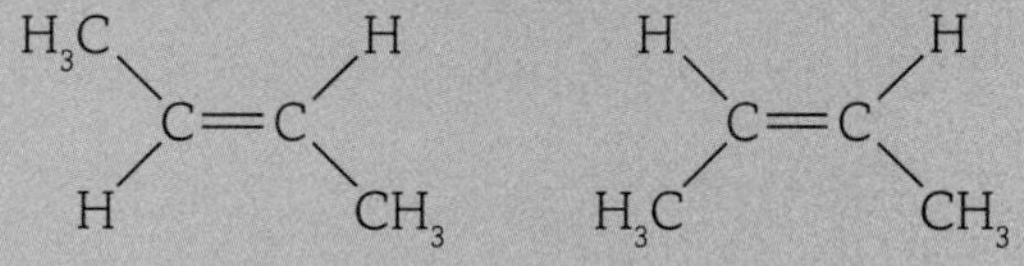

Figure 7

Answer

1. The molecules are already drawn – focus on the double bond.
2. Assign the priorities. In the first carbon atom, the carbon of the methyl group has higher relative atomic mass than hydrogen, so the methyl group has the highest priority. The same is true for the second carbon atom.
3. In the first diagram, the two highest priority groups are opposite and so the name is *E*-but-2-ene and is an example of a *trans* isomer. In the second diagram, the two highest priority groups are on the same side, so the name is *Z*-but-2-ene and is an example of a *cis* isomer.

DID YOU KNOW?

Fats form an important food group and are part of a balanced diet. Almost all naturally occurring unsaturated fats are *cis* fats, where the highest priority groups are on the same side of the C=C double bond. Human bodies can digest *cis* fats and use the products for many things, like making cell membranes.

However, in many processed foods, like margarine, there are *trans* fats. These have a more 'kinked' shape and this makes them have higher melting points. The body cannot digest these fats in the same way. *Trans* fats have had unexpected adverse health effects, such as increasing levels of cholesterol in the blood, which can lead to coronary heart disease.

Questions

1. Draw and name the five structural formulas of C_6H_{14}.
2. Draw and name the *E*/*Z* isomers of $CH_3CH{=}CHCH_2CH_3$.
3. Are these *E*/*Z* isomers an example of *cis–trans* isomerism?

Reaction mechanisms

By the end of this topic, you should be able to demonstrate and apply your knowledge and understanding of:

- the different types of covalent bond fission:
  - (i) homolytic fission (in terms of each bonding atom receiving one electron from the bonded pair, forming two radicals)
  - (ii) heterolytic fission (in terms of one bonding atom receiving both electrons from the bonded pair)
- the term radical (a species with an unpaired electron) and use of 'dots' to represent species that are radicals in mechanisms
- a 'curly arrow' as the movement of electrons showing either heterolytic fission or formation of a covalent bond
- reaction mechanisms, using diagrams, to show clearly the movement of an electron pair with 'curly arrows' and relevant dipoles

Introduction

DID YOU KNOW?

In a chemical reaction, some or all of the reactant bonds break, atoms rearrange themselves and new bonds are formed to make the products. This can all be explained using the movement of electrons.
In organic chemistry, models called **reaction mechanisms** are used to understand reactions and to make predictions about the products formed. These diagrams show the movement of electrons as **curly arrows**.

LEARNING TIP

In reaction mechanisms, each line of the arrow head represents one electron moving. So, ⇀ means one electron and → means two electrons moving. The curly arrows should start from a bond, a lone pair or a negative charge and finish on the atom they are moving to. All dipoles and any charges should be shown on the diagrams clearly.

Covalent bond fission

A covalent bond is a strong electrostatic attraction between a shared pair of electrons and the nuclei of the bonded atoms. During the first stage of a chemical reaction, activation energy affects the covalent bonds by causing them to break – this is described as undergoing fission.

Homolytic fission

Homolytic fission occurs when a covalent bond breaks and each electron goes to a different bonded atom. This generates two highly reactive, neutral species called **radicals**. Radicals have one or more, unpaired electrons, which are shown as dots.

Homolytic fission can be summarised in a general equation:

$X - Y \rightarrow X^{\bullet} + Y^{\bullet}$

For example, in a chlorine molecule, Cl_2, there is a shared pair of electrons between the atoms. When UV radiation strikes this covalent bond it causes it to break. One electron from the covalent bond goes to each chlorine atom forming two identical radicals – Figure 1 shows the mechanism of homolytic fission.

$$Cl - Cl \xrightarrow{UV} 2Cl\bullet$$

Figure 1 Homolytic fission of a chlorine molecule.

KEY DEFINITIONS

Reaction mechanisms are models that show the movement of electron pairs during a reaction.
Curly arrows model the flow of electron pairs during reaction mechanisms.
Homolytic fission happens when each bonding atom receives one electron from the bonded pair, forming two radicals.
Radicals are species with one or more unpaired electrons.
Heterolytic fission happens when one bonding atom receives both electrons from the bonding pair.

Heterolytic fission

Heterolytic fission occurs when a covalent bond breaks and both electrons go to one of the bonded atoms. This results in a positive ion (cation) and negative ion (anion) being formed.

Heterolytic fission can be summarised in a general equation:

$$X - Y \rightarrow X^+ + Y^-$$

For example, 2-bromopropane has a molecule that has a dipole. The bromine atom is more electronegative than the carbon atom, so the electrons spend more time near the bromine atom than the carbon atom. This makes the bromine atom $\delta-$ and the carbon atom $\delta+$.

Under certain conditions, the polar covalent bond will undergo heterolytic fission (Figure 2). Both electrons from the bond will go to the bromine atom, making a bromide ion and leaving a carbocation behind.

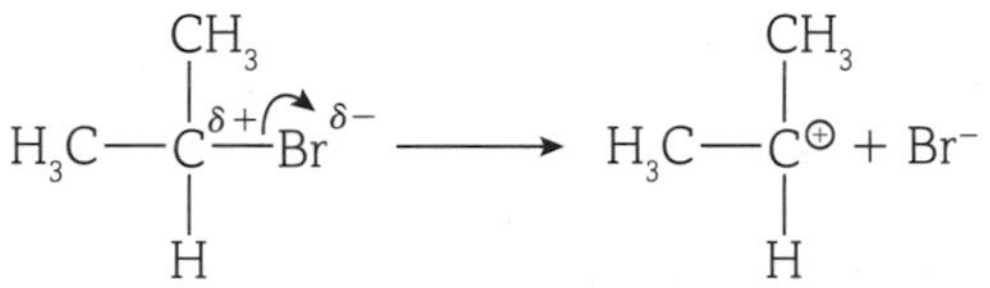

$$H_3C-\overset{CH_3}{\underset{H}{C^{\delta+}}}-Br^{\delta-} \longrightarrow H_3C-\overset{CH_3}{\underset{H}{C^{\oplus}}} + Br^-$$

Figure 2 Heterolytic fission of a polar bond.

Covalent bond formation

When two radicals or two oppositely charged ions collide, a new bond is formed.

For example, in the production of tetrachloromethane, methyl radicals are made. When two methyl radicals collide they form ethane (Figure 3).

$$CH_3^{\bullet} \; {}^{\bullet}CH_3 \longrightarrow H_3C-CH_3$$

Figure 3 The formation of ethane from two radicals.

Carbocations are very reactive intermediates (Figure 4) because they are unstable.

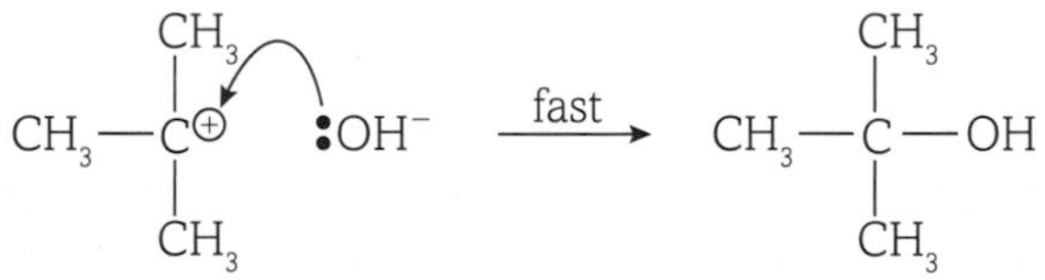

$$CH_3-\overset{CH_3}{\underset{CH_3}{C^{\oplus}}} \quad :OH^- \xrightarrow{\text{fast}} CH_3-\overset{CH_3}{\underset{CH_3}{C}}-OH$$

Figure 4 The formation of 2-methylpropan-2-ol from a carbocation.

Questions

1. Draw a reaction mechanism to show homolytic fission in the reaction between chlorine and methane to form chloromethane under UV light.
2. Draw a reaction mechanism to show the formation of tetrabromoethane from two radicals.
3. Draw a reaction mechanism to show the reaction between a carbocation and an iodide ion to form iodoethane.

6 Properties of alkanes

By the end of this topic, you should be able to demonstrate and apply your knowledge and understanding of:

- alkanes as saturated hydrocarbons containing single C–C and C–H bonds as σ-bonds (overlap of orbitals directly between the bonding atoms)
- free rotation of the σ-bond
- explanation of the tetrahedral shape and bond angle around each carbon atom in alkanes in terms of electron pair repulsion
- explanation of the variations in boiling points of alkanes with different carbon-chain length and branching, in terms of induced dipole–dipole interactions (London forces)

Introduction

The **alkanes** form a **homologous series**. They:

- are **hydrocarbons** containing only hydrogen atoms and carbon atoms
- are saturated containing only single covalent bonds
- have the general formula C_nH_{2n+2}
- have a gradation in physical properties
- have similar chemical properties.

KEY DEFINITIONS

Alkanes are a homologous series of saturated hydrocarbons.
A **homologous series** is a family of organic compounds that have the same functional group, but successive members differ by CH_2.
A **hydrocarbon** is an organic molecule that contains only carbon atoms and hydrogen atoms.

Bonding in alkanes

As carbon is in group 14 of the periodic table, its atoms have four electrons in their outer shell. These pair up with electrons on other atoms to form four covalent bonds.

Each covalent bond has direct overlap of the electron clouds from each atom making a sigma bond. Therefore C–C and C–H bonds are examples of σ-bonds. In the alkane structure, every carbon atom has a tetrahedral shape with a bond angle of 109.5° (Figure 1).

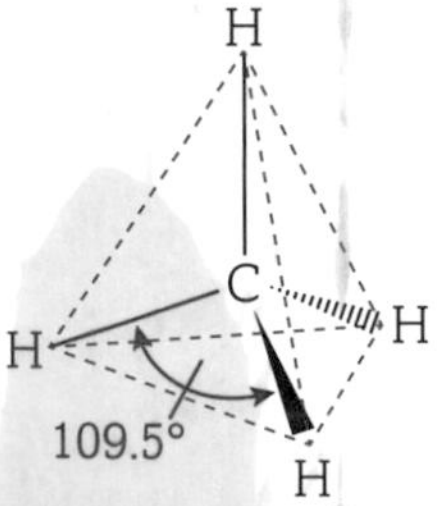

Figure 1 Dot-and-wedge diagram showing the tetrahedral shape of the first member of the alkane series.

LEARNING TIP

You will be expected to draw 3D representations of small alkane molecules to show the tetrahedral arrangement of the bonds around the carbon atoms. The dot-and-wedge notation can be useful for this.

DID YOU KNOW?

In alkanes the carbon 2s and 2p atomic orbitals make four hybrid orbitals described as sp^3. Two sp^3 orbitals in neighbouring carbon atoms overlap to produce a C–C σ-bond between the atoms. Then the remaining three sp^3 orbitals on each carbon atom overlap with 1s orbitals from three separate hydrogen atoms to produce three C–H σ-bonds.

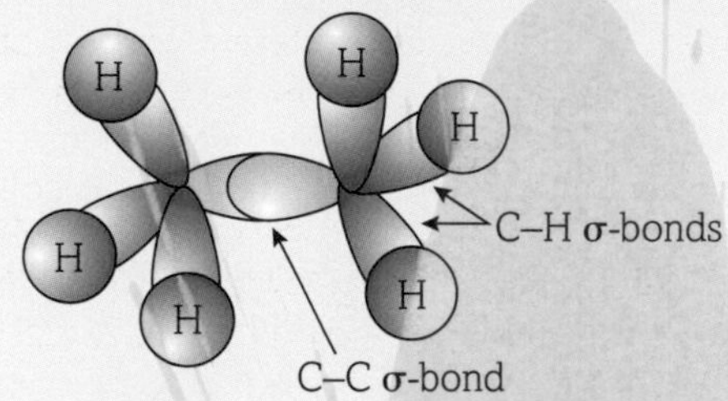

The Pauling electronegativity values of carbon and hydrogen are 2.55 and 2.20 respectively. As these values are very similar, alkane molecules are non-polar and without any significant dipoles. However, because electrons are moving round the shells all the time, occasionally there may be a lack of balance in charge distribution. This causes an instantaneous dipole, which will induce dipoles in neighbouring molecules (Figure 2). This is the weakest intermolecular force of attraction and is called an induced dipole–dipole interaction or London force.

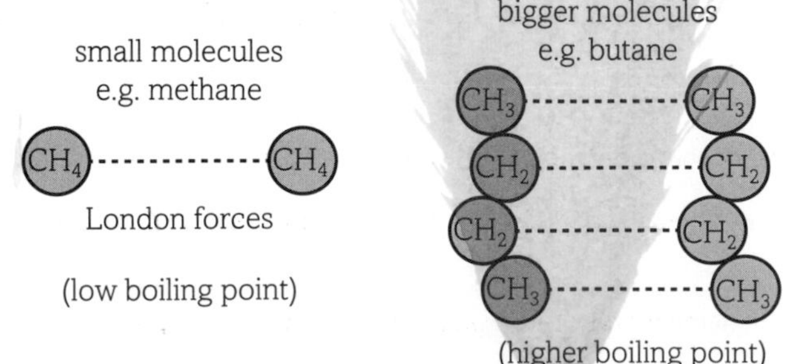

Figure 2 Induced dipole–dipole forces between alkane molecules.

Boiling points of alkanes

The boiling point of a pure substance is a precise temperature where the substance changes state from a liquid to a gas or a gas to a liquid. As boiling point changes with pressure, the standard values are given in Kelvin (K) at 100 kPa.

Carbon-chain length

As an alkane chain gets longer, its relative molecular mass increases. Larger molecules have more surface area contacts between adjacent molecules. This increases the number of induced dipole–dipole forces. So more energy is needed to overcome the intermolecular attraction in order that the alkane can change state. When the chain length is plotted against boiling point (Figure 3), you can see a positive correlation.

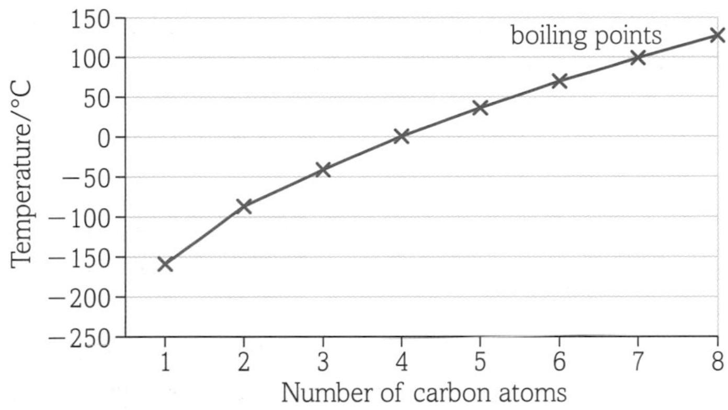

Figure 3 Showing how the boiling point changes as the chain length of an alkane changes.

DID YOU KNOW?

Many alkanes are derived from the fractional distillation of crude oil. As each fraction is a mixture of hydrocarbons they have their boiling points given as a small range of temperatures.

Branching

Structural isomers of alkanes have different boiling points. The more branched a compound is, the fewer surface area interactions there are between molecules – this is because the molecules cannot fit together as neatly.

Therefore branched molecules have fewer induced dipole–dipole attractions compared to the straight-chain isomer with the same molecular formula (Figure 4). This leads to branched molecules having a lower boiling point than the equivalent straight-chain isomer.

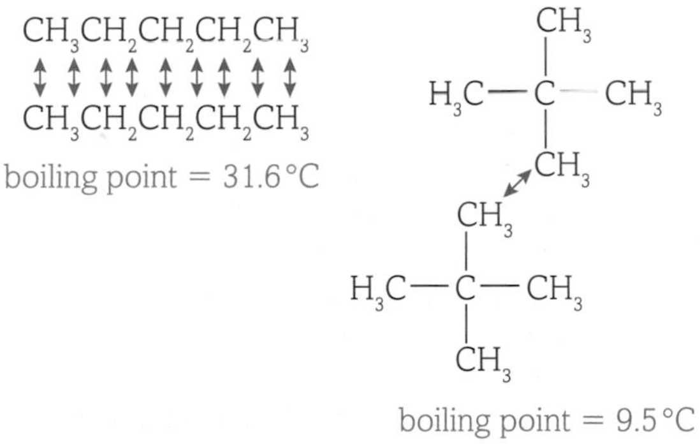

$CH_3CH_2CH_2CH_2CH_3$

$CH_3CH_2CH_2CH_2CH_3$

boiling point = 31.6 °C

CH_3 | $H_3C—C—CH_3$ | CH_3

CH_3 | $H_3C—C—CH_3$ | CH_3

boiling point = 9.5 °C

Figure 4 Straight-chain alkanes have a higher surface area between molecules than their branched-chain isomers.

Questions

1. Which has the highest boiling point – butane or methylpropane? Explain your answer.
2. Explain why there are more extensive induced dipole–dipole interactions in pentane compared to in ethane.
3. Draw the different structural isomers for C_6H_{14} and predict the order of their boiling points.

4.1

7 Reactions of alkanes

By the end of this topic, you should be able to demonstrate and apply your knowledge and understanding of:

* the low reactivity of alkanes with many reagents in terms of the high bond enthalpy and very low polarity of the σ-bonds present
* complete combustion of alkanes, as used in fuels, and the incomplete combustion of alkane fuels in a limited supply of oxygen with the resulting potential dangers from CO
* the reaction of alkanes with chlorine and bromine by radical substitution using ultraviolet radiation, including a mechanism involving homolytic fission and radical reactions in terms of initiation, propagation and termination
* the limitations of radical substitution in synthesis by the formation of a mixture of organic products, in terms of further substitution and reactions at different positions in a carbon chain

Reactivity of alkanes

Alkanes have a low reactivity with many reagents (Figure 1) – this is for two reasons:

- All the covalent bonds in alkane molecules have high bond enthalpies – a large amount of energy is required to break the bonds.
- The carbon–hydrogen σ-bonds have very low polarity because the electronegativities of carbon and hydrogen are almost the same.

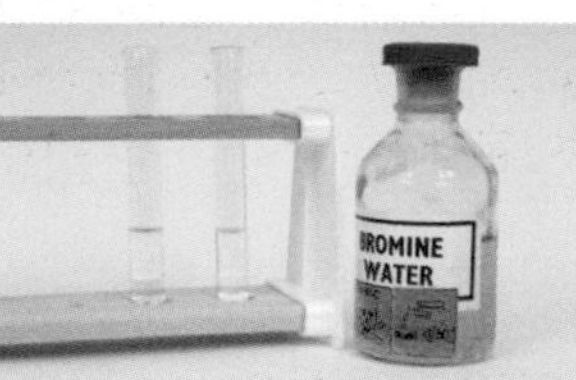

Figure 1 Alkanes such as hexane do not react with bromine in bromine water, unlike alkenes such as hexene.

Combustion

Alkanes are often used as fuels – they undergo combustion and transfer their stored chemical energy to a usable form such as thermal energy. Combustion is a rapid oxidation reaction which combines oxygen, usually from the air, with another substance.

Complete combustion

When there is a plentiful supply of air, an alkane undergoes **complete combustion**. This makes a clean, blue flame and transfers the maximum amount of thermal energy because it fully oxidises carbon and hydrogen – only carbon dioxide and water are produced.

A Bunsen burner (Figure 2) mainly burns methane, the first member of the alkane homologous series. When the air hole is open, there is a plentiful supply of oxygen and complete combustion occurs. This can be represented:

$$CH_4 + 2O_2 \rightarrow CO_2 + 2H_2O$$

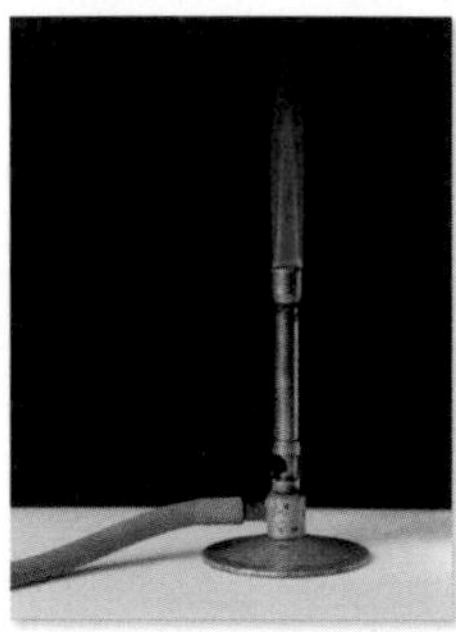

Figure 2 Complete combustion of methane in a Bunsen burner.

Incomplete combustion

When there is a limited supply of air, an alkane undergoes **incomplete combustion**. This produces carbon particles (soot) and makes a cooler, dirty yellow flame. Water and carbon dioxide are also produced, as is the colourless, odourless and toxic gas carbon monoxide (CO).

When the air hole is closed in a Bunsen burner, there is a limited supply of air and incomplete combustion occurs. As there is a mixture of products, this process can be represented by a variety of balanced chemical equations. They include:

$$2CH_4 + 2\tfrac{1}{2}O_2 \rightarrow CO + C + 4H_2O$$
$$2CH_4 + 3\tfrac{1}{2}O_2 \rightarrow CO_2 + CO + 4H_2O$$
$$3CH_4 + 4\tfrac{1}{2}O_2 \rightarrow CO_2 + CO + C + 6H_2O$$

Figure 3 Incomplete combustion of methane in a Bunsen burner.

DID YOU KNOW?

Carbon monoxide is a toxic gas and is a product of incomplete combustion. Once inside the lungs it diffuses into the blood and forms a coordinate bond to the haemoglobin in the red blood cells. This stops the molecules binding to oxygen and so reduces the oxygen-carrying capacity of the blood. Initially, carbon monoxide poisoning makes sufferers feel light-headed and breathe faster on exercising. In serious cases, victims lose consciousness and may even die.
It is important to have gas appliances checked regularly (Figure 4) and to install a carbon monoxide detector.

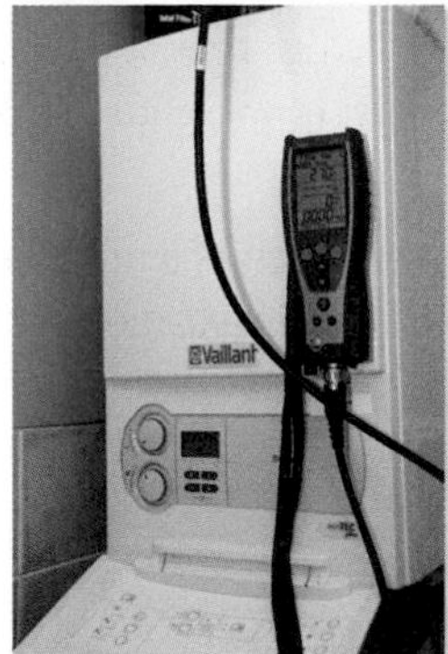

Figure 4 Gas appliances need to be checked regularly to ensure that they are not leaking carbon monoxide.

KEY DEFINITIONS

Complete combustion is oxidising a fuel in a plentiful supply of air.
Incomplete combustion is oxidising a fuel in a limited supply of air.

Radical substitution

Alkanes react with both chlorine and bromine in the presence of ultraviolet (UV) light to form halogenated organic compounds. Halogenation happens by homolytic fission of the halogen molecule, forming radicals. A hydrogen atom in the alkane is substituted by a halogen atom. If the reaction conditions are maintained, the substitution can continue until all the hydrogen atoms in the original alkane have been replaced by halogen atoms.

The radical substitution mechanism has three stages.

1. Initiation – the formation of the radicals.
2. Propagation – two repeated steps that build up the desired product in a side-reaction.
3. Termination – two radicals collide and make a stable product.

Radical substitution is very unpredictable and difficult to control. The reactions are randomly caused by a very reactive radical colliding with another species. A mixture of products is formed, which need to be separated by processes such as fractional distillation or chromatography before they can be used.

LEARNING TIP

In the propagation step there is always the same number of radical species. So if a reaction has one radical as a reactant it must have one radical in the product.

Chlorination of methane to form chloromethane

The basic steps are shown in Table 1.

Step	Equations	Comments
initiation	$Cl_2 \rightarrow 2Cl^\bullet$	Ultraviolet light or temperatures of about 300 °C are needed.
propagation	$CH_4 + Cl^\bullet \rightarrow {}^\bullet CH_3 + HCl$ ${}^\bullet CH_3 + Cl_2 \rightarrow CH_3Cl + Cl^\bullet$	• Step 1 generates an alkyl radical and hydrogen chloride. • Step 2 generates the desired product and regenerates the chlorine radical.
termination	$2Cl^\bullet \rightarrow Cl_2$ $2{}^\bullet CH_3 \rightarrow C_2H_6$ ${}^\bullet CH_3 + Cl^\bullet \rightarrow CH_3Cl$	A mixture of products is produced by the random collisions between radicals. Only one product is desirable (CH_3Cl) so this reaction has a low atom economy.

Table 1

The chloromethane molecules made may collide again with other chlorine radicals and this will cause further substitutions. So the products of the reaction of methane with chlorine in the presence of UV light could include, as well as chloromethane, varying proportions of the undesired side-products – hydrogen, chlorine, ethane, dichloromethane, trichloromethane, 1,2-dichloroethane, 1,1-dichloroethane, 1,1,2-trichloroethane, 1,1,1-trichloroethane 1,1,2,2-tetrachloroethane, 1,1,1,2-tetrachloroethane, 1,1,1,2,2-pentachloroethane and 1,1,1,2,2,2-hexachloroethane.

Bromination of ethane to form bromoethane

The basic steps are shown in Table 2.

Step	Equations
initiation	$Br_2 \rightarrow 2Br^\bullet$
propagation	$C_2H_6 + Br^\bullet \rightarrow {}^\bullet C_2H_5 + HBr$ ${}^\bullet C_2H_5 + Br_2 \rightarrow C_2H_5Br + Br^\bullet$
termination	$2Br^\bullet \rightarrow Br_2$ $2{}^\bullet C_2H_5 \rightarrow C_4H_{10}$ ${}^\bullet C_2H_5 + Br^\bullet \rightarrow C_2H_5Br$

Table 2

Questions

1. Write balanced chemical equations for the complete and incomplete combustion of camping gas (propane).
2. Explain why it is important that a gas cooker hob has a plentiful supply of air.
3. Write the equations and label the stages to show the bromination of methane to make bromomethane.
4. Write the equations and label the stages to show the chlorination of ethane to make chloroethane.
5. Explain why radical substitution does not produce a high yield of desirable product.

8 Properties of alkenes

By the end of this topic, you should be able to demonstrate and apply your knowledge and understanding of:

* alkenes as unsaturated hydrocarbons containing a C=C bond comprising a π-bond (sideways overlap of adjacent p-orbitals above and below the bonding C atoms) and a σ-bond (overlap of orbitals directly between the bonding atoms)
* explanation of the trigonal planar shape and bond angle around each carbon in the C=C alkenes in terms of electron pair repulsion
* explanation of the term *stereoisomers*

Introduction

The **alkenes** form a homologous series. They:

- are hydrocarbons, containing only hydrogen atoms and carbon atoms
- are **unsaturated** containing at least one C=C bond
- have the general formula C_nH_{2n}
- have a gradation in physical properties
- have similar chemical properties.

Bonding in alkenes

The C–H covalent bonds in alkenes are sigma bonds. All alkene molecules contain at least one C=C bond. A double covalent bond has two sections involving the overlap of electron clouds (Figure 1).

- A σ-bond is formed between two carbon atoms using the direct overlap of the electron clouds of the two atoms.
- The π-bond is formed by the electrons in the adjacent p-orbitals overlapping above and below the carbon atoms. This type of bond can only be made after a σ-bond has been formed.

The π-bond holds the atoms in position by restricting rotation around the double bond. So alkene molecules have a flat shape in the region of the double bond(s) – they are described as 'planar'.

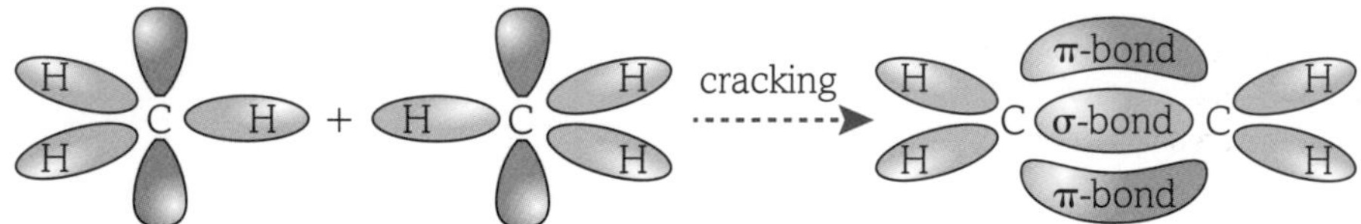

Figure 1 Bonding in ethene showing the σ- and π-bonds.

The π-bond is the reactive part of the molecule because of the high electron density around it.

Shapes of alkene molecules

Ethene is the first member of the alkene homologous series. It has the molecular formula C_2H_4 – each carbon atom in the molecule has three areas of electron density:

- two separate covalent bonds between the carbon atoms and two different hydrogen atoms – these are σ-bonds
- the covalent bonds between the two carbon atoms, made from a σ-bond and a π-bond.

Each of these electron densities repel by the same amount forming bond angles of 120°, which results in a trigonal planar shape (Figure 2).

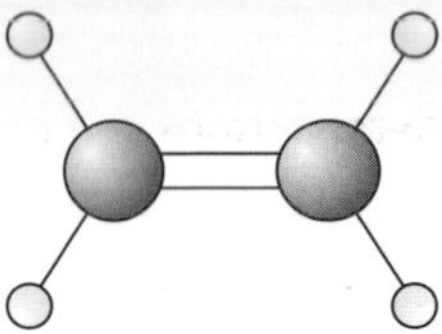

Figure 2 Ethene is a planar (flat) molecule.

Stereoisomerism

Stereoisomers are compounds that have the same structural formula but different arrangements of atoms in space.

If an alkene has different alkyl groups attached to both carbon atoms of the double bond, then stereoisomers can be formed. The restricted rotation around the C=C bond allows *E* and *Z* isomers to be generated. If two identical substituent groups are attached to each carbon atom, then a special case of the isomerism called ***cis–trans* isomerism** is possible (Figure 3). See topic 4.1.4 to remind yourself about stereoisomerism.

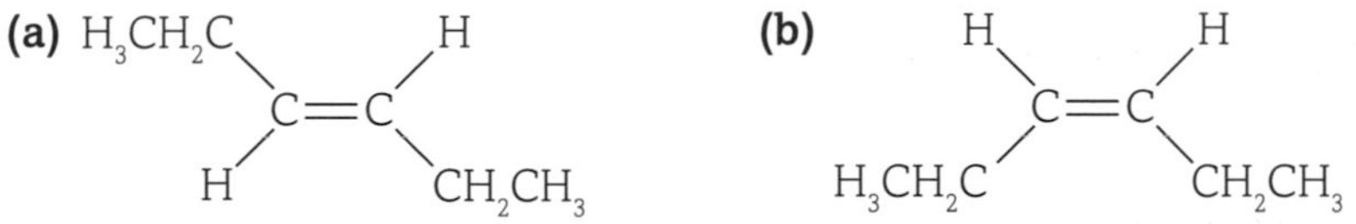

(a) H_3CH_2C H
C=C
H CH_2CH_3

(b) H H
C=C
H_3CH_2C CH_2CH_3

Figure 3 (a) *E*-hex-3-ene, also known as the *trans*-isomer; (b) *Z*-hex-3-ene, also known as the *cis*-isomer.

LEARNING TIP

Cis-trans is a special case of *E*/*Z* isomerism where the two different groups on the carbon atoms are the same.

KEY DEFINITIONS

Alkenes are a homologous series of unsaturated hydrocarbons.
Unsaturated organic compounds contain at least one multiple C=C bond.
Stereoisomers have the same structural formula but different arrangements in space.
***Cis–trans* isomerism** is a type of *E*/*Z* isomerism in which two substituent groups attached to each carbon atom in the C=C bond are the same.

Questions

1. Describe the difference between sigma bonds and pi bonds.
2. Explain why ethene cannot form stereoisomers.
3. Draw the structural formula of:
   (a) *E*-but-2-ene
   (b) the *cis*-stereoisomer of but-2-ene.

Addition reactions of alkenes

By the end of this topic, you should be able to demonstrate and apply your knowledge and understanding of:

- the reactivity of alkenes in terms of the relatively low bond enthalpy of the π-bond
- addition reactions of alkenes with:
  (i) hydrogen in the presence of a suitable catalyst, e.g. Ni, to form alkanes
  (ii) halogens to form dihaloalkanes, including the use of bromine to detect the presence of a double C=C bond as a test for unsaturation in a carbon chain
  (iii) hydrogen halides to form haloalkanes
  (vi) steam in the presence of an acid catalyst, e.g. H_3PO_4, to form alcohols
- definition and use of the term *electrophile* (an electron pair acceptor)
- the mechanism of electrophilic addition in alkenes by heterolytic fission
- use of Markownikoff's rule to predict formation of a major organic product in addition reactions of H–X to unsymmetrical alkenes, e.g. H–Br to propene, in terms of the relative stabilities of carbocation intermediates in the mechanism

Reactivity of alkenes

Alkenes are more reactive than alkanes (Figure 1) even though they are both hydrocarbons. The only difference is the presence of a C=C bond in an alkene molecule. So this must be the reason for the difference in reactivity.

```
    H   H          H         H
    |   |           \       /
H — C — C — H        C  =  C
    |   |           /       \
    H   H          H         H
```

Figure 1 Ethene is more reactive than its alkane analogue, ethane.

KEY DEFINITION

An **electrophile** is an electron-pair acceptor.

LEARNING TIP

Reagents are chemicals used in a reaction. So, concentrated nitric and sulfuric acids are the reagents but NO_2^+ is the reacting species.

A mean bond enthalpy is the average energy required for one mole of a given bond to undergo homolytic fission in the gaseous state. The mean bond enthalpy for C–C is $+347\,kJ\,mol^{-1}$, so a reasonable prediction would be that the mean bond enthalpy for C=C would be $2 \times 347\,kJ\,mol^{-1}$ or $+694\,kJ\,mol^{-1}$. However, it is reported as $+612\,kJ\,mol^{-1}$.

This shows that a π-bond needs only $(612 - 347) = 265\,kJ\,mol^{-1}$ to undergo homolytic fission and is weaker than a σ-bond. Therefore, in alkene chemical reactions the π-bonds will break first and react, leaving the σ-bond between the two carbon atoms.

The double bond is an area of high electron density which attracts electrophiles like Br_2, HBr and NO_2^+. **Electrophiles** are species that are electron-pair acceptors – examples include positive ions (cations) and molecules with a $\delta+$ region of charge.

Understanding electrophilic addition reactions

In addition reactions, two or more molecules become bonded to make one product. For alkenes, the π-bond breaks and a small molecule is added across the two carbon atoms to make a saturated organic product.

A general equation for this reaction is:

```
H         H                             H   H
 \       /                              |   |
  C  =  C      +    X—Y    ——>      H — C — C — H
 /       \                              |   |
H         H                             X   Y
```

Hydrogenation

Hydrogenation is an addition reaction in which hydrogen is added across a C=C bond.

At a temperature of 150 °C and using a suitable catalyst, like nickel, gaseous hydrogen and an alkene are mixed. They are then passed over the catalyst in the reaction chamber. This saturates the alkene and an alkane is produced:

$$H_2C{=}CH_2 + H_2 \xrightarrow[\text{Nickel}]{150\,°C} H_3C{-}CH_3$$

Halogenation

Halogenation is an addition reaction in which a halogen is added across C=C bonds – the alkene becomes saturated and a dihaloalkane is produced. This reaction can be used as a test for saturation.

When bromine solution or iodine solution is mixed with a saturated compound there is no reaction. However, when either of them are mixed with an unsaturated compound like ethene, decolorisation results because an addition reaction occurs (Figure 2).

Figure 2 An alkene decolorising bromine solution.

$$H_2C{=}CH_2 + Br_2 \longrightarrow CH_2Br{-}CH_2Br$$

We can use a **reaction mechanism** to explain how this electrophilic addition reaction occurs.

Stage	Diagram	Explanation
Starting the reaction	$CH_3CH{=}CH_2$; induced dipole $Br^{\delta+}{-}Br_{\delta-}$	Bromine has non-polar molecules. But when it gets close to the electron-rich C=C bond a dipole is induced. The $\delta+$ side of the molecule is attracted to the high electron density of the C=C bond.
Reaction intermediates	$CH_3CH{=}CH_2$ + $Br^{\delta+}{-}Br^{\delta-}$ $\longrightarrow$ $CH_3\overset{+}{C}H{-}CH_2Br$ + $:Br^-$	The electrons from the π-bond make a bond with a bromine atom. This causes the heterolytic fission of the bromine molecule. A positive charge remains on the second carbon atom. This species is very reactive and known as a carbocation.
Making 1,2-dibromopropane	$CH_3\overset{+}{C}H{-}CH_2Br$ + $:Br^-$ $\longrightarrow$ $CH_3CHBr{-}CH_2Br$	Two electrons from the bromide ion are shared with the carbocation, making a second bond and a stable product.

Table 1 The addition reaction stages of halogenation.

Hydration

Hydration is an addition reaction between a gaseous alkene and steam – it is used in industry to make alcohols. The conditions involve high temperatures and high pressures with a phosphoric acid catalyst.

$$CH_2{=}CH_2 + H_2O \xrightarrow[300\,°C,\ 65\,atm]{H_3PO_4} CH_3CH_2OH$$

Addition of hydrogen halides

This is an addition reaction in which a hydrogen halide is added across the C=C bond. This produces a class of compounds called haloalkanes. These are saturated organic compounds based on alkane chain skeletons. At least one hydrogen atom is replaced by a halogen atom.

Hydrogen halides like HCl, HBr and HI are gases at room temperature. These reactants are bubbled through the alkene to cause a reaction.

$$CH_2{=}CH_2 + HBr \longrightarrow CH_3CH_2Br$$

ethene → bromoethane

We can use a reaction mechanism to explain how this electrophilic addition reaction occurs.

Stage of the reaction	Diagram	Explanation
Starting the reaction	$CH_2{=}CH_2$ with $H^{\delta+}$–$Br^{\delta-}$ approaching	There is a large difference in electronegativity between hydrogen and bromine, so a hydrogen bromide molecule is polar. The δ+ hydrogen side of the hydrogen bromide molecule is attracted to the high electron density of the C=C bond.
Reaction intermediates	$CH_2{=}CH_2 + H^{\delta+}$–$Br^{\delta-}$ → CH_3–CH_2^+ + Br^- a carbocation is formed	The electrons in the π-bond make a bond with the hydrogen, causing heterolytic fission – electrons in the HBr molecule are given to the bromine atom making a bromide ion. A carbocation is also made.
Making bromoethane	CH_3–CH_2^+ + :Br^- → CH_3–CH_2Br	The bromide ion and the carbocation bond to make a stable product.

Table 2 The addition reaction stages of making a haloalkane.

When a hydrogen halide is added to an unsymmetrical alkene there are two possible products.

Consider the reaction of propene with hydrogen bromide. Can you see that either 1-bromopropane or 2-bromopropane could be made – however, 2-bromopropane is the major product.

We can use Markownikoff's rule to help in predicting the product. This states that when H–X is added to an unsymmetrical alkene, the hydrogen becomes attached to the carbon with the most hydrogen atoms to start with. This is because carbocations that have alkyl groups attached are more stable than those with hydrogen atoms attached.

Carbocation	Diagram
Primary (low)	H | R—C—H +
Secondary	R | R—C—H +
Tertiary (high)	R | R—C—R +

low
stability
high

Table 3 Carbocation stability.

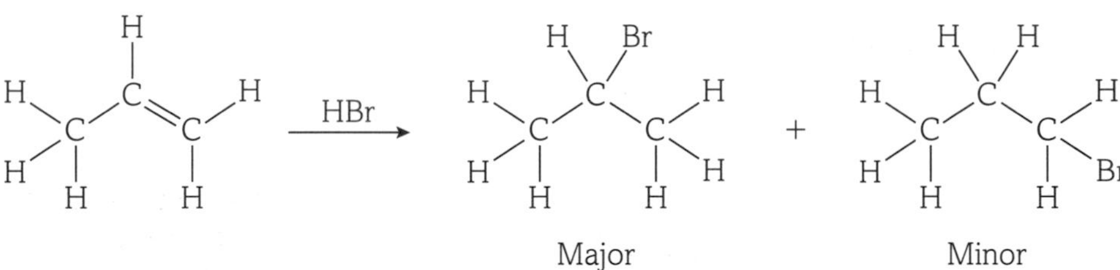

Figure 3 When HBr is added across the double bond in propene there is a mixture of products. 2-bromopropane is favoured because the carbocation intermediate is more stable than for the minor product.

Questions

1. Explain why an alkene is more reactive than an alkane with the same number of carbon atoms.
2. Draw a reaction mechanism for the reaction between ethene and bromine.
3. Draw a reaction mechanism for the reaction between ethene and hydrogen iodide.
4. Predict the major product when but-1-ene reacts with hydrogen chloride.

KEY DEFINITION

A **reaction mechanism** is a model that shows the movement of electrons in an organic reaction.

10 Addition polymerisation

By the end of this topic, you should be able to demonstrate and apply your knowledge and understanding of:

* addition polymerisation of alkenes and substituted alkenes, including:
  (i) the repeat unit of an addition polymer deduced from a given monomer
  (ii) identification of the monomer that would produce a given section of an addition polymer

Introduction

Small molecules, known as **monomers**, join together to make very long macromolecule chains. For addition polymerisation, all the monomers have a C=C double bond made up of a σ-bond and a π-bond.

During the polymerisation reaction, monomers have their π-bond broken. The electrons from each π-bond make a σ-bond with a neighbouring carbon atom on a different monomer. This connects the monomers chemically, generating a saturated polymer with long carbon chains.

The general equation for addition polymerisation is:

repeat unit

n (W)(Y)C=C(X)(Z) ⟶ [–C(W)(Y)–C(X)(Z)–]$_n$

monomer – any alkene   polymer

KEY DEFINITIONS

Monomers are small molecules that are used to make polymers.
Polymers are macromolecules made from small repeating units.
A **repeating unit** is a specific arrangement of atoms that occurs in a structure over and over again.

Polymers can be millions of atoms long and it is not useful to try to draw a full structural formula. However, they are made up of a repeating pattern atom groups – like the beads in a necklace. So, we can draw the structural formula of the **repeating unit** using square brackets to show that it is part of a polymer. The bonds must extend through the bracket lines and a small subscript n is used to show there are n repeats of this unit – n can have large values.

Common addition polymers

Table 1 shows some common addition polymers.

Name of monomer	Structural formula of monomer	Name of polymer	Repeating unit	Use of polymer
ethene	(H)(H)C=C(H)(H)	poly(ethene)	[–C(H)(H)–C(H)(H)–]$_n$	plastic bags and bottles
propene	(H)(H)C=C(CH_3)(H)	poly(propene)	[–CH_2–CH(CH_3)–]$_n$	ropes and crates
chloroethene	(H)(H)C=C(Cl)(H)	poly(chloroethene) (PVC)	[–C(H)(H)–C(Cl)(H)–]$_n$	electrical cable insulation

Name of monomer	Structural formula of monomer	Name of polymer	Repeating unit	Use of polymer
phenylethene	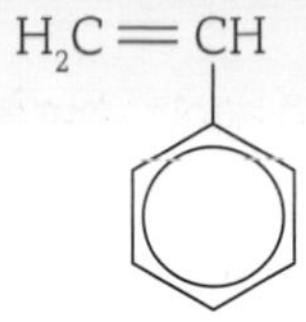 $H_2C=CH$ (with benzene ring on CH)	poly(phenylethene) (commonly known as polystyrene)	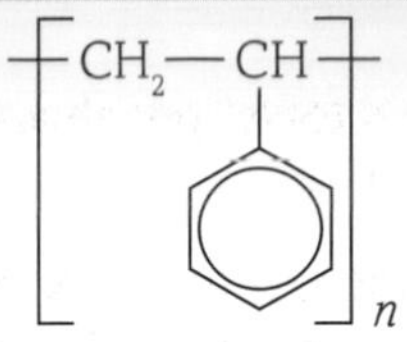 $[-CH_2-CH(C_6H_5)-]_n$	packaging
tetrafluorethene	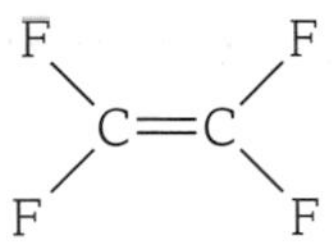 $F_2C=CF_2$	poly(tetrafluoroethene) (Teflon®)	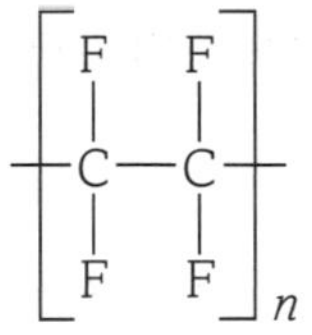 $[-CF_2-CF_2-]_n$	non-stick coatings on cooking pans

Table 1

Deducing the structures

Use the steps below to help you to generate the repeating unit of a polymer from its monomer.

- Step 1: draw the monomer structure so that the C=C bond is the focus of the diagram.
- Step 2: draw square brackets round the monomer.
- Step 3: change the C=C to C–C and draw two lines extending from each C atom through the square brackets.
- Step 4: add a subscript '*n*'.

If you are given a section of the polymer chain and are asked to deduce the monomer, look carefully at the order of the atoms. Choose one group and see where that repeats – the repeating unit is between these two points. Then use the steps above in reverse to generate the structure of the monomer.

LEARNING TIP

You need to be able to deduce the monomer or polymer structure for a given polymer or monomer.

(a) $-CH_2-CH(CN)-CH_2-CH(CN)-CH_2-CH(CN)-CH_2-CH(CN)-$

(b) $[-CH_2-CH(CN)-]_n$

(c) $H_2C=CH(CN)$
monomer

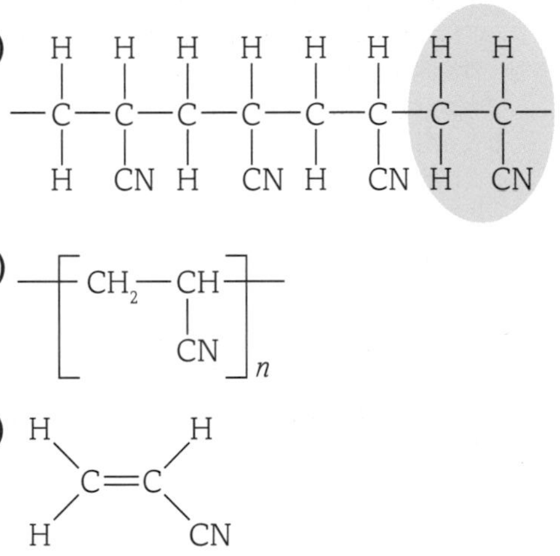

Figure 1 (a) Part of the structure of the synthetic fibre polyacrylonitrile (orlon) polymer; (b) Repeating unit for the polyacrylonitrile polymer; (c) The monomer used to make the polyacrylonitrile polymer.

Questions

1. Draw the structural formula of the monomer that makes poly(2-chloropropene).
2. Draw the repeating unit of the polymer made from 1-chloropropene.
3. Draw the structural formula of the monomer and the repeating unit of the polymer made from 1,2-diphenylethene.

Polymers – dealing with polymer waste

By the end of this topic, you should be able to demonstrate and apply your knowledge and understanding of:

- the benefits for sustainability of processing waste polymers by:
  - (i) combustion for energy production
  - (ii) use as an organic feedstock for the production of plastics and other organic chemicals
  - (iii) removal of toxic waste products, e.g. removal of HCl formed during disposal by combustion of halogenated plastics (e.g. PVC)
- the benefits to the environment of development of biodegradable and photodegradable polymers

Introduction

Synthetic polymers have become a very important part of everyday life because they are so versatile. They have many uses, ranging from clothing to materials in cars. We often regard this type of material as disposable. But addition polymers are saturated organic compounds and so are very stable. When we have finished using them, they do not get attacked by environmental conditions or by microorganisms. This means that plastics can remain chemically unchanged for hundreds of years.

There may come a time when we cannot make the plastics that we have become so reliant on. Addition polymers are made from compounds made available from the processing of crude oil. Crude oil is a non-renewable resource and we are using it faster than the natural processes in our planet can make more.

Figure 1 Currently, more than 80% of plastic waste is used once and then sent to landfill.

Polymer waste

Just over 10% of household waste is plastic. Such plastics are disposed of in different ways:

- landfill
- combustion
- combustion with electrical generation
- reusing
- recycling
- using them as organic feedstock.

Let us look at each of these in turn.

Landfill

Large holes are dug into the landscape (Figure 1). The holes are lined to stop any contaminants seeping into the water table. Rubbish is then put in the holes and compacted. When a site is full, it is capped and landscaped.

After rubbish is buried, the conditions change from normal atmospheric to anaerobic and there is often a limited supply of water. This reduces the rate of decomposition of biodegradable materials. As many plastics are non-biodegradable, the waste does not break down and can become a danger to wildlife.

Combustion

Plastics are mainly organic and can be burnt (Figure 2). As they contain large amounts of carbon, their combustion releases carbon dioxide. Carbon dioxide is a greenhouse gas and is linked to climate change.

Depending on the plastic, other toxic or polluting gases, like HCl, can be made. These gases can be removed by using gas scrubbers, where a base such as CaO neutralises the acidic gas.

Figure 2 Gas scrubber on an incinerator.

As plastics have a high calorific value, they can be burnt in power stations (Figure 3). The chemical energy transferred can be used to drive turbines and generate electricity.

Figure 3 Some power stations are run on waste.

Reusing

Some plastics can be reused for the same function many times (Figure 4). For example, a drinks bottle can be washed out and refilled rather than using new bottles every time.

Figure 4 Some plastic products can be used many times.

Recycling

For recycling to take place, plastics must be sorted into different types (Figure 5). This is expensive because it is either labour-intensive or uses costly high technology to automate the process.

Plastics are cleaned, melted down and reshaped into new products. There is a limited market for recycled plastic – companies are concerned about quality and whether or not there could be contamination from its previous use.

Figure 5 Only about 7% of plastic is recycled currently.

Organic feedstock

After waste plastic has been sorted into different types, a series of chemical reactions can be used to break the plastic polymers up into small organic molecules. This allows the recovered chemicals to be used in other industrial reactions.

Biodegradable and photodegradable polymers

Biodegradable polymers

Much research has been done on plastics that can be attacked by microorganisms and environmental conditions so that they break down chemically into harmless, or even useful, substances. Sometimes a **biodegradable** polymer like plant starch is mixed with an addition polymer like poly(ethene) (Figure 6). Given time, the starch part breaks down making the polymer chains smaller and the material biodegrades. However, there is growing concern that these small pieces of addition polymers still cause hazards to the ecosystem.

Figure 6 Plastic carrier bags made from an addition polymer and plant starch mix.

If only the biodegradable polymer is used, the new material is called a **bioplastic**. Plant starch can be used to make bin bags. If the starch is derived from sustainable farming methods then the material could be classified as 'carbon neutral'.

To be fully compostable (Figure 7) a plastic must decompose landfill material about as quickly as compost would form from grass clippings and other green waste. The only products should be carbon dioxide, water, inorganic compounds and biomass.

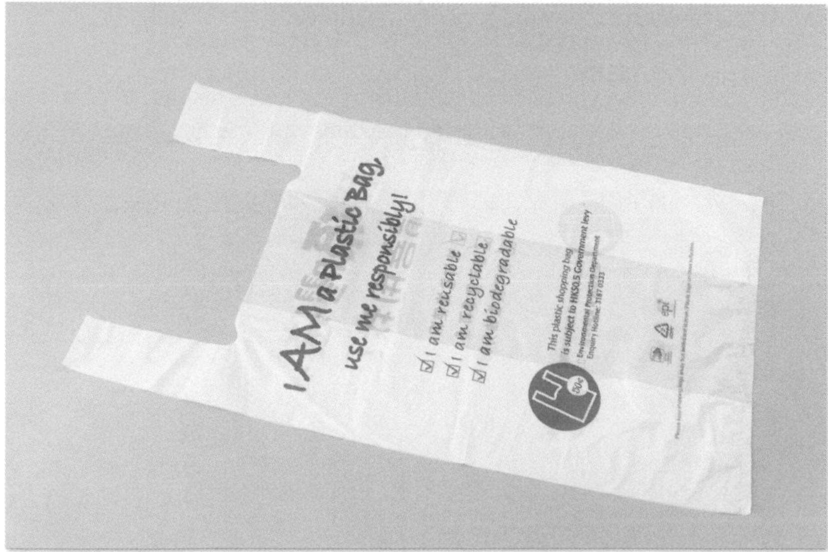

Figure 7 Compostable green waste bag.

A substance like poly(lactic acid) will decompose in about 180 days and is currently used for making cold-drink cups and to wrap fresh fruit and vegetables.

Photodegradable polymers

Photodegradable plastics break down chemically using energy with wavelengths similar to light. They are either addition polymers with bonds within their structure that are weakened by absorption of light or they have an additive that is affected by light, which then weakens bonds in the polymer. Once the polymer has been exposed to light, it begins to break down and it is not possible to stop the process. However, photodegradable plastics in landfill may not be exposed to sufficient light to degrade.

KEY DEFINITIONS

Biodegradable materials are affected by the action of microorganisms and environmental conditions, leading to decomposition.
A **bioplastic** is a material made from a renewable source that is biodegradable.

LEARNING TIP

Make sure you can evaluate the use of plastics, bearing in mind disposal.

DID YOU KNOW?

In the 1980s multi-pack drinks cans were held together with plastic rings around the necks of the cans. When the can holders were discarded, the non-biodegradable plastic remained in the environment for a long time and presented a major problem as marine litter. Wildlife such as birds got trapped in them and died (Figure 8). Since the 1990s, when legislation was passed, this packaging has been made using photodegradable low-density poly(ethene) and so is more environmentally friendly.

Figure 8 Bird caught in plastic packaging.

Questions

1. Explain why generating electricity from waste plastic is better for the environment than disposing of plastics in landfill.

2. Explain why bioplastics are better than polyethene/cornstarch-mixed plastic.

THINKING BIGGER

TOWARDS A GREENER ENVIRONMENT

In the extract below, examples of the role of *catalysts* in the modern petroleum industry are considered. The questions that follow will also enable you to consider aspects of organic chemistry and quantitative chemistry covered so far.

HETEROGENEOUS AND HOMOGENEOUS CATALYSTS

Heterogeneous catalysts are solids that catalyse the reactions between liquid or gaseous reactants. (Note that the reaction itself usually occurs at the surface of the catalyst.) The catalytically active solid is typically coated onto a high surface area support to ensure the maximum exposure to the gas or liquid reaction mixture. These catalysts, usually transition metals and their compounds, are used in circa 85% of industrial processes because they are easy to separate from the products at the end of the reaction. Examples include the Haber–Bosch process for the production of NH_3, catalytic cracking and the hydrogenation of vegetable oils.

A heterogeneous catalyst provides a lower energy path via a sequence that involves adsorption of reactant molecules upon an active site in the surface. The molecules become chemisorbed on the active site, their bonds are disrupted and rearrangements take place to form the activated complex; desorption then follows to release the product molecule(s) into the gas, or liquid, phase. The active site is once again vacant to repeat the process. Figure 1 shows adsorbed ethene and adsorbed hydrogen upon the active sites of a nickel catalyst. Following adsorption and reaction, the product ethane desorbs back into the gas phase, leaving the active site vacant for the next reactant molecules. This reaction is the basis of the hydrogenation of unsaturated vegetable oils in the manufacture of margarine.

Clearly the adsorption is an important step and the precise structure of the surface is vital in providing the active site. Defects in the solid surface will create different types of site, which may have different or no catalytic properties. The active sites are characterised by having a critical geometry associated with the compounds that adsorb onto the catalyst surface. The geometry of an active site may also be determined by the structure of the underlying support. Thus changing the support can have a profound effect upon activity and may even redirect the course of a reaction.

An active site may be made more active by introducing other atoms or compounds. These are known as promoters, which alone are relatively inactive. For example, if a small amount of cobalt is added to the desulfurisation catalyst molybdenum disulfide there is a marked increase in activity. Poisons have the counter effect and build up on the catalyst surface during its life and often mark the end of the catalyst's life. A platinum hydrogenation catalyst, for example, is poisoned by sulfur-containing compounds.

The understanding of surface structure therefore plays a big role in the development of new catalysts and so it is here that much of the research effort is directed. Where only a small amount of a catalyst surface is productive there is clearly scope for improvement.

Several industrial processes use homogeneous catalysts, which are in the same phase (liquid or gas) as the reactants and products. Although more difficult to separate at the end of a reaction, homogeneous catalysts are often more active and selective than heterogeneous catalysts and tend to work at lower temperatures. This is because all the metal ions of a dissolved catalyst are potentially active sites for the reaction, whereas in a solid only those atoms at the surface are accessible to the reactant molecules. Examples include the acid catalysed esterification of carboxylic acids and alcohols and the gas-phase catalysed reaction of ozone destruction in the stratosphere in which chlorine free radicals, from CFCs, act as catalysts for the reaction.

Low sulfur fuels: desulfurisation catalysis

Petroleum-derived fuels contain a small amount of sulfur. Unless removed this sulfur persists throughout the refining processes and ends up in the petrol or diesel. Pressure to reduce atmospheric sulfur has driven the development of catalytic desulfurisation. One of the problems was that much of the sulfur present was in compounds such as the thiophenes, which are stable and resistant to breakdown.

The catalyst molybdenum disulfide coated on an alumina support provided one solution. Cobalt is added as a promoter, suggesting that the active site is a molybdenum-cobalt sulfide arrangement. In the catalytic reaction (see Scheme 1), which is essentially a hydrogenation sequence, the adsorbed thiophene molecule is hydrogenated and its aromatic stability destroyed. This enables the C—S bond to break and release the sulfur as hydrogen sulfide. This is an interesting example of a catalyst performing different types of reactions: hydrogenation, elimination and isomerisation.

- Extract taken from: 'Catalysts for a green industry' http://www.rsc.org/education/eic/issues/2009July/catalyst-green-chemistry-research-industry.asp

Where else will I encounter these themes?

1.1 | 2.1 | 2.2

Figure 1

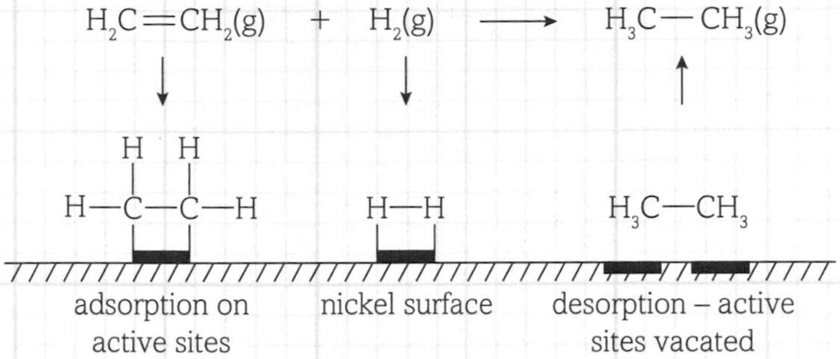

$$H_2C{=}CH_2(g) + H_2(g) \longrightarrow H_3C{-}CH_3(g)$$

H–C–C–H (H H) H–H $H_3C{-}CH_3$

adsorption on active sites | nickel surface | desorption – active sites vacated

Figure 2 Adsorption and reaction in heterogenous catalysis.

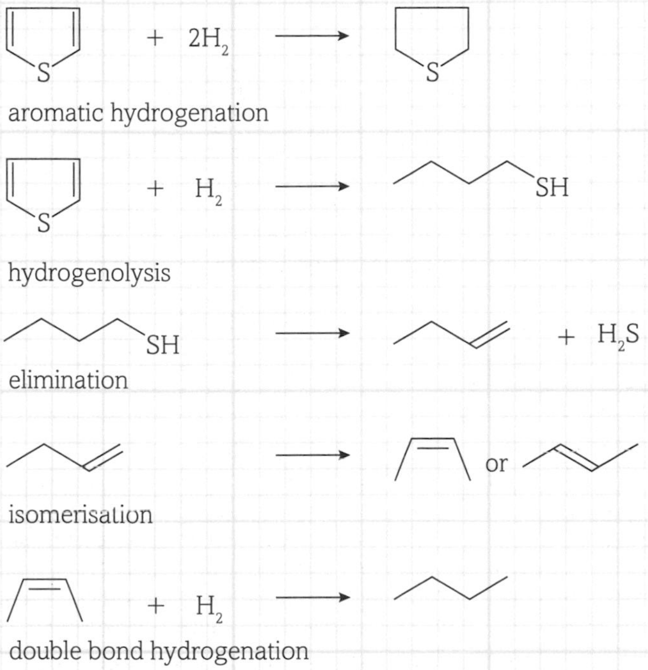

+ $2H_2$ ⟶ (S)

aromatic hydrogenation

+ H_2 ⟶ SH

hydrogenolysis

SH ⟶ + H_2S

elimination

⟶ or

isomerisation

+ H_2 ⟶

double bond hydrogenation

Figure 3 Desulfurisation of thiophenic compounds from petroleum.

DID YOU KNOW?

Hydrogen sulfide (H_2S) is highly toxic but, luckily, most humans can detect it at concentrations of less than 0.5 parts per billion (or ppb; that's 1 molecule in 2×10^9 air molecules!). It also smells like rotten eggs so we get plenty of warning before the level of 800 000 ppb, which can be fatal, is reached!

Let us start by considering the nature of the writing in the article.

1. In the second extract on 'low sulfur fuels' a good deal of scientific literacy is required. Imagine that you are required to convince the general public that removing the sulfur from fuels is worth the extra cost. Design a pamphlet to present the argument as strongly as possible.

Think about how illustrations can elicit very emotive responses. How might you use illustrations in your pamphlet?

Now we will look at the chemistry in, or connected to, this article. Do not worry if you are not ready to give answers to these questions yet. You may like to return to the questions once you have covered other topics later in the book. Use the timeline at the bottom of the page to help you put this work in context with what you have already learned and what is ahead in your course.

2. Use the text to explain the difference between homogeneous and heterogeneous catalysts. What are the pros and cons of each type?

3. a. Work out the molecular formula of thiophene (shown below).

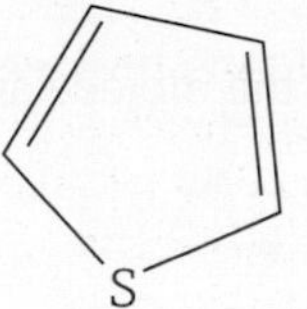

Figure 4

b. Calculate the percentage mass of sulfur in the molecule.

c. Write a balanced equation for the complete combustion of thiophene. (You can assume the oxidised product of sulfur is SO_2 only.)

d. During the elimination (Figure 3) part of the reaction sequence, butan-1-thiol is converted into two products. Name them.

e. The isomerisation process (Figure 3) gives rise to two stereoisomers. Explain what is meant by a geometric isomer and name both.

f. Why can the double bond hydrogenation reaction (Figure 3) be considered to have 100% atom economy?

Activity

You may wonder why sulfur appears in fossil fuels at all! The chemistry of sulfur gives it some special properties and, apart from carbon, hydrogen, oxygen and nitrogen, it is the only other element present in the building blocks of all proteins; amino acids. Prepare a 5 min presentation to the class on the importance of sulfur in proteins: Your presentation should include (1) which amino acids contain sulfur, (2) why sulfur is so important in protein structure and (3) what the consequences of a diet low in sulfur can be (no more than four slides). (Links with Biology.)

4.1 Practice questions

1. What is the correct name for the molecule with the formula $CH_3C(CH_3)_2OH$? [1]
   A. methylethanol
   B. 1,1,dimethylethanol
   C. 2-methylpropan-2-ol
   D. hydroxybutane

2. Which of the following statements about $CH_3CH(CH_3)CH_2\ CH_2CH_3$ is **not** true? [1]
   A. It is an aliphatic hydrocarbon.
   B. It is a structural isomer of 2,2 dimethyl butane.
   C. Its empirical formula is C_3H_7.
   D. It must be an unsaturated hydrocarbon.

3. How many isomers have the molecular formula $C_5H_{12}O$ **and** are alcohols? [1]
   A. 3
   B. 4
   C. 8
   D. 10

4. Look at the reaction step shown below.
   $CH_4 + Cl\bullet \rightarrow CH_3\bullet + HCl$
   Which of the statements below is true? [1]
   A. It is a nucleophilic substitution reaction.
   B. A methyl radical is generated.
   C. A methyl radical is consumed.
   D. It is an example of a termination step.

[Total: 4]

5. Crude oil is a source of alkanes.
   (a) Fractional distillation is used to separate useful hydrocarbons found in crude oil.
   Explain, in terms of intermolecular forces, how fractional distillation works. [2]
   (b) The petroleum industry processes straight-chained alkanes into cycloalkanes such as cyclopentane and cyclohexane.

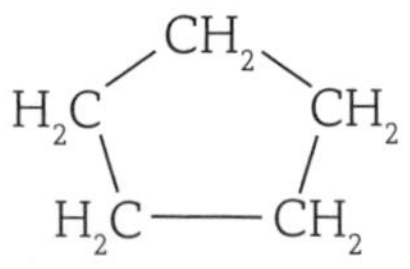

CH_2, H_2C, CH_2, H_2C—CH_2

cyclopentane

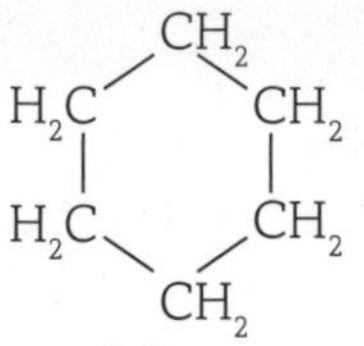

CH_2, H_2C, CH_2, H_2C, CH_2, CH_2

cyclohexane

   (i) Deduce the general formula of a **cycloalkane**. [1]
   (ii) Construct the equation to show the formation of cyclohexane from hexane. [1]
   (iii) Suggest why the petroleum industry processes hexane into cyclohexane. [1]

   (c) The flowchart below shows some of the organic compounds that could be made starting from cyclohexane.

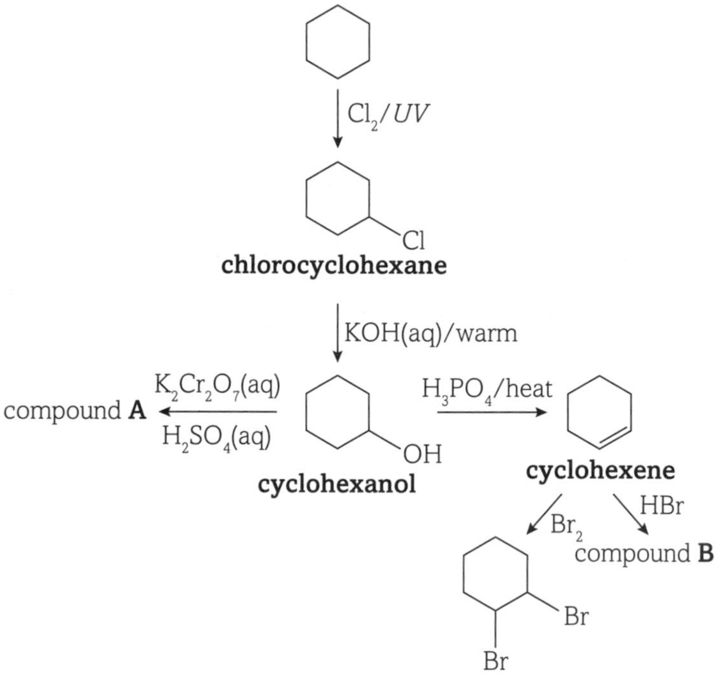

Cl_2/UV

Cl

chlorocyclohexane

KOH(aq)/warm

compound **A** ← $K_2Cr_2O_7$(aq) / H_2SO_4(aq) — **cyclohexanol** (OH) — H_3PO_4/heat → **cyclohexene**

cyclohexene — Br_2 ↙ ; ↘ HBr compound **B**

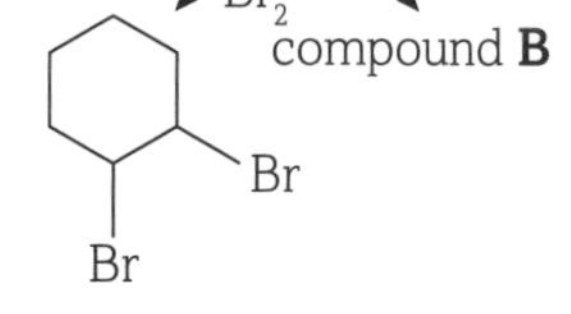

   (i) Explain why cyclohexene is described as *unsaturated* and as a *hydrocarbon*. [2]
   (ii) The reaction between chlorine and cyclohexane is an example of radical substitution.
   State **one** problem of using this reaction to prepare a sample of chlorocyclohexane. [1]
   (iii) The formation of cyclohexanol from chlorocyclohexane involves the reaction of a nucleophile, the hydroxide ion.
   Suggest what feature of the hydroxide ion makes it able to act as a nucleophile. [1]

(iv) Using the flowchart, draw the structures of compound **A** and compound **B**. [2]

(v) Describe, using the 'curly arrow model', the mechanism for the reaction between Br_2 and cyclohexene. Show relevant dipoles and charges. [4]

$H_2C—CH_2$
H_2C CH_2 ⟶
C=C
H H
Br
|
Br

[Total: 15]

[Q1, F322 Jan 2011]

6. (a) Draw out the skeletal formulae of:
   (i) butane [2]
   (ii) propan-1-ol. [2]
   (b) Calculate the relative molecular masses of both compounds in g mol^{-1}. [1]
   (c) (i) Look up the boiling points of the two compounds in your data booklet. [1]
   (ii) Describe and explain the difference in boiling points of the two compounds in terms of intermolecular forces. [4]

[Total: 10]

7. Analysis of molecule D shows it to have the following composition by mass to 1 d.p.:

   24.7% carbon, 2.1% hydrogen 73.2% chlorine.

   (a) (i) Calculate its empirical formula. [2]
   (ii) Given that its molecular mass is 97 g mol^{-1} calculate its molecular formula. [1]
   (b) (i) A sample of D decolourises bromine in the absence of sunlight but does **not** form stereoisomers. Draw the displayed formula of D and name it. [3]
   (ii) Draw the repeating unit of the polymer formed by D. [2]
   (iii) Highlight **two** problems associated with the disposal of the polymer formed from molecule D [4]

[Total: 12]

8. In a reaction between methane and bromine in the presence of ultraviolet light a mixture of products are formed including:

   bromomethane, dibromomethane, hydrogen bromide, ethane and bromoethane.

   (a) Draw the displayed formula for each of the molecules named above. [5]
   (b) With reference to homolytic fission mechanisms explain how each of the above molecules can form. [5]

[Total: 10]

MODULE 4

Core organic chemistry

CHAPTER 4.2

ALCOHOLS, HALOALKANES AND ANALYSIS

Introduction

Scientists like to sort chemicals into groups with similar properties and features. This helps chemists to classify new discoveries and predict properties of new materials. Researchers look at the structure of a chemical and focus on the parts where chemical reactions are likely to happen. This is then used to classify the chemicals.

One family of chemicals is the alcohols, which all contain an –OH functional group. Ethanol is probably the most recognised alcohol. It can be made in nature by fermentation in over ripe fruit which has fallen from the tree. The naturally occurring yeasts use sugars to respire producing ethanol and carbon dioxide. Archaeologists have found evidence which suggests that fermentation was first harnessed to make alcoholic drinks as early as the Stone Age. It is also possible that, historically, beer was used in London as a safe water supply because the low concentration of alcohol ensured the harmful microbes were not able to survive.

Haloalkanes contain halogen atoms and are an important class of compounds used as refrigerants, fire extinguishers and an important feed stock for the chemical industry. However, they are not without controversy as many are greenhouse gases and may contribute to climate change. The use of chlorofluorocarbons is now heavily regulated as it is widely accepted that their use has lead to ozone depletion which in turn can increase the occurrence of cancer and cataracts in humans.

Analytical chemistry is an important area of the science which allows you to identify chemicals such as those found in toxicology in medicine or crime investigations. The information from different techniques can also be used to infer the structure of an unknown chemical.

All the maths you need

To unlock the puzzles of this chapter you need the following maths:

- Be able to use general formula to generate molecular formula
- The units of measurement (*e.g. the mole*)
- Use of standard form and ordinary form (*e.g. representing the amount of an atom*)
- Using an appropriate amount of standard figures (*e.g. in multi-step calculations*)
- Changing the subject of an equation (*e.g. finding a molecular mass*)
- Substituting numerical values into algebraic equations (*e.g. using* $n = \frac{m}{M_r}$)

What have I studied before?

- How to draw structural formula
- Defining homologous series
- Describing how ethanol can be produced
- How to recognise and draw isomers

What will I study later?

- How more analytical techniques can be used to differentiate between organic molecules
- More complex reaction mechanisms

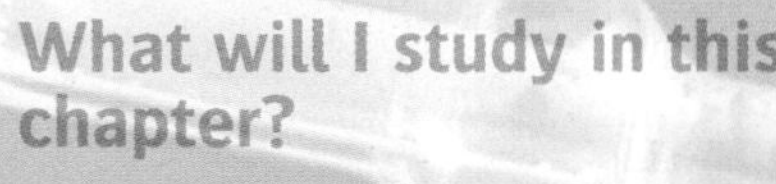

What will I study in this chapter?

- How to use IUPAC naming conventions
- To recognise classes of chemicals and list their key features
- To be able to generate the structural formula for a variety of types of isomerism
- To use recognise and generate different representations of the same molecule
- To use arrow pushing to model reaction mechanisms
- To be able to interpret analytic data including spectra
- To explain how organic synthesis can be used to interconvert organic chemicals

4.2  Properties of alcohols

By the end of this topic, you should be able to demonstrate and apply your knowledge and understanding of:

* an explanation of the polarity of alcohols and, in terms of hydrogen bonding, of the water solubility and the relatively low volatility of alcohols compared with alkanes
* classification of alcohols in the primary, secondary and tertiary alcohols

Introduction

Alcohols form a homologous series that:

- are saturated, containing only single covalent bonds
- have the general formula, $C_nH_{2n+1}OH$
- have a gradation in physical properties
- have similar chemical properties.

Classifying alcohols

The –OH functional group is the characteristic part of an alcohol molecule. This functional group can be in different positions on the parent carbon chain. The position is indicated by using a number, as shown in Figure 1.

(a)

```
    H  H  H
    |  |  |
 H—C—C—C—O—H
    |  |  |
    H  H  H
```

(b)

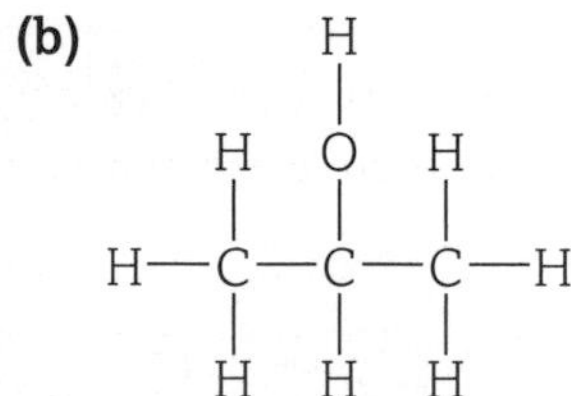

Figure 1 (a) In propan-1-ol, the functional group is on the first carbon atom; (b) In propan-2-ol, the functional group is on the second carbon atom.

Alcohols can be classified as:

- primary
- secondary
- tertiary.

Primary alcohols

A **primary alcohol** has the alcohol group attached to an end of a chain (Figure 2).

```
    H  H
    |  |
 H—C—C—O—H
    |  |
    H  H
```

Figure 2 Ethanol is an example of a primary alcohol.

Secondary alcohols

A **secondary alcohol** has the alcohol group attached to a carbon atom with two alkyl chains and one hydrogen atom (Figure 3).

```
          H
          |
    H  H  O  H
    |  |  |  |
 H—C—C—C—C—H
    |  |  |  |
    H  H  H  H
```

Figure 3 Butan-2-ol is an example of a secondary alcohol.

Tertiary alcohols

A **tertiary alcohol** has the alcohol group attached to a carbon atom with three alkyl chains attached.

```
          H
          |
        H—C—H
    H  H  |  H
    |  |  |  |
 H—C—C—C—C—H
    |  |  |  |
    H  H  O  H
          |
          H
```

Figure 4 2-methylbutan-2-ol is an example of a tertiary alcohol.

KEY DEFINITIONS

A **primary alcohol** has the functional group attached to a carbon atom with no more than one alkyl group.
A **secondary alcohol** has the functional group attached to a carbon atom with two alkyl groups.
A **tertiary alcohol** has the functional group attached to a carbon atom with three alkyl groups.

Changing state

Boiling point

The boiling points of alcohols increase as the chain length increases. This is for the same reason as for alkanes – as the molecules get longer, there are more surface area contacts (Figure 5) and so stronger induced dipole–dipole intermolecular forces. This results in more energy being needed to overcome these attractive forces, and therefore a higher boiling point.

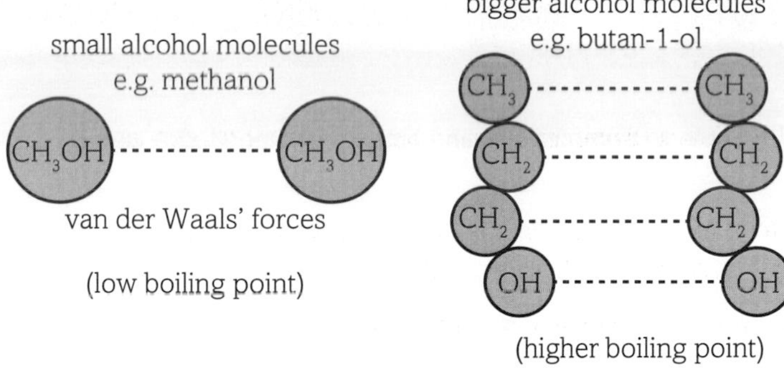

Figure 5 The longer the molecule, the more surface area contacts.

The boiling points of alcohols are higher than the corresponding alkanes (Figure 6). This is because there are hydrogen bonds between the alcohol molecules (Figure 7). This is the strongest type of intermolecular force of attraction – about 10% of the strength of a covalent bond. Therefore, more energy is needed to overcome this force and the boiling point of an alcohol is higher than that of the corresponding alkane.

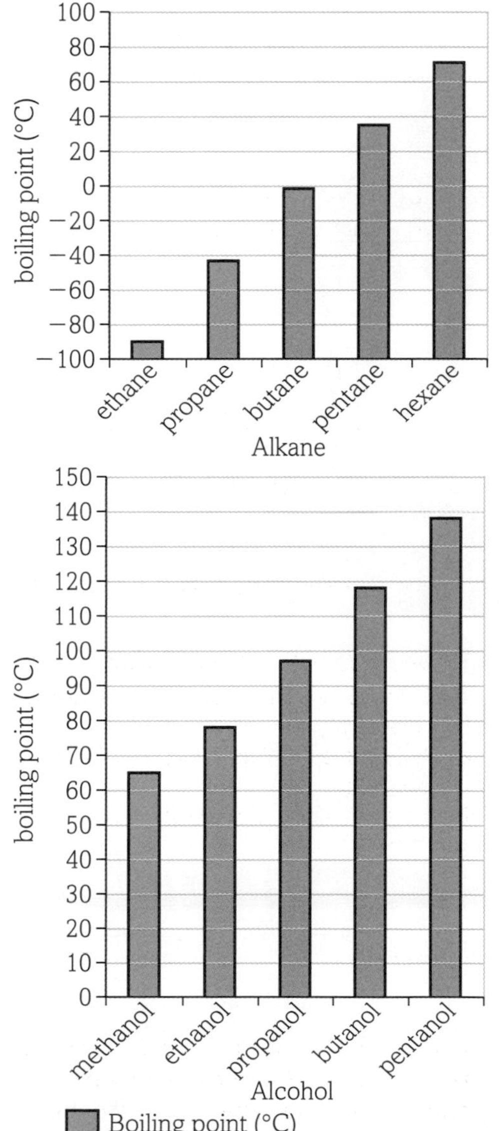

Figure 6 A comparison of boiling points for the first five members of the alkanes and of the primary alcohols.

hydrogen bond

Figure 7 Hydrogen bonding between ethanol molecules.

Evaporation

A substance that is volatile evaporates easily at room temperature and pressure. Volatility increases as the boiling point decreases. Alcohols have hydrogen bonds and this makes them less volatile than a corresponding alkane.

Solubility

Water molecules are polar and so is the alcohol functional group. This means that methanol, ethanol and propanol are soluble in water (Figure 8), their molecules form hydrogen bonds with the molecules. This is sometimes described as miscibility.

hydrogen bond

Figure 8 Hydrogen bonding between ethanol and water molecules.

As alkyl chain length increases, the solubility of the alcohol decreases. This is because the aliphatic chain cannot form hydrogen bonds and this becomes the larger part of the molecule.

LEARNING TIP

Remember that dipole induced dipole forces can also be called London forces.

Questions

1 Name and classify these alcohols:
(a) CH_3OH
(b) $CH_3CH(OH)CH_3$
(c) $CH_3C(CH_3)(OH)CH_3$

2 State and explain the trend in boiling points of alcohols as the chain length increases.

3 Explain why methanol will dissolve in water but hexan-1-ol is insoluble in water.

4.2

2 Oxidation of alcohols

By the end of this topic, you should be able to demonstrate and apply your knowledge and understanding of:

* combustion of alcohols
* oxidation of alcohols by an oxidising agent, e.g. $Cr_2O_7^{2-}/H^+$ (i.e. $K_2Cr_2O_7/H_2SO_4$) including,
  (i) the oxidation of primary alcohols to form aldehydes and carboxylic acids; and the control of the oxidation product using different reaction conditions
  (ii) the oxidation of secondary alcohols to form ketones
  (iii) the resistance to oxidation of tertiary alcohols

Combustion

Alcohols can be used as fuels by undergoing combustion to transfer their stored chemical energy in a usable form such as thermal energy. Combustion is a rapid oxidation reaction that combines oxygen, usually from the air, with another substance.

When an alcohol is completely burnt the products are carbon dioxide and water – for example:

$$C_2H_5OH(l) + 3O_2(g) \rightarrow 2CO_2(g) + 3H_2O(l)$$

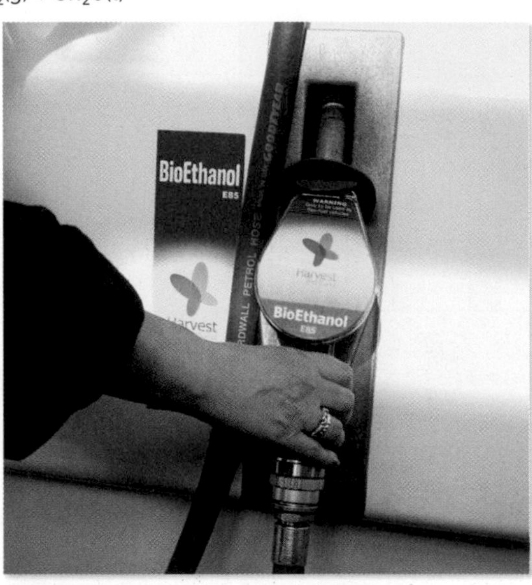

Figure 1 Ethanol can be used as a fuel for cars.

Overview of oxidation

Primary and secondary alcohols can be oxidised easily; tertiary alcohols cannot. Tertiary alcohols are very resistant to oxidation and are not oxidised by common oxidising agents like acidified potassium dichromate (VI) – see Table 1.

Class of alcohol	Reaction conditions	Product	Observation
primary	gentle heating	aldehyde	colour change from orange to green
primary	stronger heating under reflux with excess potassium dichromate (VI)	carboxylic acid	colour change from orange to green
secondary	heat under reflux	ketone	colour change from orange to green
tertiary		no reaction	remains orange

Table 1 Oxidation of primary, secondary and tertiary alcohols using acidified potassium dichromate (VI).

DID YOU KNOW?

A reagent is a chemical that takes part in a reaction. The reagents used for the oxidation of alcohols are a strong acid like sulfuric acid (H_2SO_4) and an oxidising agent like potassium dichromate(VI) ($K_2Cr_2O_7$). The species actually involved in the reaction are H^+ and $Cr_2O_7^{2-}$.

Primary alcohols

Oxidising agents provide oxygen to other substances, so they are often simplified to [O] in equations.

When you gently heat a primary alcohol like ethanol with acidified potassium dichromate (VI), the alcohol group loses its hydrogen atom and is partially oxidised to an aldehyde (Figures 2 and 3).

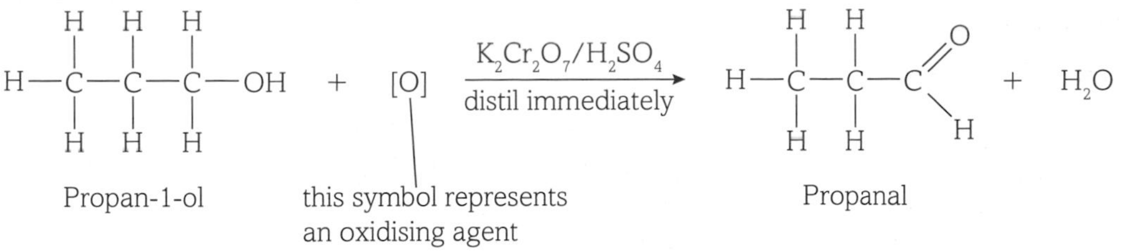

$$\mathrm{H{-}CH_2{-}CH_2{-}CH_2{-}OH} + [O] \xrightarrow[\text{distil immediately}]{K_2Cr_2O_7/H_2SO_4} \mathrm{H{-}CH_2{-}CH_2{-}CHO} + H_2O$$

Propan-1-ol — [O]: this symbol represents an oxidising agent — Propanal

Figure 2 Oxidation of propan-1-ol to propanal.

Figure 3 Oxidation of propan-1-ol to propanal with acidified potassium dichromate (VI) involves a colour change.

The aldehyde must be distilled immediately to prevent any further reaction (Figure 4). If a primary alcohol is heated strongly with excess acidified potassium dichromate (VI) then full oxidation occurs to form a carboxylic acid.

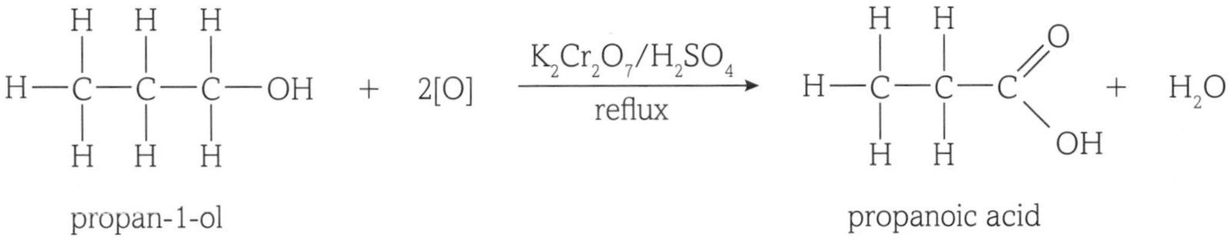

$$\mathrm{H{-}CH_2{-}CH_2{-}CH_2{-}OH} + 2[O] \xrightarrow[\text{reflux}]{K_2Cr_2O_7/H_2SO_4} \mathrm{H{-}CH_2{-}CH_2{-}COOH} + H_2O$$

propan-1-ol — propanoic acid

Figure 4 Oxidation of propan-1-ol produces propanoic acid.

To fully oxidise a primary alcohol on purpose, you must use a **reflux** set-up.

KEY DEFINITION

Reflux is the constant boiling and condensing of a reaction mixture. This ensures that the reaction goes to completion as fully as possible without losing reactants or products as vapour to the air.

Secondary alcohols

Secondary alcohols will oxidise to form a ketone (Figure 5) using oxidising agents like acidified potassium dichromate (VI). Ketones will not undergo any further reaction.

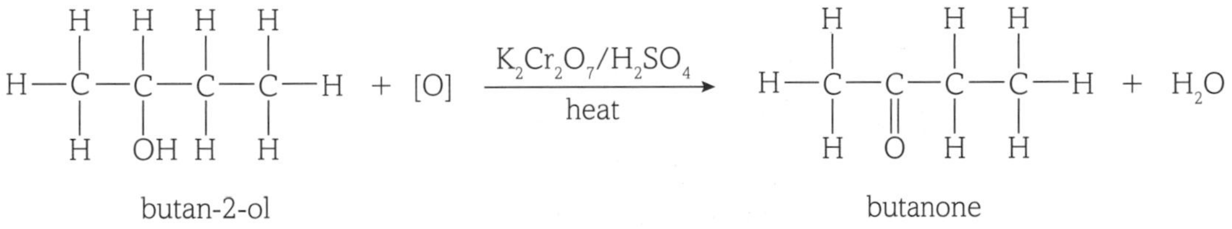

$$\mathrm{H{-}CH_2{-}CH(OH){-}CH_2{-}CH_2{-}H} + [O] \xrightarrow[\text{heat}]{K_2Cr_2O_7/H_2SO_4} \mathrm{H{-}CH_2{-}C({=}O){-}CH_2{-}CH_2{-}H} + H_2O$$

butan-2-ol — butanone

Figure 5 Oxidation of butan-2-ol to butanone.

DID YOU KNOW?

Using simple laboratory tests you can classify an unknown alcohol. First you heat the test alcohol with excess acidified potassium dichromate (VI) under reflux.

- If there is no colour change to green – the test alcohol is tertiary.
- If there is a colour change to green, then distil the product and test it with Fehling's solution or Tollens' reagent.
  - A second colour change indicates that the test alcohol is secondary.
  - No second colour change indicates that the test alcohol is primary.

Questions

1. Write balanced equations for the complete combustion of:
   (a) propan-1-ol
   (b) hexan-2-ol.

2. Draw the skeletal formula of the organic product formed when excess acidified potassium dichromate (VI) is heated under reflux with:
   (a) 2-methylhexan-1-ol
   (b) 3-methylpentan-2-ol
   (c) 2-methylpropan-2-ol.

3. Define the reflux process and explain why it is used in organic synthesis.

Other reactions of alcohols

By the end of this topic, you should be able to demonstrate and apply your knowledge and understanding of:

- **elimination of H_2O from alcohols in the presence of an acid catalyst (e.g. H_3PO_4 or H_2SO_4) and heat to form alkenes**
- **substitution with halide ions in the presence of acid (e.g. $NaBr/H_2SO_4$) to form haloalkanes**

Esterification

Esterification is the chemical reaction used to make an ester. Esters are organic compounds that contain a –COOR functional group, where R is an alkyl chain. This group of compounds are widely used in the manufacture of foods because they have a pleasant, fruity smell. Small-molecule esters are used in industry as powerful solvents.

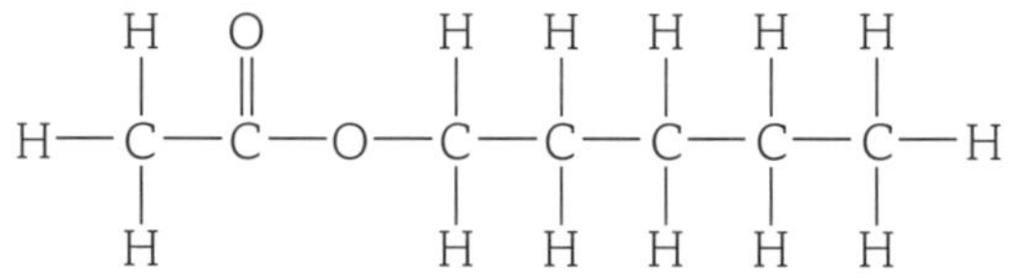

```
   H  O     H  H  H  H  H
   |  ||    |  |  |  |  |
H—C—C—O—C—C—C—C—C—H
   |        |  |  |  |  |
   H        H  H  H  H  H
```

Figure 1 Pentyl ethanoate is an ester that smells like bananas.

Esterification of alcohols

A common way of making an ester is by reacting an alcohol with a carboxylic acid in the presence of an acid catalyst. This is a reversible reaction and has a slow rate. During the reaction, the O–H bond in the alcohol and the C–O bond in the carboxylic acid are broken. New bonds are made between the H and O–H to form water. The remaining two species bond to form the ester.

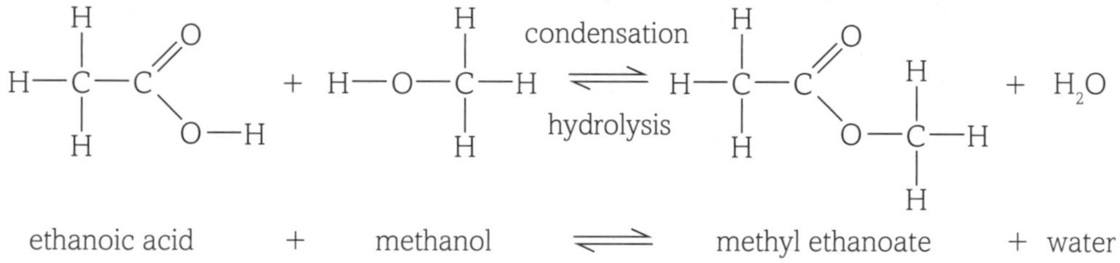

```
   H                    H   condensation     H
   |    O               |                    |    O     H
H—C—C//       +  H—O—C—H   ⇌        H—C—C//      |        +  H2O
   |    \               |   hydrolysis       |    \   O—C—H
   H     O—H            H                    H         |
                                                       H
```

ethanoic acid + methanol ⇌ methyl ethanoate + water

Figure 2 Esterification reaction to form methyl ethanoate.

On a small laboratory scale, a few cubic centimetres of the liquid alcohol and carboxylic acid are mixed in a test tube. A few drops of concentrated sulfuric acid are added to act as a catalyst. To increase the rate of reaction, the test tube is then held in a hot (80 °C) water bath. The classic sweet smell of an ester is noticeable after a few minutes.

As only small-molecule esters are soluble in water, many are visible as an oily layer floating on top of the reaction mixture.

When working on a large scale, the reaction mixture can be continually distilled to collect the ester product.

> **LEARNING TIP**
>
> You do not need to know about esterification of alcohols for the AS level course. However, if you take the full OCR Chemistry A level, you will cover this topic in more detail in the second year of your course.

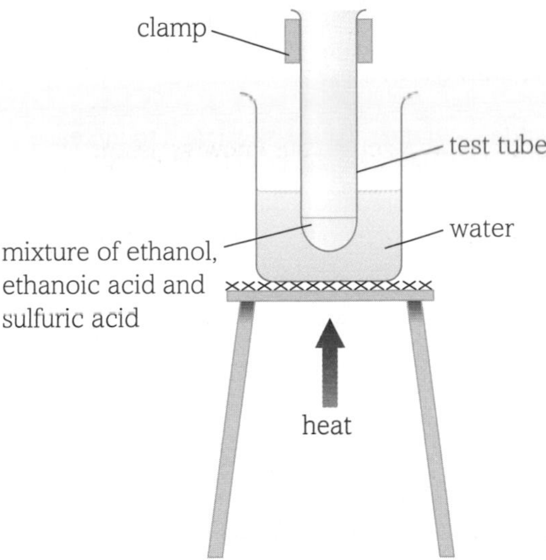

Figure 3 Laboratory-scale preparation of an ester.

Figure 4 This ester is floating in a layer on top of the reaction mixture.

Naming esters (Figure 5) involves using two words to indicate the IUPAC name:

- the first part is the name of the alkyl chain that was attached to the alcohol group
- the second part is the stem of the name of the carboxylic acid with the suffix -oate. For example, when methanol reacts with hexanoic acid, the resulting ester is called methyl hexanoate.

(a)

```
      O
     //      H   H
H—C          |   |
     \   O—C—C—H
             |   |
             H   H
```

(b)

```
    H
    |     O
H—C—C//      H
    |   \     |
    H    O—C—H
              |
              H
```

Figure 5 (a) When ethanol reacts with methanoic acid, ethyl methanoate is made; (b) When methanol reacts with ethanoic acid, methyl ethanoate is made.

Dehydration

Dehydration is a chemical reaction in which water is lost from an organic compound (Figure 6). This is a type of **elimination** reaction.

When alcohols are heated with a strong acid (like concentrated sulfuric acid), water is eliminated to make an alkene. This happens when an –OH group from one carbon atom and an H atom on an adjacent carbon atom lost to form a water molecule. Next a π-bond forms between the two adjacent carbon atoms.

$$\mathrm{H{-}CH_2{-}CH_2{-}OH} \xrightarrow[170\,^\circ\mathrm{C}]{\mathrm{H_2SO_4}} \mathrm{H_2C{=}CH_2} + \mathrm{H_2O}$$

Figure 6 The dehydration of ethanol.

LEARNING TIP

You do not need to know the mechanism for dehydration of ethanol.

On the laboratory scale (Figure 7), the alcohol is heated under reflux with concentrated sulfuric acid (H_2SO_4) or concentrated phosphoric acid (H_3PO_4) for about 40 minutes. The concentrated strong acid is a catalyst in the reaction.

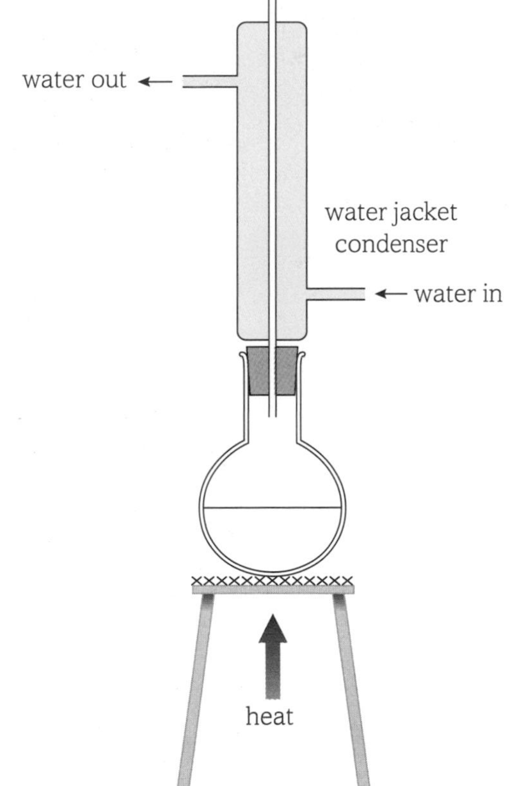

Figure 7 Laboratory-scale dehydration of an alcohol.

Halide substitution

Halide ions like Cl^-, Br^- and I^- will react with alcohols. The halide takes the place of, or substitutes, the alcohol group to form a haloalkane. The general equation is:

$$ROH + HX \rightarrow RX + H_2O$$

Although the halide is the species directly involved in the reaction, the reagent would be a hydrogen halide like HCl. For substitution with bromide (Figure 8), a salt like sodium bromide is usually used, which reacts with the sulfuric acid to make HBr in situ.

An acid catalyst such as concentrated sulfuric acid is added and the mixture is warmed to increase the rate of reaction. However, when 'iodide' is reacted then phosphoric acid is used. This is because sulfuric acid oxidises iodide ions to iodine – so the yield of the desired iodoalkane is very low.

You do not need to know the mechanism for this reaction.

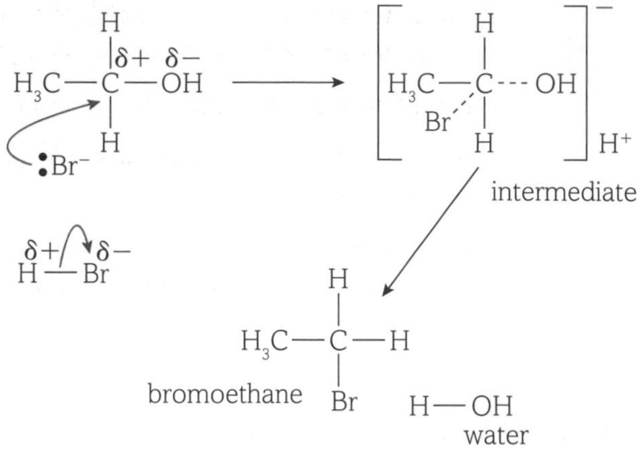

Figure 8 Bromide ions reacting with ethanol to make bromoethane.

KEY DEFINITIONS

Dehydration is a chemical reaction in which water molecules are eliminated from an organic compound.

Elimination is an organic reaction in which one reactant forms two products. Usually a small molecule like water is released.

Questions

1. Name the products of these reactions:
   (a) ethanol and propanoic acid
   (b) propan-1-ol and methanoic acid
   (c) propan-2-ol and propanoic acid.

2. Butan-2-ol can be dehydrated, using concentrated phosphoric acid, to produce three possible alkenes that are isomers. Draw the structural formulae and name the alkenes formed.

3. Write a balanced equation for the formation of a haloalkane from a reaction mixture containing sodium bromide, ethanol and concentrated sulfuric acid.

4. Name the product of the reaction between sodium bromide and propan-1-ol in the presence of an acid catalyst.

Haloalkanes

By the end of this topic, you should be able to demonstrate and apply your knowledge and understanding of:

- hydrolysis of haloalkanes in a substitution reaction:
  - (i) by aqueous alkali
  - (ii) by water in the presence of $AgNO_3$ and ethanol to compare experimentally the rates of hydrolysis of different carbon–halogen bonds
- definition and use of the term *nucleophile* (an electron pair donor)
- the mechanism of nucleophilic substitution in the hydrolysis of primary haloalkanes with aqueous alkali
- explanation of the trend in the rates of hydrolysis of primary haloalkanes in terms of the bond enthalpies of carbon–halogen bonds (C–F, C–Cl, C–Br and C–I)

What are haloalkanes?

Haloalkanes are saturated organic compounds that contain carbon atoms and at least one halogen atom.

A **primary haloalkane** (Figure 1) has the halogen atom at the end of a chain.

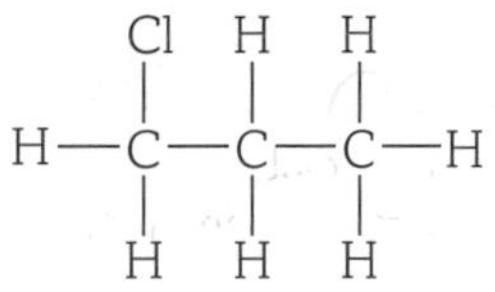

Cl H H
H—C—C—C—H
H H H

Figure 1 1-chloropropane is a primary haloalkane.

Reactivity

There is a large difference in the electronegativities of carbon and all halogens. This makes the carbon–halogen bond in haloalkanes polar (Figure 2). The electrons in the C–X bonds spend more time near the halogen atom than the carbon atom. This results in a $\delta+$ charge on the carbon atom and a $\delta-$ charge on the halogen atom.

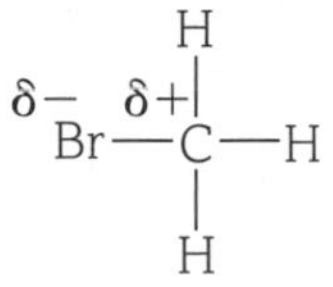

H
δ− δ+
Br—C—H
H

Figure 2 Polarity in bromomethane.

The $\delta+$ carbon atom is attacked by atoms, molecules or ions (Figure 3) that have a partial or full negative charge, like H_2O, OH^- and NH_3. These species are called **nucleophiles** and they are electron-pair donors.

> **DID YOU KNOW?**
> As nucleophiles are electron-pair donors, they are also Lewis bases.

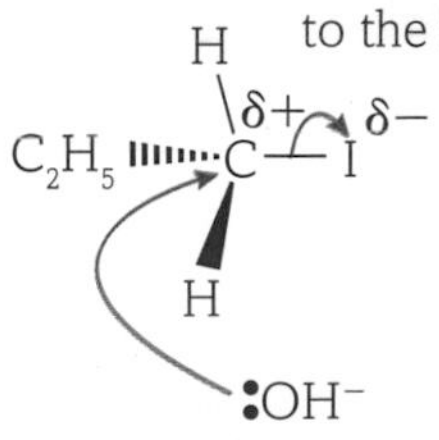

The carbon–iodine bond breaks as the two electrons in the covalent bond move to the iodine, forming an iodide ion, I^-.

H
δ+ δ−
C_2H_5 C—I
H
:OH^-

Two electrons from the hydroxide ion, OH^-, form a covalent bond with the carbon atom.

Figure 3 Nucleophile attacking a haloalkane.

Two factors affect reactivity in this type of reaction.

- Strength of the C–X bond – going down the halogen group in the periodic table, the atoms get larger. This means that their bonding electrons are further away from the nuclei and feel stronger shielding (Figure 4). This results in lower mean bond enthalpies and the C–X bonds are easier to break. The higher the mean bond enthalpy, the less likely that the halogen will react in a nucleophilic substitution reaction.

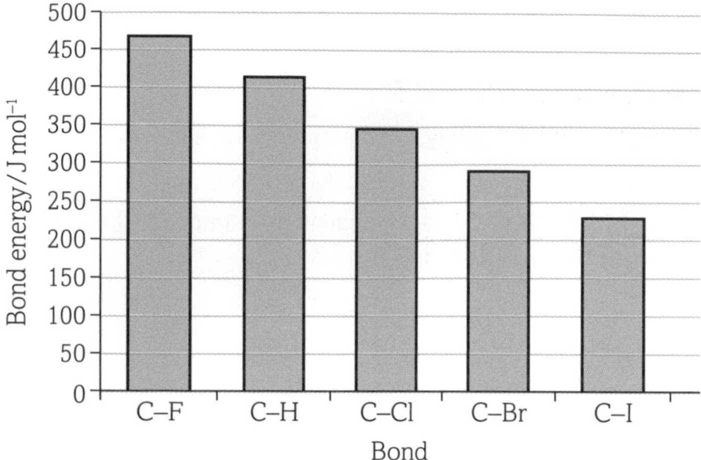

Figure 4 How C–X bond enthalpy changes.

- The polarity of the C–X bond: fluorine is the most electronegative element in group 17 so electronegativity decreases going down the group. This means that the polarity of the C–X bond decreases going down the group and it is the molecules with the highest polarity that are more likely to react.

KEY DEFINITIONS

A **primary haloalkane** has the halogen atom on the end of a chain.
Nucleophiles are electron-pair donors.
Nucleophilic substitution is a chemical reaction in which an atom or group of atoms is exchanged for a nucleophile.
Hydrolysis is a chemical reaction in which water is a reactant. There are alkali hydrolysis reactions where $-OH^-$ is the reacting species.

Nucleophilic substitution

Nucleophilic substitution is a reaction where an atom or group of atoms is exchanged for a nucleophile. The nucleophile is attracted to the partial positive charge on the carbon atom and donates its lone pair of electrons to form a new covalent bond.

What is hydrolysis?

In haloalkane reactions **hydrolysis** is a nucleophilic substitution reaction. Water is often the reactant that provides the hydroxide ions that act as nucleophiles. If water is used (Figure 5), one of the H–O bonds in a molecule undergoes heterolytic fission to produce the nucleophile OH^-. This species attacks the electron-deficient carbon atom in the haloalkane molecules. Once the nucleophile has bonded to the haloalkane, the C–X bond undergoes heterolytic fission to produce a halide ion.

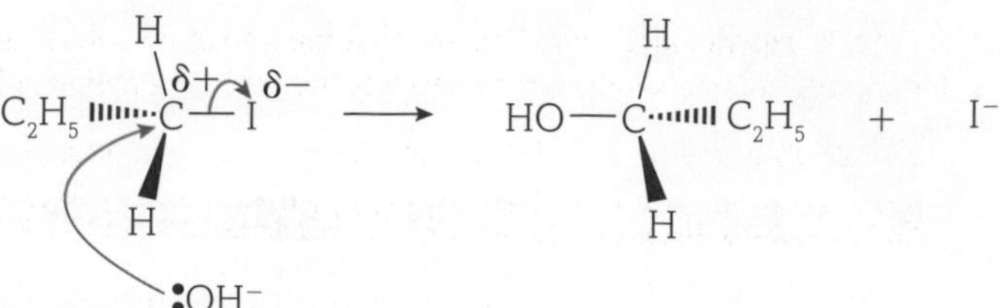

Figure 5 Hydrolysis of 1-iodopropane.

Measuring the rate of hydrolysis

The rate of a chemical reaction is the change in concentration of a reactant or product in a given time.

One of the products of hydrolysis is halide ions, X^-. These form a coloured precipitate of a silver halide when acidified silver nitrate is added.

INVESTIGATION

Fluoride ions react with silver nitrate solution to make silver fluoride. However, silver fluoride is very soluble and therefore no precipitate is seen. The silver nitrate test cannot be used to determine the presence of fluoride ions in a compound.

Haloalkane	Precipitate formed with $AgNO_3$	Colour
chloroalkane	AgCl(s)	white
bromoalkane	AgBr(s)	cream
iodoalkane	AgI(s)	yellow

Table 1 Halide ion test colours.

By measuring how long it takes for a precipitate to form, the rate of reaction can be monitored. To investigate the rate of hydrolysis, the independent, dependent and control variables need to be considered.

- The *independent variable* is the factor you want to investigate. In this experiment the halogen present in the haloalkane is the independent variable.
- The *dependent variable* is measured in the experiment – here it is the time taken for the precipitate to form.
- The *control variables* are kept the same so that the results can be compared. These include the length of alkyl chain, the amount (moles) of haloalkane used , the concentrations of the reactants, the volume of the reactants and the temperature maintained using a water bath.

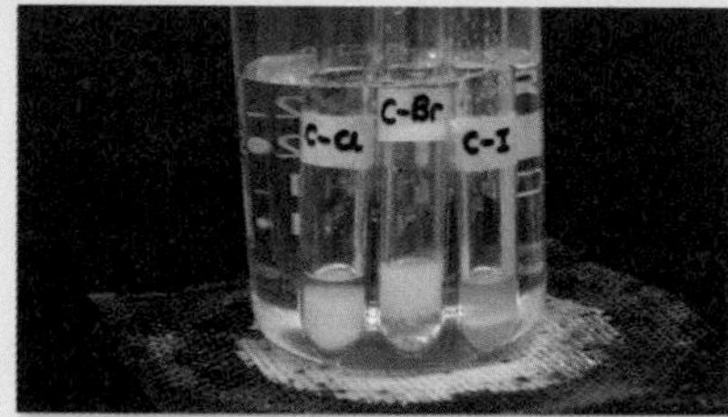

Figure 6 Investigating the rate of hydrolysis for a primary haloalkane.

Here is the outline method for this investigation.

- Using a measuring cylinder, add 1 cm³ of each haloalkane to different test tubes and place them in a water bath set at 50 °C.
- Half-fill three separate test tubes with ethanol, water and aqueous silver nitrate and place them in the same water bath.
- When all liquids have reached the temperature of the water bath, add 1 cm³ of ethanol, water and aqueous silver nitrate to each haloalkane and time how long it takes for the precipitate to appear.

The results of this experiment (Figure 7) show that the rate of hydrolysis is fastest for primary iodoalkanes and slowest for primary chloroalkanes. We can predict the position of primary fluoroalkanes.

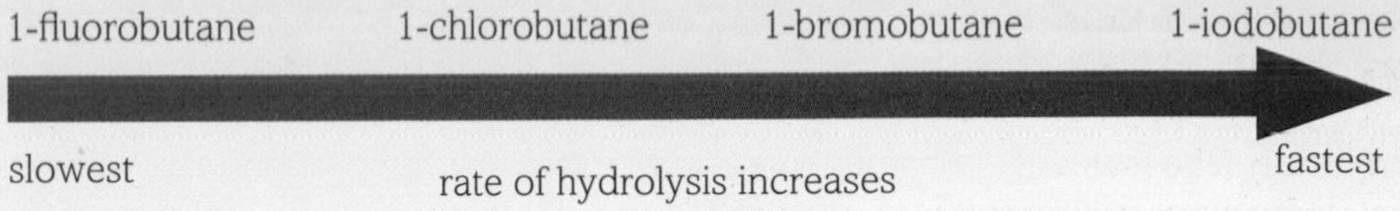

1-fluorobutane 1-chlorobutane 1-bromobutane 1-iodobutane

slowest rate of hydrolysis increases fastest

Figure 7 Rates of hydrolysis.

Hydrolysis and bond enthalpies of carbon–halogen bonds

As the reactivity of haloalkanes in nucleophilic substitution reactions depends on both polarity and bond enthalpy, we can conclude that bond enthalpy is the most important factor in the hydrolysis of primary haloalkanes.

The C–Cl bond is the most polar in the investigation described above and therefore should be the most attractive to nucleophilic attack, but because of the high C–Cl bond enthalpy chloroalkanes actually have the lowest rate of hydrolysis.

The least polar, and therefore least attractive, C–I group has a higher rate of hydrolysis because the bond is weaker and easier to break.

Hydrolysis using aqueous alkali

When a haloalkane is heated under reflux with an aqueous solution containing hydroxide ions, an alcohol is made. This is a nucleophilic substitution reaction (Figure 8) and also an example of hydrolysis. The rate of the reaction is faster than that using water.

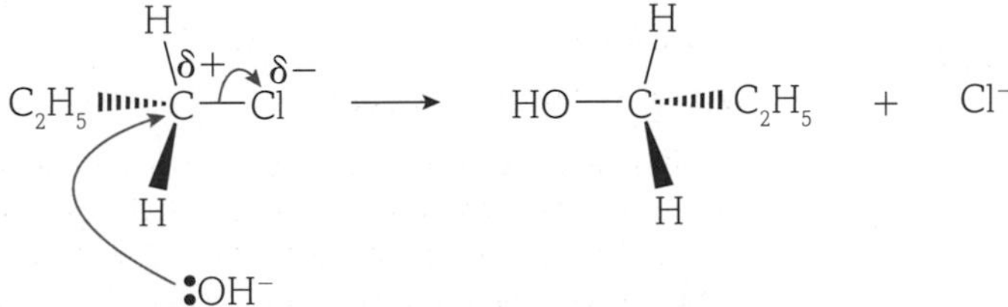

Figure 8 Mechanism to show hydrolysis of 1-chloropropane with sodium hydroxide.

Questions

1. Explain why a carbon–chlorine bond is more polar than a carbon–bromine bond.
2. Explain why haloalkanes are susceptible to nucleophilic attack.
3. Write the mechanism to show the hydrolysis of 1-chloroethane.
4. Explain how the rate of hydrolysis of primary haloalkanes is affected by the halogen used.

5 Haloalkanes and the environment

By the end of this topic, you should be able to demonstrate and apply your knowledge and understanding of:

* production of halogen radicals by the action of ultraviolet (uv) radiation on CFCs in the upper atmosphere and the resulting catalysed breakdown of the Earth's protective ozone layer, including equations to represent:
  (i) the production of halogen radicals
  (ii) the catalysed breakdown of ozone by Cl• and other radicals e.g. •NO

Chlorofluorocarbons

CFCs are a class of organic compounds that contain chlorine atoms and fluorine atoms. As they are inert, in the past they were used widely because they do not react with other substances. This property made them very useful in fire-fighting equipment and they are still used today in fire-safety equipment in the aviation industry. CFCs also had domestic uses such as refrigerants (Figure 1) because they are non-toxic.

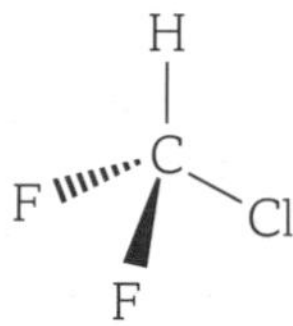

Figure 1 R-22 ($CHClF_2$) – a CFC used as a refrigerant.

CFCs were first synthesised in the early 1900s for use as refrigerant gases. They were later used as blowing agents (to make bubbles in materials) to make plastics like polystyrene because they did not react with the polymer, and also in fire extinguishers (Figure 2) because they are inert and smother fires.

Figure 2 Aircraft 'halon' fire extinguishers use green recyclable cylinders. Halon hasn't been manufactured since 1994 due to strict legislation to try to protect the ozone layer.

CFCs and the atmosphere

CFCs are gases at room temperature and after use they disperse into the atmosphere. As the gases diffuse up through the layers of the atmosphere (Figure 3) – they eventually become exposed to ultraviolet (UV) radiation in the stratosphere.

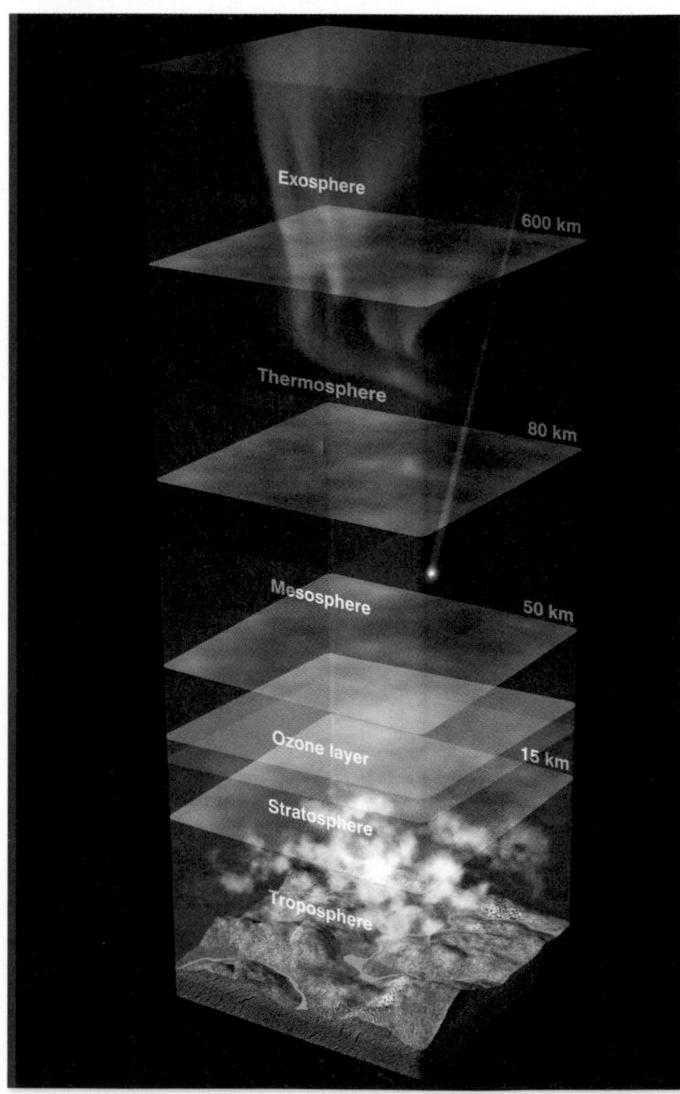

Figure 3 Layers of the atmosphere.

The C–Cl bonds in $C_2F_2Cl_2$ undergo homolytic fission due to the UV light:

Initiation step: $C_2F_2Cl_2 \rightarrow C_2F_2Cl^\bullet + Cl^\bullet$

The Earth's **ozone layer** is in the stratosphere, about 20 km from the Earth's surface. This is a layer of O_3 molecules (Figure 4). Ozone is toxic to humans in the troposphere and a key contributor to photochemical smog. In the stratosphere it protects us from harmful UV radiation from the Sun – it also stabilises temperatures.

The ozone layer absorbs all UV-C (200–283 nm) radiation and most of the UV-B (280–320 nm) and so protects organisms from genetic damage. Prolonged exposure to UV-B radiation can lead to cell mutations, which can cause skin cancer and cataracts. UV-A (320–400 nm) radiation is not absorbed by the ozone layer – it has lower energy and does not pose health concerns.

Figure 4 Ozone, O_3, is an allotrope of diatomic oxygen, O_2.

Stratospheric ozone is decomposed naturally by UV radiation and more ozone is made in a natural process.

The chlorine **radicals** from CFCs can catalyse the decomposition of ozone to make diatomic oxygen. The chlorine radical attacks ozone molecules and turns them into diatomic oxygen:

First propagation step: $Cl^{\bullet} + O_3 \rightarrow {}^{\bullet}ClO + O_2$

Second propagation step: ${}^{\bullet}ClO + O_3 \rightarrow Cl^{\bullet} + 2O_2$

Overall: $2O_3 \rightarrow 3O_2$

The ozone layer is shrinking and in some places there is no ozone in the upper stratosphere and lower troposphere. Scientists have used satellites and ground stations (Figure 5) to measure an average of 4% reduction in ozone levels per decade since the 1970s.

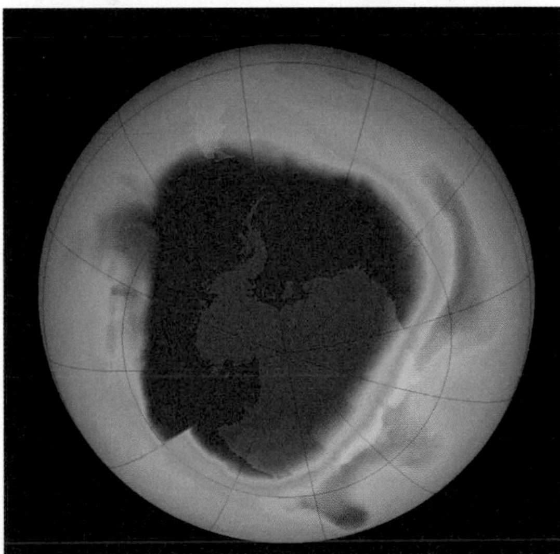

Figure 5 Satellite image of ozone concentrations in the stratosphere and troposphere – the blue part is the ozone 'hole'.

However, the Earth's polar regions show a larger and seasonal decline. Unfortunately, CFCs have been shown to remain in the atmosphere for longer than 50 years – it is estimated that a single CFC molecule can destroy 100 000 ozone molecules.

KEY DEFINITIONS

The **ozone layer** is an area of high concentration of ozone, O_3, in the stratosphere.

A **radical** is a highly reactive species with one or more unpaired electrons.

Other radicals and ozone

In the high temperatures of combustion engines, nitrogen and oxygen can react to make nitrogen oxides, known as NO*x*. This can also occur naturally when lightning is produced. Nitrogen monoxide can decompose ozone:

${}^{\bullet}NO + O_3 \rightarrow {}^{\bullet}NO_2 + O_2$

${}^{\bullet}NO_2 + O \rightarrow {}^{\bullet}NO + O_2$

Overall: $O_3 + O \rightarrow 2O_2$

LEARNING TIP

Single dots on symbols or formulas of substances indicate a single unpaired electron or a radical.

DID YOU KNOW?

The Montreal Protocol

In 1985, worldwide monitoring of atmospheric gases showed that there was a 'hole' in the ozone layer. Since then, a lot of research has been completed and scientists now understand that CFCs not only deplete the ozone layer but are also greenhouse gases and have been linked to global warming.

The Montreal Protocol began in 1985. It controls tightly the manufacture, use and disposal of CFCs. Since 2000, no new CFCs have been made and there is a timetable for a phased removal of them from use. CFCs are now being replaced by hydrocarbons, hydrofluorocarbons (HFCs) and hydrochlorofluorocarbons (HCFCs). However, these chemicals are now causing scientists concern because of mounting evidence that these too are very potent greenhouses gases.

The Montreal Protocol is informed by the scientific community and has made amendments as new evidence has arisen. It is widely considered to be the most successful global environmental agreement ever signed.

In 2003, scientists concluded that the depletion of the ozone layer is slowing down. This observation has been attributed to the reduced use of CFCs.

Questions

1. State the effects of ozone depletion.
2. Using balanced equations, explain how a CFC molecule can lead to ozone depletion.
3. Outline what the Montreal Protocol is and how its success could be measured.

Practical skills for organic synthesis

By the end of this topic, you should be able to demonstrate and apply your knowledge and understanding of:

* use of the techniques and procedures for:
  - (i) use of Quickfit apparatus including for distillation and heating under reflux
  - (ii) preparation and purification of an organic liquid including:
    - use of a separating funnel to remove an organic layer from an aqueous layer
    - drying with an anhydrous salt (e.g. $MgSO_4$, $CaCl_2$)
    - redistillation

Quickfit apparatus

Organic chemistry reactions are often carried out using Quickfit apparatus (Figure 1). This is a selection of heat-resistant glassware with connectors that can be easily fixed in a variety of arrangements.

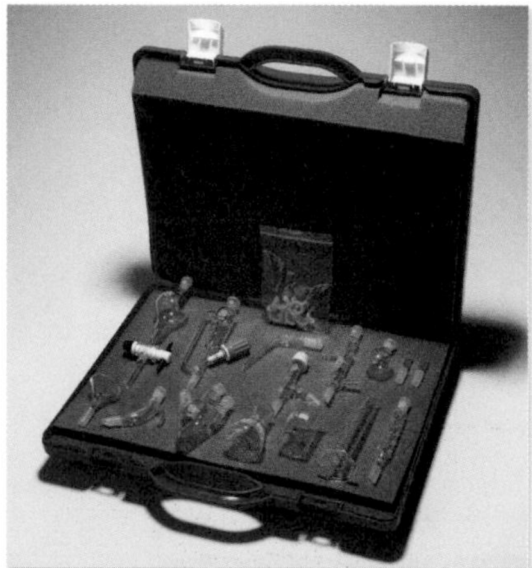

Figure 1 Quickfit apparatus.

Before you connect Quickfit apparatus, grease the joints slightly (Figure 2). Using a small amount of petroleum jelly on your finger wipe it inside the ground glass joints. You can also use plastic connectors to fix the equipment in place.

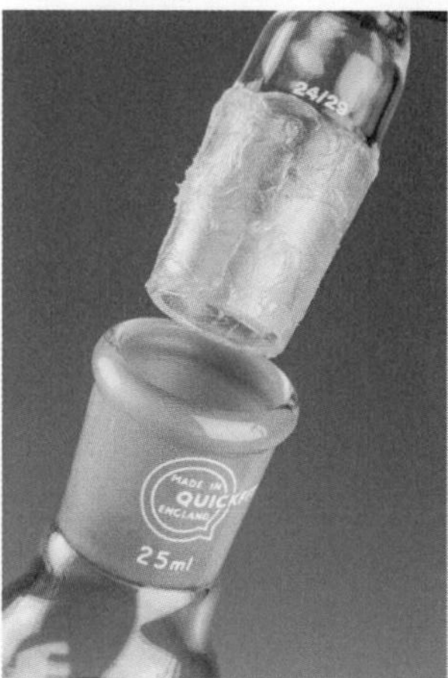

Figure 2 Quickfit joints must be secured.

KEY DEFINITIONS

Distillation is a technique used to separate miscible liquids or solutions.
Reflux is a technique used to stop reaction mixtures boiling away into the air.

Distillation

Distillation is a separating technique and is often used to collect a product from an organic synthesis reaction. For a distillation you need:

- a round-bottomed flask for the mixture
- a heating mantle
- a thermometer holder
- a thermometer
- a condenser
- a connector
- a collecting flask.

Once you have collected all your apparatus, lay it out on the bench (Figure 3). Then use clamp stands and bosses to secure each piece of equipment as you build up your apparatus.

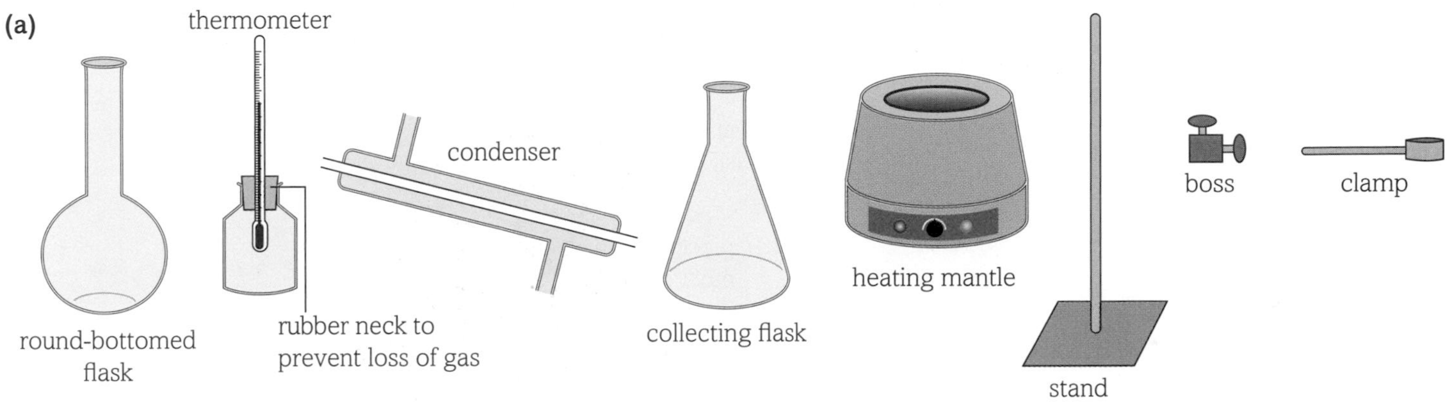

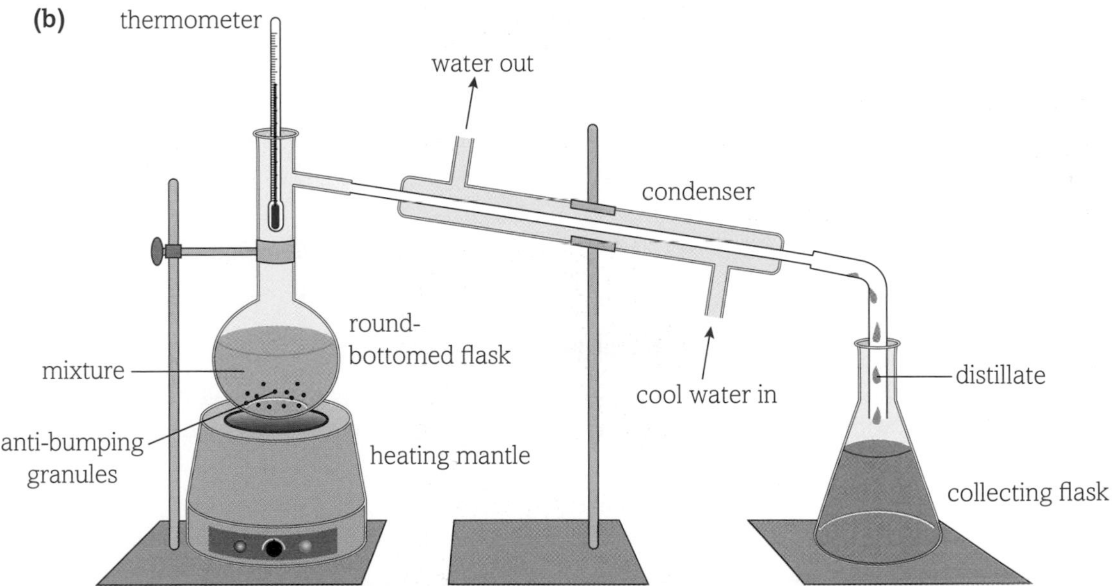

Figure 3 (a) Laying out the apparatus for distillation; (b) Quickfit apparatus for distillation.

Connect the condenser to the water supply by carefully attaching the bottom condenser hose to a water tap; put the top rubber hose into the sink and turn on the water tap to give a gentle flow.

Pour the organic mixture into the round-bottomed flask with a few anti-bumping granules to ensure smooth boiling.

To complete the distillation, heat the mixture using a heating mantle (or a Bunsen burner with a tripod and gauze). For more gentle heating, a water bath can be used on top of the tripod and gauze.

LEARNING TIP

As the product is collected, note the temperature on the thermometer – this may be useful in identifying the product.

Heating under reflux

Reflux is a technique used in organic synthesis to ensure that a reaction mixture does not boil away. To heat under reflux (Figure 4) you need:

- a pear-shaped flask
- a heating mantle
- a condenser.

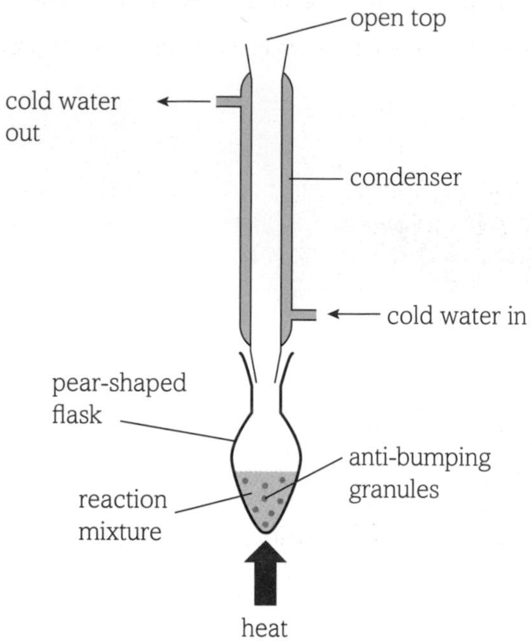

Figure 4 Quickfit apparatus for reflux.

Preparing and purifying an organic liquid

Using a separating funnel

Many organic liquids have an oily consistency and are immiscible in water. The organic layer will float on the water. This allows the product to be removed as an organic layer, which is less dense than the aqueous layer.

One method of separating immiscible liquids is to use a **separating funnel** (Figure 5).

1. Mount an iron ring on a clamp stand and put the separating funnel in it.
2. Remove the stopper and make sure that the tap at the bottom is closed.
3. Carefully pour the mixture into the funnel so that the funnel is no more than half full. Wash out the reaction vessel with water and add this to the mixture in the funnel – there should still be some air in the funnel. Put the stopper back.
4. Take the funnel out of the ring and invert it. Open the tap to equalise the pressure. Turn the tap back to closed. Gently shake the mixture in the funnel and equalise the pressure as required. Repeat until you no longer hear a 'whistle'.
5. Replace the funnel in the iron ring and give the mixture time to separate into layers.
6. Remove the stopper, put a beaker under the spout and open the tap. Collect the lower (water) layer in a beaker. Turn the tap off. As the organic product is in the upper layer, this aqueous layer can be discarded.
7. Using a clean, dry beaker open the tap and collect the desired organic product.
8. Shake the liquid with a small amount of drying agent and pour the final dry product into a clean, dry container.

Using drying agents

Inorganic salts (Figure 6) such as magnesium sulfate ($MgSO_4$) and calcium chloride ($CaCl_2$) absorb water slowly to become hydrated. The anhydrous salts can be used to dry other chemicals.

- Add a few spatulas of the drying agent to the organic product.
- If the drying agent clumps together, add some more.
- When the drying agent remains free-moving the organic product is dry.
- Use gravity filtration and collect the filtrate, which is the dry organic product.

Figure 6 Example of a drying agent.

Figure 5 A separating funnel can be used to separate immiscible liquids.

KEY DEFINITIONS

A **separating funnel** is a piece of equipment used to separate immiscible liquids.
Redistillation is the purification of a liquid using multiple distillations.

Redistillation

Redistillation is the purification of a liquid by performing multiple distillations. For instance, when ethanol is made by fermentation, distillation is used to remove the ethanol from the mixture of water, yeast and any remaining sugar. This distillate is usually distilled again to purify the ethanol further.

Questions

1. Describe how to separate a dry sample of paraffin from a paraffin/water mixture.
2. Why must you always put the top on a bottle of anhydrous magnesium sulfate that is to be used as a drying agent?
3. Explain why a separating funnel cannot be used to separate ethanol and water.

Synthetic routes in organic synthesis

By the end of this topic, you should be able to demonstrate and apply your knowledge and understanding of:

* an organic molecule containing several functional groups:
  (i) identification of the individual functional groups
  (ii) prediction of properties and reactions
* two-stage synthetic routes for preparing organic compounds

Laboratory tests for functional groups

Some simple practical tests (Table 1) can be used in a laboratory to show the presence of certain **functional groups**.

Functional group	Reagents	Observations
unsaturated hydrocarbon	add a few drops of bromine water to the sample and shake	bromine water decolourises
haloalkane	silver nitrate, ethanol and water	• white precipitate indicates chloro- • cream precipitate indicates bromo- • yellow precipitate indicates iodo-
carbonyl	• acidified potassium dichromate (VI) • Fehling's solution • Tollens' reagent	• ketones – no change • aldehydes – orange turns green • ketones – no change • aldehydes – dark red precipitate • ketones – no silver mirror • aldehydes – silver mirror
carboxylic acid	• universal indicator or pH probe • reactive metal, e.g. Mg • metal carbonate, e.g. $CaCO_3$	• pH of a weak acid • effervescence (hydrogen) • effervescence (carbon dioxide)
alcohol	warm with an equal volume of carboxylic acid and a few drops of H_2SO_4	sweet smell of an ester after a short time

Table 1 Tests to show the presence of some functional groups.

KEY DEFINITION

A **functional group** is a group of atoms responsible for the characteristic reactions of a compound.

More than one functional group

Many organic compounds have more than one functional group.

Aspirin

Aspirin (Figure 1) is an analgesic (painkiller) and was originally obtained from willow bark. It has a number of functional groups including a carboxylic acid and an aromatic ring.

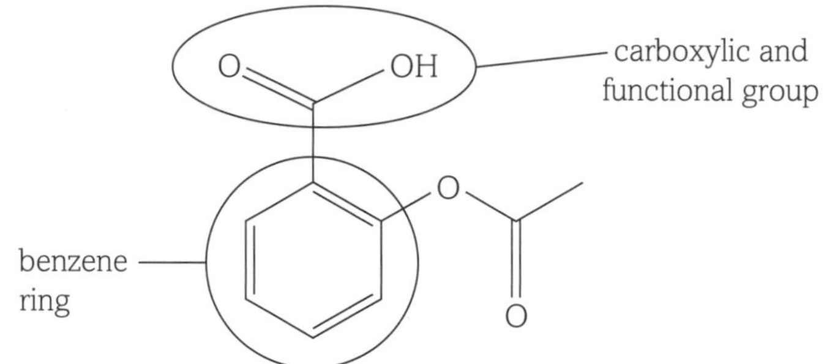

Figure 1 Skeletal formula of aspirin.

Niacin

Vitamins help our bodies to work more efficiently. Many vitamins are complex organic molecules with several functional groups.

Vitamin B3, also known as niacin, has the IUPAC name pyridine-3-carboxylic acid (Figure 2). It is needed for healthy hair, eyes, skin and liver. This organic molecule contains nitrogen in a stable structure known as a pyridine ring and a carboxylic acid functional group.

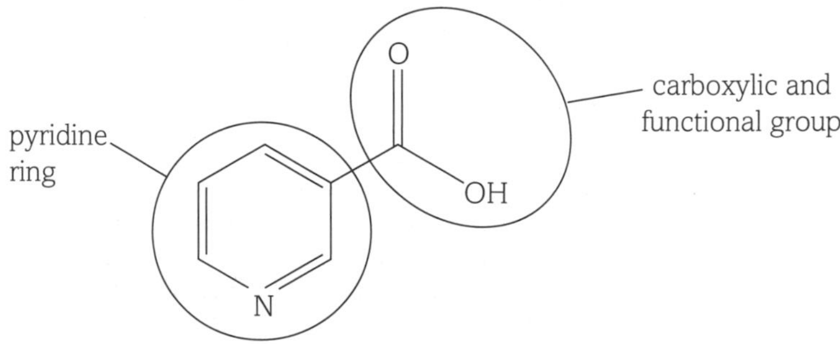

Figure 2 Skeletal structure of niacin.

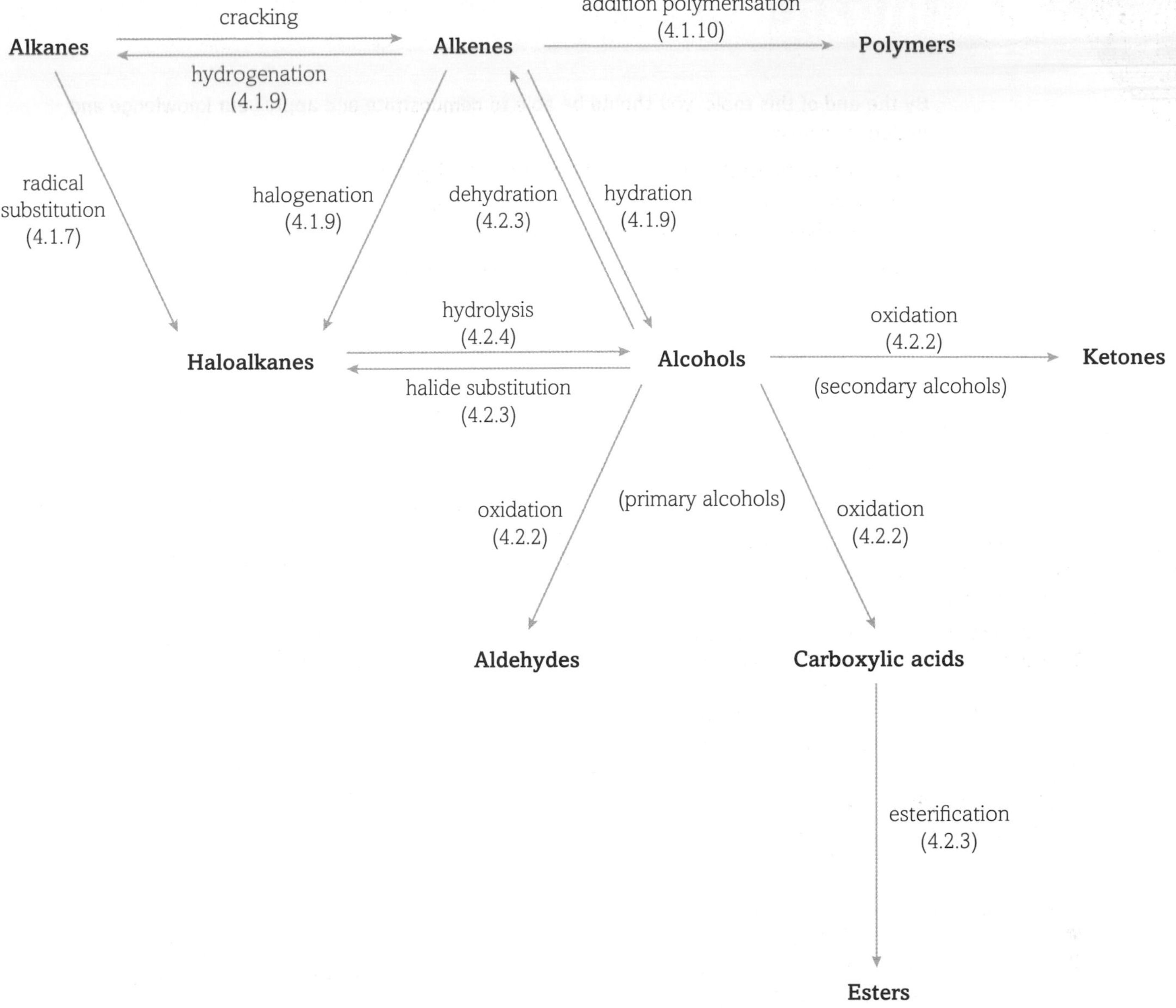

Figure 3 Reaction routes to make some organic compounds.

Synthetic routes

It is often common to use more than one reaction to generate a target molecule for a particular organic reactant. Figure 3 summarises some reaction routes showing how reactions could be sequenced to produce certain organic compounds.

DID YOU KNOW?

Organic chemists and chemical engineers design synthesis routes to generate target molecules. When the synthetic route is being designed, it is important that the atom economy, yield, reaction conditions, availability and the cost of starting materials are considered. All these factors need to be balanced so that a profit can be made.

Questions

1. Name the type of reaction that produces chloroethane from ethene. Illustrate your answer with a balanced chemical equation.
2. Describe the laboratory test(s) needed to label two liquids as chloroethane or bromoethene.

4.2

8 Infrared spectroscopy

By the end of this topic, you should be able to demonstrate and apply your knowledge and understanding of:

* infrared radiation (IR) causes covalent bonds to vibrate more and absorb energy
* absorption of infrared radiation by atmospheric gases containing C=O, O–H and C–H bonds (e.g. H_2O, CO_2 and CH_4), the suspected link to global warming and resulting changes to energy usage
* use of infrared spectroscopy to monitor gases causing air pollution (e.g. CO and NO from car emissions) and in modern breathalysers to measure ethanol in the breath

Infrared radiation and molecules

All molecules absorb infrared radiation. This absorbed energy makes covalent bonds vibrate more, with either a stretching or bending motion as shown in Figure 1.

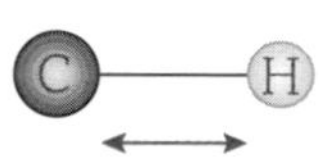

The C–H bond stretches when it absorbs infrared radiation.

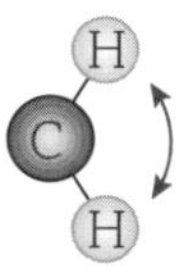

The C–H bond bends when it absorbs infrared radiation.

Figure 1 When molecules absorb infrared radiation, it causes them to vibrate more in stretching or bending motions.

Every bond vibrates at its own unique frequency. The amount of vibration depends on:

- the bond strength
- the bond length
- the mass of each atom involved in the bond.

Most bonds vibrate at a frequency between 300 and 4000 cm^{-1} – this is in the infrared part of the electromagnetic spectrum.

The absorbed energies can be displayed as an infrared spectrum. By analysing such a spectrum, we can determine details about a compound's chemical structure. In particular, the spectrum indicates the presence of functional groups in the compound under investigation.

In a modern infrared spectrometer, a beam of infrared radiation is passed through a sample of the material under investigation. The beam contains the full range of frequencies in the infrared region. The molecules absorb some of these frequencies and the emerging beam is analysed to plot a graph of transmittance against frequency – this is the *infrared spectrum* of the molecule. The frequency is measured using wavenumbers, with units of cm^{-1}.

DID YOU KNOW?

Infrared radiation and global warming

Molecules don't just absorb radiation inside an infrared spectrometer – it happens all around us. For example, C=O, O–H and C–H bonds are very good at absorbing infrared radiation. These bonds are present in many common greenhouse gases such as CO_2, H_2O and CH_4. They are all released every time we burn fuels such as coal or petrol. As we release more and more of these gases into the atmosphere, more and more infrared radiation can be absorbed, which eventually leads to artificial warming of the planet – known as global warming.

This suspected link between the absorption of such gases and global warming has led to many governments introducing policies that encourage the use of technologies and energy resources that do not release greenhouse gases into the environment.

What does an infrared spectrum look like?

A compound's spectrum may show a number of troughs, surprisingly called *peaks*. Each peak represents the absorbance of energy from infrared radiation that causes the vibration of a particular bond in the molecules under investigation. Figure 2 shows an example of an infrared spectrum.

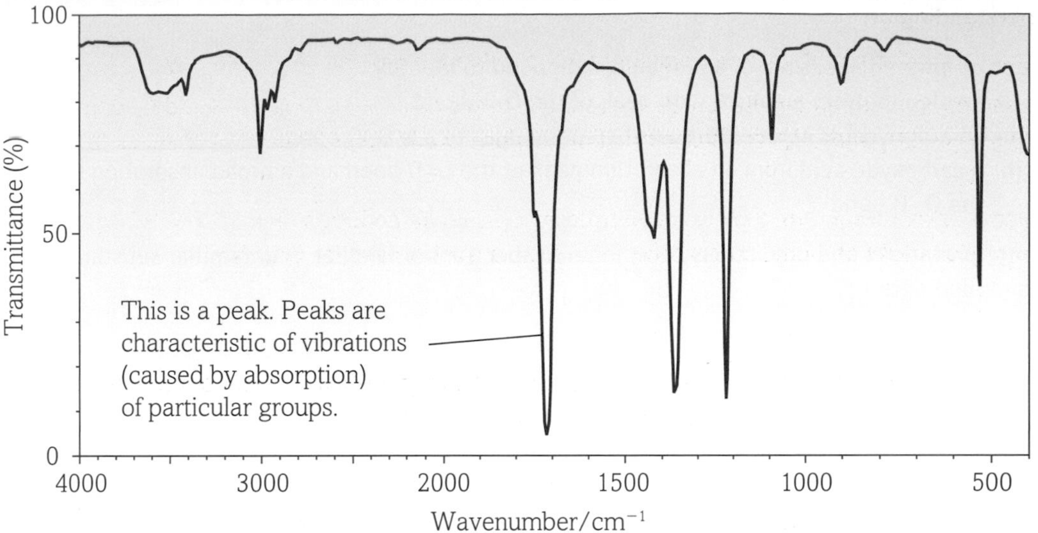

Figure 2 An infrared spectrum showing absorption peaks.

LEARNING TIP

An infrared spectrum is a bit like a graph. It is produced by a computer that analyses radiation that passes through completely uninterrupted (i.e. not having passed through any substances) and compares this with the radiation that has been passed through the substance being investigated.

Absorption is shown by peaks (the dips!). These show where absorption of the radiation has occurred. These parts of the spectrum indicate to chemists the bonds present in the substance.

When it comes to analysing spectra, you may find it helpful to annotate the actual diagram – for example, by labelling individual peaks.

DID YOU KNOW?

Applications of infrared spectroscopy

Infrared spectroscopy has many everyday uses. It is used extensively in forensic science – for example, to analyse paint fragments from vehicles in hit-and-run offences. Other uses that rely on infrared spectrometry include:

- monitoring the degree of unsaturation in polymers
- quality control in perfume manufacture
- drug analysis.

Infrared spectrometers are also used as one of the main methods for testing the breath of suspected drunken drivers for ethanol. The characteristic peak for an O–H bond identifies the presence of alcohol. The level of absorption is related to concentration of alcohol in the blood. This relationship is described by a principle known as the Beer–Lambert law.

Sometimes other substances present in the breath may produce similar spectra. So most modern breathalysers combine infrared spectroscopy with other technologies, such as chemical reactions to identify alcohol and fuel cells where reactions occur that allow more specific results to be obtained on the concentration of alcohol present.

Because many known substances contain bonds that will vibrate characteristically when infrared radiation is passed through them, this kind of spectroscopy can have many applications. Air pollutants, such as CO and NO, and their concentrations can be monitored – for example, to study the effect of traffic on air quality. Infrared spectroscopy technologies can be used to detect toxic leaks from industrial plants or to monitor changes in the different levels of the atmosphere.

Questions

1. Explain how covalent bonds are affected by infrared radiation.
2. Describe how infrared spectroscopy can help to identify whether or not someone is a drink driver.

Infrared spectroscopy: functional groups

By the end of this topic, you should be able to demonstrate and apply your knowledge and understanding of:

* use of infrared spectrum of an organic compound to identify:
  (i) an alcohol from an absorption peak of the O–H bond
  (ii) an aldehyde or ketone from an absorption peak of the C=O bond
  (iii) a carboxylic acid from an absorption peak of the C=O bond and a broad absorption peak of the O–H bond
* interpretations and predictions of an infrared spectrum of familiar or unfamiliar substances using supplied data

Identifying functional groups

Typical data on the spectra of functional groups is available in reference sheets. These show the typical wavenumbers for peaks caused by certain bonds when they absorb infrared radiation.

The values given in Table 1 show the peaks you will need to be familiar with in order to identify different types of molecules:

- alcohols
- aldehydes
- ketones
- carboxylic acids.

Bond	Functional group	Wavenumber/cm^{-1}
C=O	aldehydes, ketones, carboxylic acids	1640–1750
O–H	carboxylic acids	2500–3300 (very broad)
O–H	alcohols	3200–3550 (broad)

Table 1 Characteristic absorption ranges for bonds found in alcohols, aldehydes, ketones and carboxylic acids.

You also need to be aware that most organic compounds produce a peak at approximately 3000 cm^{-1} due to absorption by C–H bonds. You need to be careful that you are not tempted into thinking that this is an O–H absorption peak.

Identifying alcohols

The infrared spectrum of methanol, CH_3OH, is shown in Figure 1.

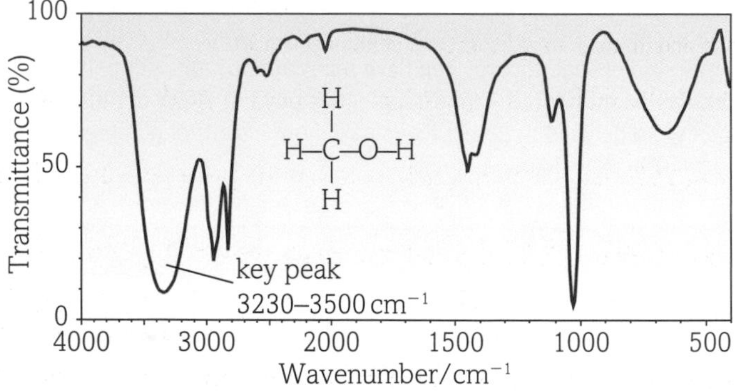

Figure 1 Infrared spectrum of methanol.

The peak at 3230–3500 cm^{-1} represents an O–H group in alcohols.

If this peak is present, the substance being analysed contains an O–H bond as part of an alcohol group.

LEARNING TIP

If peaks appear that do not match with any wavenumbers for groups you will need to know about you can ignore them, for example the peak just after 1000 cm^{-1} in Figure 1.

Identifying aldehydes and ketones

The spectrum of the aldehyde propanal, CH_3CH_2CHO, is shown in Figure 2.

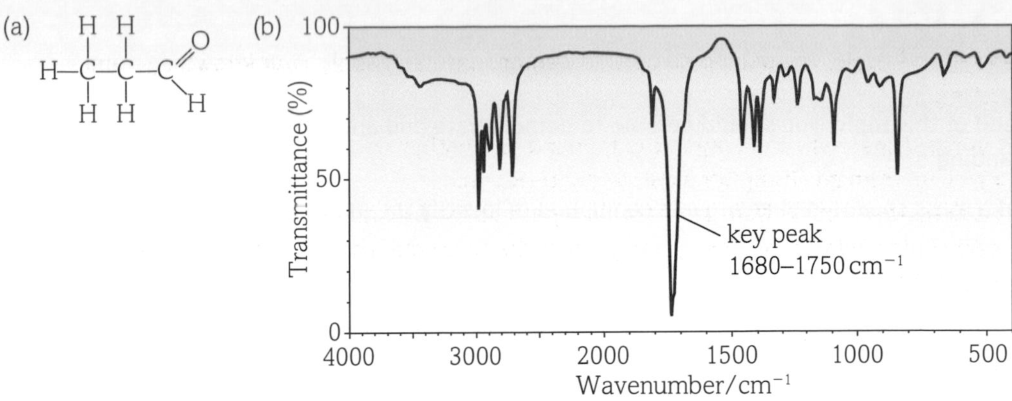

Figure 2 (a) The structure and (b) the infrared spectrum of propanal.

The peak at 1680–1750 cm^{-1} represents a C=O bond in both aldehydes and ketones. If this peak is present, the substance being analysed contains a C=O bond.

You need to take care if you analyse a spectra that contains a C=O peak – they are present in aldehydes, ketones and carboxylic acids. You can see if the spectra is from an aldehyde or a ketone, because there will be a C=O peak, but *no* peak from an O–H group. You will see in the next section why an O–H peak and a C=O peak together would indicate a carboxylic acid.

Identifying carboxylic acids

The spectrum of the carboxylic acid propanoic acid, CH_3CH_2COOH, is shown in Figure 3.

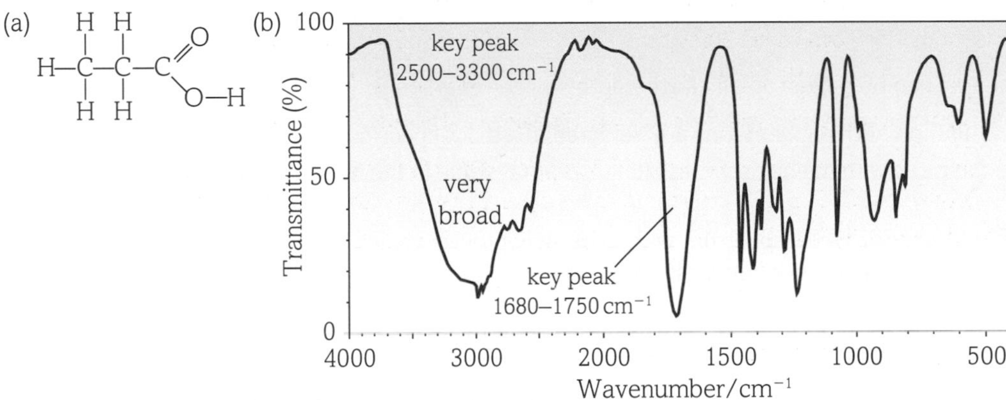

Figure 3 (a) The structure and (b) the infrared spectrum of propanoic acid.

For a substance to be identified from its spectra as a carboxylic acid, it must contain two characteristic peaks:

- a very broad peak at 2500–3300 cm^{-1}, which indicates the presence of the O–H group in a carboxylic acid
- a strong, sharp peak at 1680–1750 cm^{-1} that represents the C=O group in a carboxylic acid.

Questions

1. An organic compound has an absorption peak in the infrared spectrum at 1700 cm^{-1}. There is no peak at a wavenumber higher than 3000 cm^{-1}. What functional group must be present in the compound?

2. State the approximate wavenumbers for the key infrared absorptions for these compounds:
   (a) pentan-2-ol
   (b) 2-hydroxypentanal
   (c) butanoic acid.

4.2

Mass spectrometry in organic chemistry

By the end of this topic, you should be able to demonstrate and apply your knowledge and understanding of:

* use of a mass spectrum of an organic compound to identify the molecular ion peak and hence to determine molecular mass

Mass spectrometry and molecules

When an organic compound is vaporised and driven through through a mass spectrometer, some molecules lose an electron each and are ionised. The resulting positive ion is called the molecular ion and is given the symbol M^+.

- The mass of the lost electron is negligible.
- The molecular ion has a molecular mass equal to the relative molecular mass of the compound.
- Molecular ions can be detected and analysed.

Excess energy from the ionisation process can be transferred to the molecular ion, making it vibrate. This causes bonds to weaken and the molecular ion can split into pieces by fragmentation. Fragmentation means that the original molecules break up into positive fragment ions and other neutral species.

Fragmentation is unpredictable because it can happen anywhere within the molecule. A possible fragmentation of the ethanol molecular ion would be:

$$C_2H_5OH^+ \rightarrow CH_3 + CH_2OH^+$$

Fragment ions can be broken up still further into smaller fragments.

The molecular ion and fragment ions are detected in a mass spectrometer. The molecular ion, M^+, produces the peak with the highest mass/change (m/z) value in the mass spectrum. High-resolution mass

spectrometry can produce a spectrum with extremely precise values for M^+, making accurate molecular mass determination relatively easy.

LEARNING TIP

A **molecular ion, M^+**, is the positive ion formed in mass spectrometry when a molecule loses an electron.

Determining molecular mass

The molecular mass of a compound can be determined using mass spectrometry by locating the M^+ peak. The mass spectrum of ethanol is shown in Figure 1.

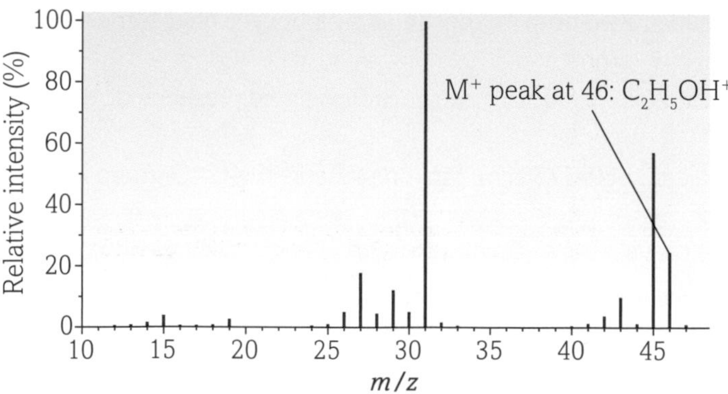

Figure 1 Mass spectrum of ethanol.

The molecular ion peak is located at an m/z ratio of 46. This indicates that the relative molecular mass of ethanol is 46. The other peaks in the spectrum are a result of fragmentation.

LEARNING TIP

In many mass spectra, you can see a small peak that is one unit beyond the molecular ion, M^+, peak – this is called the M+1 peak. It arises from the presence of the isotope carbon-13, which makes up 1.11% of all carbon atoms. Don't be fooled into thinking that this is the M^+ ion.

Fragmentation patterns

Mass spectrometry can be used to determine the structure of an unknown compound and, in many cases, give its precise identity.

- Although the molecular ion peaks of two isomers will have the same *m/z* value, the fragmentation patterns will be different.
- Each organic compound produces a unique mass spectrum, which can be used as a fingerprint for identification.

The mass spectra of isomers will have the same M^+ peak, but the fragmentation patterns will be different for each. An example of this is shown in Figure 2.

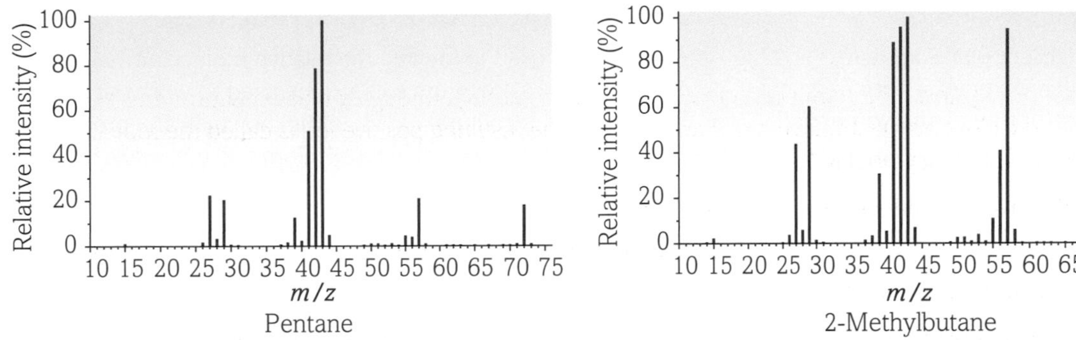

Figure 2 Pentane and 2-methylbutane have the same M^+ peak because they are isomers and have the same relative molecular mass. However, their fragmentation patterns are very different.

Questions

1. Write balanced equations showing the ionisation of the following compounds to form their molecular ions. In each case, state the likely *m/z* value for the molecular ion peak:
   (a) propan-1-ol
   (b) butane
   (c) octane.

2. In the spectra of compounds A and B in Figure 3, identify the *m/z* value for each molecular ion peak.

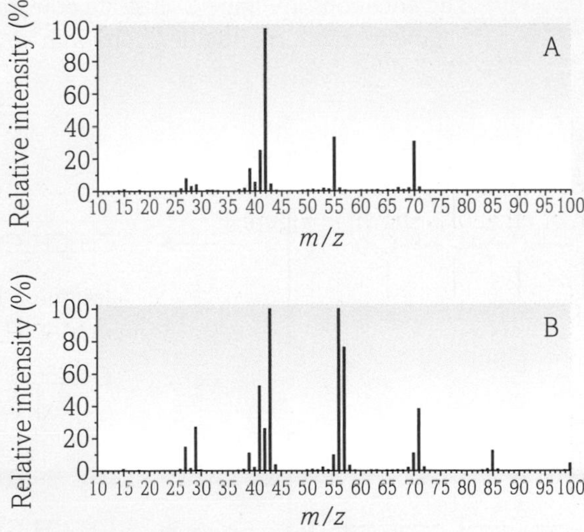

Figure 3

Mass spectrometry: fragmentation patterns

By the end of this topic, you should be able to demonstrate and apply your knowledge and understanding of:

- analysis of fragmentation peaks in a mass spectrum to identify parts of structures

Identifying fragment ions

When looking at a mass spectrum, fragment peaks appear alongside the more important molecular ion peak. These fragment peaks can give information about the structure of the compound.

Even with simple compounds, it is often impossible to identify every peak in a mass spectrum. However, there are a number of common peaks that can be identified. Common peaks for fragment ions are shown in Table 1.

$\frac{m}{z}$ value	Possible identity of fragment ion
15	CH_3^+
17	OH^+ (from alcohols)
29	$C_2H_5^+$
43	$C_3H_7^+$
57	$C_4H_9^+$

Table 1 *m/z* values for common fragment ions.

LEARNING TIP

It is important to realise that some fragments may have the same *m/z* values as listed in Table 1 – but they won't necessarily be due to the fragments listed. For example, if a molecule has a fragment peak at 29, but couldn't possibly have a $C_2H_5^+$ fragment, then it must be due to a different fragment.

Identification of organic structures

As well as being used to identify the relative molecular mass of a compound, a mass spectrum can also be used to work out some of the molecules' structural detail. An unknown alkane produced the mass spectrum shown in Figure 1. Some of the peaks have been labelled with the *m/z* ratio and with possible fragments.

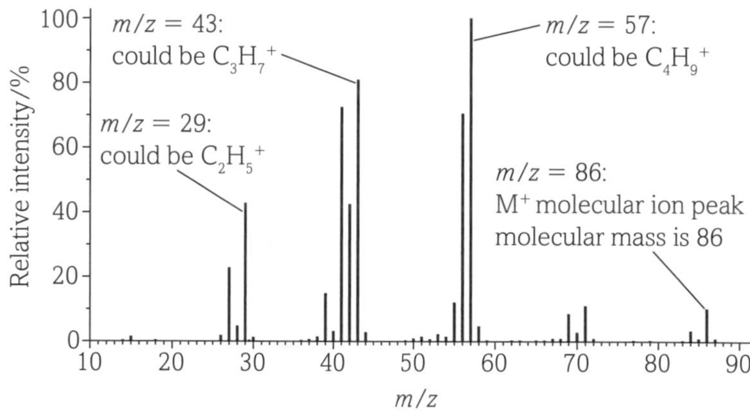

Figure 1 Mass spectrum for an unknown alkane.

The mass spectrum shown in Figure 1 was produced from hexane. The equations in Figure 2 illustrate how the molecular ion could fragmented to form fragment ions with *m/z* values of 57 and 43.

[H–C(H)(H)–C(H)(H)–C(H)(H)–C(H)(H)–C(H)(H)–C(H)(H)–H]$^+$ (bond breaks between the fourth and fifth carbon) → $CH_3CH_2CH_2CH_2^+$ + C_2H_5

fragment ion $m/z = 57$ — Detected and shown as peak in spectrum. — neutral species

[H–C(H)(H)–C(H)(H)–C(H)(H)–C(H)(H)–C(H)(H)–C(H)(H)–H]$^+$ (bond breaks between the third and fourth carbon) → $CH_3CH_2CH_2^+$ + C_3H_7

fragment ion $m/z = 43$ — Detected and shown as peak in spectrum. — neutral species

Figure 2 Equations explaining how some of the peaks arise in the mass spectrum for hexane.

LEARNING TIP

It is a good habit to make notes on spectra as you analyse them. Remember to explain that you have done this in any questions about spectra, so it is obvious where your answers are.

Questions

1. The spectrum in Figure 3 was produced from the alkene called pent-1-ene.

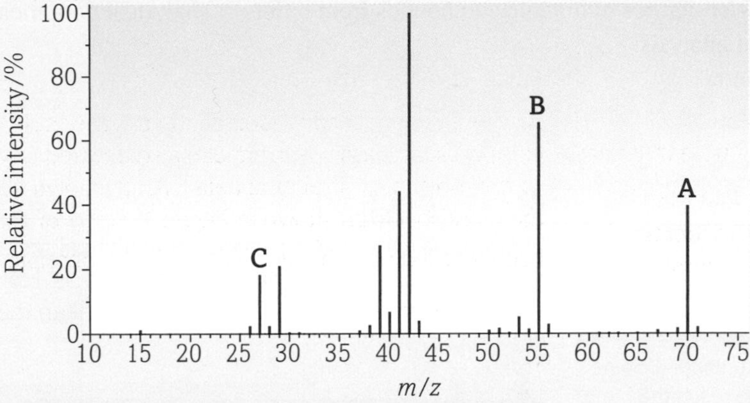

Figure 3 Mass spectrum of pent-1-ene.

Suggest the ions responsible for the peaks labelled A, B and C.

2. Figure 4 shows the mass spectrum of an alkane.

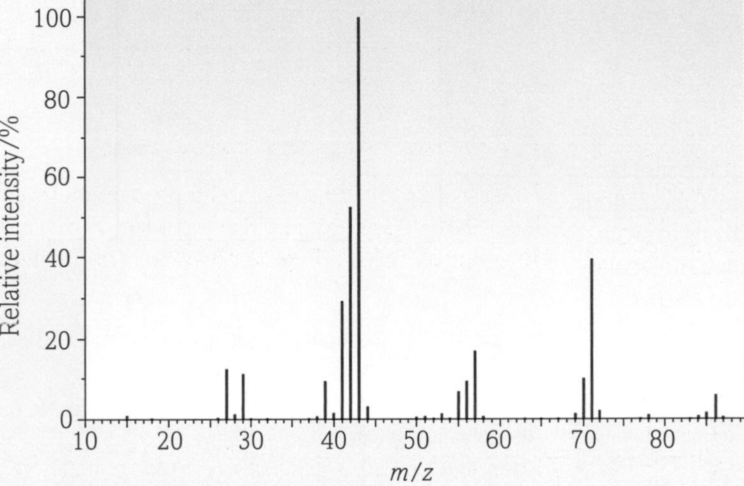

Figure 4 Mass spectrum of an alkane.

(a) Use the spectrum to identify the molecular ion peak, and therefore the relative molecular mass and the molecular formula of the compound.

(b) Molecules of this alkane are branched with one methyl group attached to the main chain at carbon 2. Draw the skeletal formula of the alkane.

(c) Draw structures for possible fragment ions represented by the peaks with *m/z* values of 43, 57 and 71.

12 Combined techniques

By the end of this topic, you should be able to demonstrate and apply your knowledge and understanding of:

- **interpretations and predictions of an infrared spectrum of familiar or unfamiliar substances using supplied data**
- **deduction the structures of organic compounds from different analytical data including:**
  (i) elemental analysis
  (ii) mass spectra
  (iii) IR spectra

Analytical techniques are rarely used singly. Often, data from a number of different techniques are used to determine which compound is being studied.

DID YOU KNOW?

Each analytical technique provides information about a compound. Some techniques may provide different data that can be used to reach the same conclusion.
For example, a carbonyl fragment may be seen on a mass spectrum and also as an absorbance on the infrared spectrum that would correspond to this functional group – this is useful confirmation.
Each technique produces its own part of a jigsaw puzzle. By using data from multiple techniques, a whole picture can be built that allows scientists to identify a compound and determine its structure.

Elemental analysis

Elemental analysis gives scientists information about the numbers and types of element present in a compound sample. Calculations can be carried out using data from composition analysis to work out the empirical formula of a compound. The relative molecular mass of the compound and its molecular formula can also be generated.

WORKED EXAMPLE 1

Consider an organic compound that is found to contain 52.1% carbon, 13.0% hydrogen and the remainder oxygen. The molecular mass of the compound is 46. What is the empirical formula of this compound?

Answer

Step 1: Imagine that you have 100 g of the compound – there would be 52.1 g of carbon, 13.0 g of hydrogen and 100 – (52.1 + 13) = 34.9 g of oxygen

Step 2: C : H : O

$$\frac{52.1}{12.0} : \frac{13.0}{1.0} : \frac{34.9}{16.0}$$

4.3 : 13.0 : 2.2

Step 3: $$\frac{4.3}{2.2} : \frac{13.0}{2.2} : \frac{2.2}{2.2}$$

2.0 : 5.9 : 1

Step 4: 2 : 6 : 1 (round for whole numbers)

Step 5: The empirical formula is C_2H_6O

Mass spectra

Using mass spectra is a destructive analytical technique because the sample cannot be reclaimed after analysis. However, it uses very small quantities and it is a relatively cheap method of analysis.

WORKED EXAMPLE 2

An organic compound produced the mass spectrum in Figure 1. Use this and the empirical formula, C_2H_6O, to work out its molecular formula.

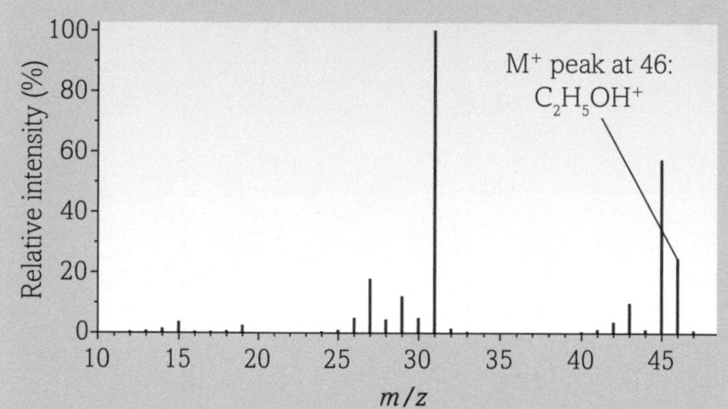

Figure 1 Mass spectrum of substance with C_2H_6O as the empirical formula.

Answer

Look at the mass value of the molecular ion peak, 46.0. The molecular formula can be calculated as follows.

Step 1: (2 × 12.0) + (6 × 1.0) + (1 × 16.0) = 46.0

Step 2: $\frac{46.0}{46.0} = 1$

Step 3: The molecular formula is C_2H_6O

In this example, the molecular formula and empirical formula are the same.

LEARNING TIP

All of the worked examples are data for the same chemical. This shows how each technique gives extra clues to the structure.

IR spectra

WORKED EXAMPLE 3

A compound with molecular formula C_2H_6O has a number of isomers. Give the displayed formula of the isomer that would generate the spectra shown in Figures 2 and 3.

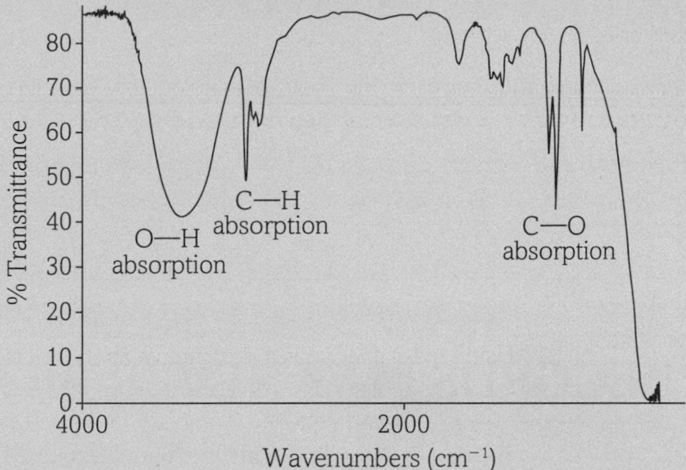

Figure 2 IR spectrum for one isomer.

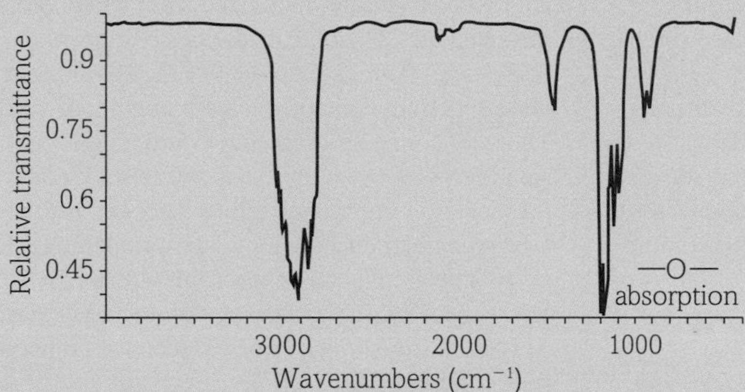

Figure 3 IR spectrum for one isomer.

Answer

Start by drawing the displayed formulae of all the possible structural isomers.

```
    H  H
    |  |
H—C—C—O—H
    |  |
    H  H
```

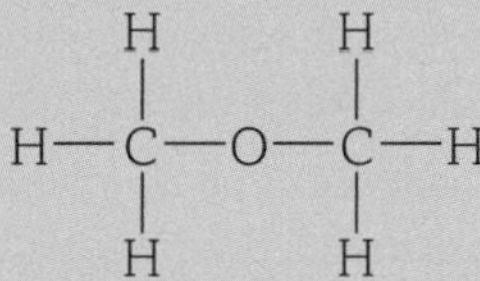

Figure 4 Displayed formula of ethanol.

Figure 5 Displayed formula of dimethylether.

Use the structures to consider the key features that differentiate the structures. You should then use the OCR Chemistry data sheet to suggest the approximate absorptions that would be expected in the infrared spectra.

- For ethanol, there should be an –OH broad absorption at 3200–3550 cm^{-1}.
- For dimethylether, there should be an –O– absorption at 1000–1300 cm^{-1}. This could also be due to an –OH group. But this isomer would have no absorption at 3200–3550 cm^{-1}.

So, Figure 2 is the IR spectrum for ethanol and Figure 3 is the IR spectrum for dimethylether.

Questions

1 Suggest how the mass spectrum of dimethylether would be (a) similar and (b) different to the mass spectrum of ethanol.

2 Explain why scientists often use more than one analytical technique.

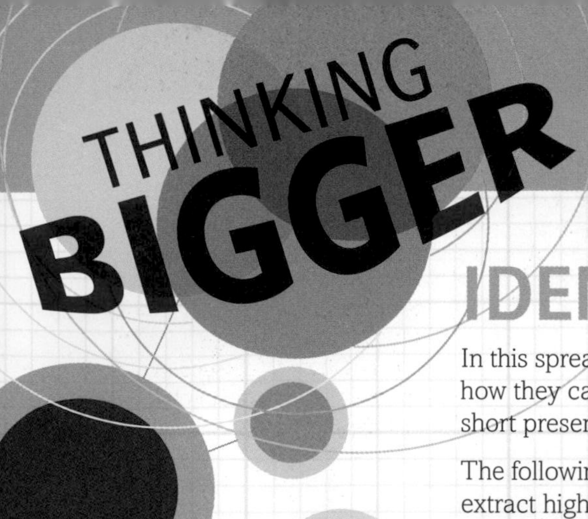

IDENTIFYING STEROIDS

In this spread you will interpret organic structures and consider various analytical techniques and how they can be used to identify different steroids. The follow-up activity will enable you to give a short presentation of the analysis of a molecule chosen from an online database.

The following two extracts come from an article entitled 'Five rings good four rings bad'. The first extract highlights the problem of drug misuse in sport and the second shows how analytical chemists are involved in drug detection.

WHY SOME ATHLETIC RECORDS MAY NEVER BE BEATEN

Extract 1

During the 1990s nandrolone became a favourite anabolic steroid for bodybuilders who wanted to build muscle. This steroid is present naturally in minute (ng) quantities and is detected (in the urine) as its breakdown product, 19-norandrosterone.
Several athletes, including the British sprinter Linford Christie have tested positive for nandrolone, as well as footballers Christophe Dugarry and Edgar Davids, the Czech tennis star Petr Korda and Pakistani Test cricketer Shoaib Akhtar. In some cases the levels of norandrosterone were up to 50 times the maximum 'natural' amount.

In a surprising development, scientists at Aberdeen University found that a combination of allowed dietary supplements, such as creatine, together with hard cardiovascular exercise can lead to norandrosterone levels over five times the legal maximum. It is unlikely, however, that this could account for all the cases.

As chemists got better at identifying steroids, people sought new ways around this. Patrick Arnold, an American chemistry graduate and a bodybuilder, studied the steroid literature, focusing on molecules that had been reported in the past but which were never commercialised. He became involved with the Bay Area Laboratory Cooperative (BALCO), producing 'designer' anabolic steroids that had never come to market and were not covered by routine drug testing. The three main molecules were desoxymethyltestosterone (Madol); norbolethone and tetrahydrogestrinone (THG). Madol and norbolethone had been synthesised in the 1960s and investigated, but neither had been marketed. THG had never been made before.

Extract 2: How Don Caitlin's team cracked THG

On 13 June 2003, Don Catlin received a methanolic solution of an unknown steroid, recovered from a hypodermic syringe. He ran standard GC-MS tests on the solution, and synthesised several derivatives. Attempts to identify the steroid failed because the mass spectrum contained a large number of unidentifiable peaks. The only compound that they could identify at this stage was a small amount of another anabolic steroid, norbolethone, evidently present as an impurity. Catlin suspected that the 'unknown' shared a common carbon skeleton with norbolethone. However, they noted a peak in its mass spectrum with $m/z = 312$, and thought this was the molecular ion. Accurate mass measurement gave 312.2080, from which they deduced the compound had the molecular formula $C_{21}H_{28}O_2$.

When they compared the mass spectrum of the unknown with other steroids, it became clear that it shared features with gestrinone and trenbolone. All three compounds had the same fragments with m/z values at 211 and below present, so Catlin deduced that they contained the same A, B and C rings.

Furthermore, when the MS of the unknown was compared with gestrinone, the fragments with m/z above 240 occur 4 Da higher in the unknown, suggesting that it was gestrinone with four additional hydrogen atoms. A possibility was that the terminal alkyne group in gestrinone had been reduced to an ethyl group.

Having tentatively identified the unknown steroid as tetrahydrogestrinone, the team then prepared an authentic sample of THG by catalytic hydrogenation of gestrinone. This required careful control of conditions (0 °C) to prevent hydrogenation of C=C double bonds (see equation).The retention time and mass spectra of the synthetic THG matched the unknown material exactly.

http://www.rsc.org/education/eic/issues/2010Mar/FiveRingsGoodFourRingsBad.asp

Where else will I encounter these themes?

1.1 | 2.1 | 2.2

The different groups of questions below will ask you to think about this article in different ways, including considering the way the author has presented the science for the audience, the writing itself, as well as the ways in which the article is linked to science you have studied in this topic and that you will go on to study later, as well as more general ideas from chemistry and other sciences you may be studying.

The first question is about the nature of the scientific writing itself rather than the science being communicated.

DID YOU KNOW?

The test for human growth hormone was not introduced until 1985. Two years earlier, the Czechoslovakian Jarmila Kratochvilova set an 800 m record that still stands. It was 1:53.28, and is almost three seconds quicker than the best Kelly Holmes managed to run.

1. **Who do you think is the intended audience for this article?**
2. **Why do you think value judgements are avoided by the author even though the article considers a very emotive issue?**

Questions 3–6 ask questions about the chemistry in, or connected to, this article. The timeline below tells you about other chapters in this book that are relevant to these questions. You may like to review these sections before giving your answers and attempting the activity..

3. **Calculate the relative molecular masses of the molecules Nandrolone and 19-Norandrosterone to 1 d.p. (see Figure 1).**

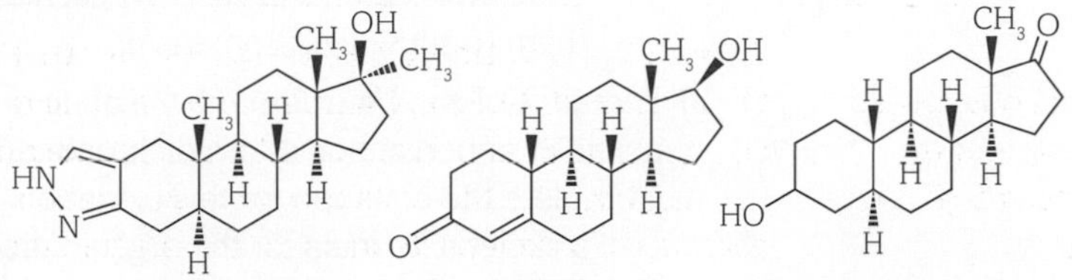

(8) Stanozolol *(9)* Nandrolone *(10)* 19-Norandrosterone

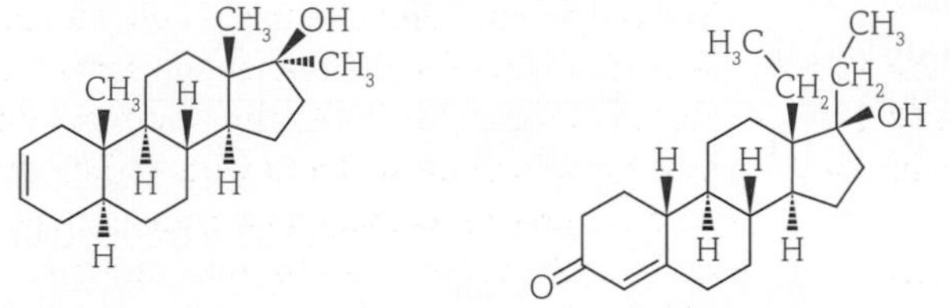

(11) Desoxymethyltestosterone (Madol) *(12)* Norbolethone

Figure 1

4. **The analytical techniques of IR spectroscopy and mass spectrometry can be used in identifying unknown molecules. Suggest which of these two techniques would be more useful in distinguishing the two molecules in question 3. You should be prepared to justify your choice.**
5. **In the following phrase – 'In some cases the levels of norandrosterone were up to 50 times the maximum 'natural' amount.' – the word natural appears in inverted commas. Why do you think the authors chose to do this?**

Activity

The free to access database at http://webbook.nist.gov/chemistry/ allows you to search for a range of organic compounds and related data.

– use the database to find the mass spectrum of one of the following compounds:

(i) chlorobenzene (ii) bromoethane (iii) ethylamine (iv) cyclohexane.

Prepare a 3–5 minute presentation showing the mass spectrum of your chosen molecule and identifying the most important peaks. Your presentation should include:

– a picture of the mass spectrum of your chosen molecule

– an identification of the main fragment and isotopic abundance peaks of your molecule explaining how each peak is formed.

4.2 Practice questions

1. Which of the following statements, about possible unsaturated molecules with the molecular formula C_4H_8, is true? [1]
   A. They can form position isomers only.
   B. They can form stereoisomers only.
   C. They can form stereo and positional isomers.
   D. They cannot form isomers.

2. Analysis of compound Q, containing carbon and hydrogen only, shows it to have an empirical formula of CH_2. Compound Q does not decolorise bromine water. What could compound Q be? [1]
   A. Hexane
   B. *E*-Hex-1-ene
   C. *Z*-Hex-1-ene
   D. Cyclohexane

3. Compound H has a boiling point of 195 °C, is only slightly soluble in water, and will turn acidified potassium dichromate (VI) from orange to green when heated under reflux. What could compound H be? [1]
   A. Octan-1-ol
   B. Propan-2-ol
   C. Hexan-3-one
   D. 2-methylheptan-2-ol

4. Mass spectrum analysis of cyclohexanol gives a very small M+1 peak due to the natural presence of ^{13}C isotopes. What will the m/z value be for this ion? [1]
   A. 101
   B. 102
   C. 95
   D. 96

[Total: 4]

5. Pentan-2-ol, shown below, is a secondary alcohol.

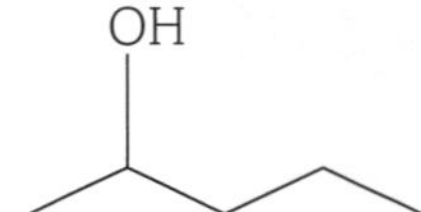

(a) Pentan-2-ol can be converted into three alkenes, **A**, **B** and **C**, by the elimination of water.
- Two of the alkenes, **A** and **B**, are stereoisomers.
- The third alkene, **C**, is a structural isomer of both **A** and **B**.

This elimination often uses a catalyst.
(i) What is a suitable catalyst for this reaction? [1]
(ii) Construct an equation, using molecular formulae, for the elimination of water from pentan-2-ol. [1]
(iii) Explain what is meant by the terms *structural isomers* and *stereoisomers.* [4]
(iv) Draw the structures of:
- stereoisomer **A** [1]
- stereoisomer **B** [1]
- isomer **C**. [1]

(v) Stereoisomers **A** and **B** show *E/Z* isomerism. State **two** features of these molecules that enable them to show *E/Z* isomerism. [2]

(b) Pentan-2-ol can be oxidised by heating under reflux with acidified aqueous potassium dichromate(VI).
Complete the equation for this oxidation.
Use a skeletal formula for the organic product.
Use [O] to represent the oxidising agent. [2]

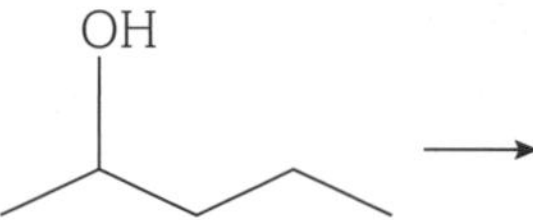

(c) Pentan-1-ol can also be oxidised but it gives two different products.
Complete the flowchart below to show the structures of the two organic products formed. [2]

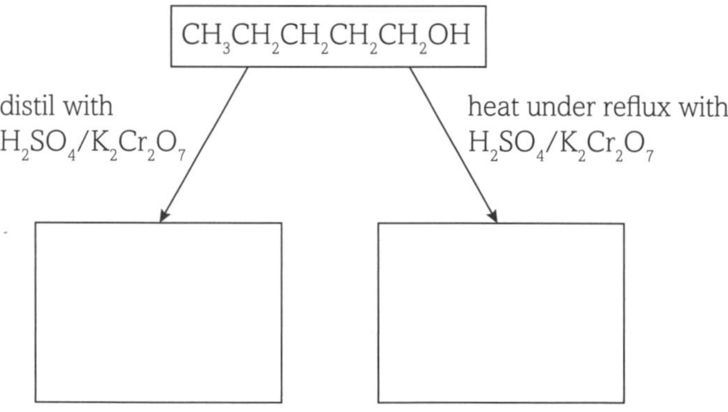

[Total: 15]
[Q3, F322 Jan 2013]

6. Outline a method that you could use to convert cyclohexanol into cyclohexene. Your method should include:
   - the apparatus, reagents and conditions needed for the conversion
   - how you would go about purifying the product and drying it
   - which instrumental technique you would use to check that your product is pure. [6]

[Total: 6]

7. This question concerns the haloalkane 2-chlorobutane.
   (a) Draw out the displayed formula of 2-chlorobutane. 2-chlorobutane reacts with potassium hydroxide in ethanolic solution by a nucleophilic substitution mechanism. [2]
   (b) Draw the mechanism for this reaction clearly indicating the nucleophile and naming the organic product formed. [4]
   (c) Explain why the organic product can be categorised as a **secondary alcohol**. [2]
   (d) Suggest how the rate of this substitution reaction can be followed. [2]
   (e) Write an equation for the complete combustion of the secondary alcohol you have identified above. [2]

[Total: 12]

8. (a) Suggest how infrared spectroscopy could be used to assess the purity of cyclohexene. Give details of particular absorbance frequencies that you would consider. [4]
   (b) The boiling points of cyclohexanol and cyclohexene are given in the table below.

Substance	Molar mass in $g\,mol^{-1}$	Boiling point in °C
Cyclohexene	82.2	83
Cyclohexanol	100.2	161

   Explain the differences in boiling points using the data in the table above and your knowledge of the different types of intermolecular force. [4]
   (c) The mass spectrum of cyclohexene gives a peak that corresponds to a $m/z = 83$ value. Explain the existence of this peak. What information does it suggest about the molecule? [4]

[Total: 12]

Maths skills

In order to be able to develop your skills, knowledge and understanding in Chemistry, you will need to have developed your mathematical skills in a number of key areas. This section gives more explanation and examples of some key mathematical concepts you need to understand. Further examples relevant to your AS/A level Chemistry studies are given throughout the book.

Arithmetic and numerical computation

Using standard form

Dealing with very large or small numbers can be difficult. For example, Avogadro's constant is an important value in Chemistry which is approximately equal to 602 000 000 000 000 000 000 000 mol^{-1}. To make such numbers easier to handle, we can write them in the format $a \times 10^b$. This is called standard form. Using standard form, Avogadro's constant can be written as $6.02 \times 10^{23}\ mol^{-1}$.

To change a number from decimal form to standard form:

- Count the number of positions you need to move the decimal point by until it is directly to the right of the first number which is not zero.
- This number is the index number that tells you how many multiples of 10 you need. If the original number was a decimal, your index number must be negative.

Here are some examples:

Decimal notation	Standard form notation
0.000 000 012	1.2×10^{-8}
15	1.5×10^{1}
1000	1×10^{3}
3 700 000	3.7×10^{6}

Using ratios, fractions and percentages

Ratios, fractions and percentages help you to express one quantity in relation to another with precision. Ratios compare like quantities using the same units. Fractions and percentages are important mathematical tools for calculating proportions.

Ratios

A ratio is used to compare quantities. You can simplify ratios by dividing each side by a common factor. For example 12 : 4 can be simplified to 3 : 1 by dividing each side by 4.

EXAMPLE

Divide 180 into the ratio 3 : 2

Our strategy is to work out the total number of parts then divide 180 by the number of parts to find the value of one part.

total number of parts = 3 + 2 = 5

value of one part = 180 ÷ 5 = 36

answer = 3 × 36 : 2 × 36 = 72 : 108

Check your answer by making sure the parts add up to 180 : 72 + 108 = 180

EXAMPLE

An excess of magnesium is added to 0.2 dm^3 of 1 mol dm^{-3} dilute hydrochloric acid. The equation for the reaction is:

$$Mg + 2HCl \rightarrow MgCl_2 + H_2$$

How many moles of hydrogen are formed?

Since there is an excess of magnesium, we know that all of the hydrochloric acid will react. We can use the following equation to calculate the number of moles of hydrochloric acid:

$$\text{concentration in mol dm}^{-3} = \frac{\text{amount in moles}}{\text{volume in dm}^{-3}}$$

$$1\ \text{mol dm}^{-3} = \frac{\text{amount in moles of HCl}}{0.2\ \text{dm}^{-3}}$$

amount of HCl reacted = 0.2 mol

The ratio for $HCl : H_2$ in this reaction is 2 : 1, so when two moles of HCl react, one mole of H_2 is formed.

$$\begin{aligned}\text{number of moles of H}_2\text{ formed} &= 0.5 \times \text{number of moles of HCl reacted}\\ &= 0.5 \times 0.2\ \text{mol}\\ &= 0.1\ \text{mol}\end{aligned}$$

Fractions

When using fractions, make sure you know the key strategies for the four operators:

To add or subtract fractions, find the lowest common multiple (LCM) and then use the golden rule of fractions. The golden rule states that a fraction remains unchanged if the numerator and denominator are multiplied or divided by the same number.

EXAMPLE

$\frac{1}{2} + \frac{1}{5} = \frac{5}{10} + \frac{2}{10} = \frac{7}{10}$

To multiply fractions together, simply multiply the numerators together and multiply the denominators together.

EXAMPLE

$\frac{2}{7} \times \frac{4}{9} = \frac{8}{63}$

To divide fractions, simply invert or flip the second fraction and multiply.

EXAMPLE

$\frac{2}{3} \div \frac{7}{9} = \frac{2}{3} \times \frac{9}{7} = \frac{18}{21} = \frac{6}{7}$

Percentages

When using percentages, it is useful to recall the different types of percentage questions.

To increase a value by a given percentage, use a percentage multiplier.

EXAMPLE

Increase 30 mg by 23%.

If we increase by 23%, our new value will be 123% of the original value. We therefore multiply by 1.23.

answer = 30 × 1.23 = 36.9 mg

To decrease a value by a given percentage, you need to focus on the part that is left over after the decrease.

EXAMPLE

Decrease 30 mg by 23%.

If we decrease by 23%, our new value will be 100 − 23 = 77% of the original value. We therefore multiply by 0.77.

answer = 30 × 0.77 = 23.1 mg

To calculate a percentage increase, use the following equation:

$$\text{percentage change} = \frac{\text{difference between values}}{\text{original value}} \times 100$$

To calculate percentage decrease, use the same equation but remember that your answer should be negative.

EXAMPLE

The volume of a solution increased from 40 ml to 50 ml. Calculate the percentage increase.

change in volume = 10 ml

$$\text{percentage increase} = \frac{10}{40} \times 100 = 25\%.$$

You often have to combine these percentage calculations with balanced equations and process yields to work out the compositions of reactions.

EXAMPLE

In a reaction with known yield of 88% and 28 g of product is required, calculate the amount of reactant you will need.

$$\text{amount of reactant} = \frac{28}{0.88} = 31.8\text{ g}$$

Reaction yield calculations combine with balanced equations.

EXAMPLE

$CaCO_3 \rightarrow CaO + CO_2$

140 g of calcium carbonate produces 64 g of calcium oxide. Calculate the percentage yield.

Molar mass of $CaCO_3$ = 100 g mol^{-1}, so 140 g = 1.400 moles of reactant.

Molar mass of CaO = 56 g mol^{-1}, so 64 g = 1.143 moles of product.

$$\text{so, percentage yield} = \frac{1.143}{1.400} = 0.816 = 81.6\%$$

Algebra

Changing the subject of an equation

It can be very helpful to rearrange an equation to express the variable that you are interested in in terms of other variables. Always remember that any operation that you apply to one side of the equation must also be applied to the other side.

EXAMPLE

A sample of 2.5 mol of a substance has a mass of 37.5 g. What is the molar mass of the substance?

The equation for calculating moles is $n = \frac{m}{M}$ where m = mass in grams, M = molar mass and n = amount of substance in moles.

If we wished to rearrange this equation to make M the subject, we would first multiply each side by M to obtain:

$$nM = m$$

Now to obtain the formula in terms of n, we divide each side by M to obtain:

$$M = \frac{m}{n}$$

We can now simply substitute in the values for the question:

$m = 37.5$ g

$n = 2.5$ mol

$$M = \frac{m}{n} = \frac{37.5\text{ g}}{2.5\text{ mol}} = 15\text{ g mol}^{-1}$$

Handling data

Using significant figures

Often when you do a calculation, your answer will have many more figures than you need. Using an appropriate number of significant figures will help you to interpret results in a meaningful way.

Remember the 'rules' for significant figures:

- The first significant figure is the first figure which is not zero.
- Digits 1–9 are always significant.
- Zeros which come after the first significant figure are significant unless the number has already been rounded.

Here are some examples:

45 678	5×10^4	4.6×10^4	4.57×10^4
45 000	5×10^4	4.5×10^4	4.50×10^4
0.002 755	3×10^{-3}	2.8×10^{-3}	2.76×10^{-3}

Applying your skills

You will often find that you need to use more than one maths technique to answer a question. In this section, we will look at four example questions and consider which maths skills are required and how to apply them.

EXAMPLE

Gallium, $A_r = 69.73$, has two stable isotopes, both of which have been used in medicine as well as research by physicists. The scientists who first analysed samples of gallium found that they consisted of 60.11% $Ga^{68.93}$ as well as one other isotope.

Determine the mass number of the other isotope in the sample of gallium.

In order to answer this question, we must substitute the numerical values provided above into the correct equation. The equation needed is that concerning the calculation of relative atomic mass, A_r, and is shown below. This is the hardest equation to work with in A level Chemistry, as its rearrangement requires several steps. However, if a student can rearrange this, they can rearrange anything at this level.

So the first step is recognising that we need to use this equation; we know this because the question asks us to deal with isotopes and this equation deals exclusively with isotopes. Next we look at the numerical values given in the question and start to substitute them into their correct place.

$$A_r = \frac{((\text{mass number A} \times \% \text{ abundance A}) + (\text{mass number B} \times \% \text{ abundance B}))}{100}$$

Therefore we can begin to substitute in the numerical values.

$$69.73 = \frac{((68.93 \times 60.11\%) + (\text{mass number B} \times \% \text{ abundance B}))}{100}$$

Additional information needed is the % abundance of *B*. This can be found from the information in the question.

$$\% \text{ abundance of } A + \% \text{ abundance of } B = 100\%$$

and so rearranging this equation gives

$$\% \text{ abundance of } B = 100 - \% \text{ abundance of A}$$

$$100\% - 60.11\% = 39.89\%$$

We now have all of the information needed for the complete equation to be rearranged.

$$69.73 = \frac{((68.93 \times 60.11\%) + (\text{mass number B} \times 39.89\%))}{100}$$

We must now change the subject of the equation and then solve it to give the mass number of isotope *B*. To change the subject of the equation we move everything on that side of the equation, which we have a numerical value for, to the opposite side. When we do this we reverse the operation (e.g. if on the right hand side it was divided by 100, on the left hand side it will be multiplied by 100; where before we had a plus, now we have a minus etc.).
This gives us the following rearranged equation:

$$\text{mass number } B = \frac{((69.73 \times 100\%) - (68.93 \times 60.11\%))}{39.89\%}$$

$$= 70.94$$

We can check theoretically what this answer tells us and compare it to the information given in the question. The A_r of Ga is 69.73, which is closer to $Ga^{68.93}$ than $Ga^{70.94}$. We know that this is the case already because $Ga^{68.93}$ has a % abundance of 60.11% compared to 39.89% of $Ga^{70.94}$. This means that there is more $Ga^{68.93}$. The answer to the equation tells us the mass number of the other isotope.

We can also check approximately that this is correct. Imagine there was 50% of each isotope of Ga. This would give an A_r of 69.94. You can calculate this easily using the relative atomic mass equation involving isotopes.

$$A_r = \frac{((68.93 \times 50\%) + (70.94 \times 50\%))}{100}$$

$$= 69.94$$

Alternatively we can calculate this without using the above equation but just by finding the mid-point between $Ga^{68.93}$ and $Ga^{70.94}$.

$$\frac{(70.94 - 68.93)}{2} = 1.005$$

so the mid-point is $68.93 + 1.005 = 69.94$

The relative atomic mass, A_r, given in the original question was 69.73, which is lower than the value obtained if the % abundances were 50%. Therefore we know we are on the right lines and have the correct answer.

EXAMPLE

10 g of Potassium was added to excess water, producing potassium hydroxide and hydrogen gas, the latter of which was set alight due to the exothermic nature of the reaction.
Calculate the volume of $H_2(g)$ produced in dm^3.

This type of question is very common. The key is to be familiar with the three different types of moles equations and when to use them. You will also usually have to work with ratios in order to find the correct number of moles of a reactant or product, and this tends to be where lots of mistakes are made.
The relevance of this question to the wider world is that, as chemists, we must know how much product we are going to make. This makes sense both for economic reasons, and in the case above where we are producing a highly flammable gas, for our safety as well!

Moles equations

(a) $\text{moles} = \frac{\text{mass}}{\text{molar mass}}$
This equation tends to be used when dealing with solids or masses.

(b) moles = volume × concentration.
This equation can be used when dealing with liquids or titrations.

(c) $\text{moles} = \text{volume in } \frac{dm^3}{24} \text{ or volume in } \frac{cm^3}{24\,000}$.
This equation is used for gases.
If you are confident with mathematics then you can also combine the above equations and rearrange them to solve the missing value.

$$\frac{\text{mass}}{\text{molar mass}} = \text{volume} \times \text{concentration}$$

$$\frac{\text{mass}}{\text{molar mass}} = \text{volume in } \frac{dm^3}{24}$$

Using the information above:

Step 1: Calculate the number of moles of potassium used in the reaction.
In order to answer this question we first need to know which equation to use. We are given a mass, and know we are dealing with solids, so we should use equation (a). Knowing this, it is now a simple case of substituting in the numerical values and solving the equation.

$$\text{moles} = \frac{\text{mass}}{\text{molar mass}}$$

$$\text{moles} = \frac{10\text{ g}}{39\text{ g mol}^{-1}} = 0.256\text{ mol}$$

Step 2: Calculate the number of moles of $H_2(g)$.

To find the number of moles of $H_2(g)$ requires us to work with ratios. First we write the symbol equation and then make sure it is balanced; sometimes this will already be done for you or it could be part of a previous question. Either way, balancing the symbol equation is key to determining the ratio of reactants to products etc. From this, we can now determine the ratio of *K* to H_2, which tells us how many moles of *K* there are compared to H_2.

$$2K(s) + 2H_2O(l) \rightarrow 2KOH(aq) + H_2(g)$$

$K : H_2 = 2 : 1$ therefore 0.256 mol : 0.128 mol

Step 3: Calculate the volume of $H_2(g)$ produced in dm^3.

The final part of the question requires us to calculate the volume of H_2 produced. To do this we must decide which moles equation to use. As we are dealing with gases it leaves us with one obvious choice.

$$\text{moles} = \frac{\text{volume in } dm^3}{24} \text{ or moles} = \frac{\text{volume in } cm^3}{24\,000}$$

Always check what units are required as you could lose valuable marks otherwise (this is usually stated either in the question or next to the space provided to write the answer). This question has asked for the answer to be shown in dm^3. Rearranging this equation, substituting in the numerical values before solving provides us with the answer.

$$\text{volume in } dm^3 = 0.128 \text{ mol} \times 24 = 3.077 \text{ dm}^3$$

It is always good to check your answer following a long question like this. The best way is to work backwards, and make sure we get 10 g of *K* as the end point.

EXAMPLE

In the United Kingdom there are currently 16 operational nuclear reactors located at 9 power stations. However, as the country's energy demands increase this number could rise. Recent surveys have found public opinion is not in favour of this increase, due to risk of leaks and radiation exposure following an explosion. One possible radioactive contaminant following an explosion is radioactive iodine, I^*, which is taken up by a person's thyroid gland. To counter this, potassium iodide, KI, is given to patients. This blocks the uptake of radioactive iodine by the thyroid gland by swamping the body with non-radioactive iodine instead.

The current recommended dose of iodine, I_2, is 130 mg. 1 mg = 1×10^{-3} g Calculate the mass of potassium iodide required to obtain the recommended dose of iodine. Show your answer in mg.

This style of question requires you to think carefully about what the question is asking. Whilst there are only two steps to finding the correct answer, this can still prove troublesome. This question is made harder by the fact that we are using iodine, which is I_2, rather than just I^-. This means the molar mass is $2 \times 126.6 \text{ g mol}^{-1}$.

We must first determine the number of moles of iodine required, using the basic moles equation to do so. We know it is this equation because we are dealing with solids and masses. Therefore the equation to solve is

$$\text{moles} = \frac{\text{mass}}{\text{molar mass}}$$

$$\frac{0.130 \text{ g}}{253.8 \text{ g mol}^{-1}} = 5.122 \times 10^{-4} \text{ mol}$$

If we know how many moles of iodine, I_2, we need, we can work out the mass of potassium iodide, KI, by simply changing the subject of the equation and then solving it.

$$\text{mass} = \text{moles} \times \text{molar mass}$$

However, the question has asked for the mass to be shown in units of mg. This means we must multiply the answer by 1000. When answering any question is it better to leave the rounding to the end, in order to avoid rounding errors.

$$(5.122 \times 10^{-4} \text{ mol} \times 165.9 \text{ g mol}^{-1}) \times 1000 = 84.98 \text{ mg}$$

EXAMPLE

Phosphoric acid is known as a polyprotic acid, which means it can donate more than one proton. Polyprotic acids can be represented as HxA. In this formula, x represents the number of H^+ ions that can be replaced by metal ions to form salts, with 'A' representing 'acid'.

Two students carried out a titration to find the value of x in the simplified formula of phosphoric acid, HxA. In the titration, 25.00 cm^3 of 0.05 mol dm^{-3} of phosphoric acid, HxA, reacted exactly with 18.75 cm^3 of 0.200 mol dm^{-3} sodium hydroxide, NaOH. A solution of sodium phosphate was produced.

(a) Calculate the amount, in mol, of HxA used.
(b) Calculate the amount, in mol, of NaOH used.
(c) Deduce the value for x in the simplified formula for phosphoric acid, HxA.

This is another question that is actually quite easy to answer but can lead to confusion. The first two parts of the question simply require you to calculate the number of moles of acid and base, respectively, before using the ratio to determine the value of *x*.

(a) moles = volume × concentration
$0.025 \text{ dm}^3 \times 0.05 \text{ mol dm}^{-3} = 1.25 \times 10^{-3}$ mol

(b) $0.01875 \text{ dm}^3 \times 0.200 \text{ mol dm}^{-3} = 3.75 \times 10^{-3}$ mol

(c) In order to answer this question we must find the ratio of NaOH to HxA. This is where mistakes are made, as some students divide the numbers the wrong way round. It is best to always divide the big number by the small number, as this will tell you how much larger the first number is compared to the second.

1.25×10^{-3} mol : 3.75×10^{-3} mol

$x = 3.75 \times 10^{-3} \text{ mol} / 1.25 \times 10^{-3}$ mol

answer = 3

This means that HxA is H_3A. This is correct as the actual formula of phosphoric acid is H_3PO_4. There is a special importance in this titration, in that chemists often want to know exactly how much reagent needs to be combined in order for there to be nothing left over. Knowing that phosphoric acid has 3 protons which it can donate is important, because we now know how many of its protons will react in a reaction.

Preparing for your exams

Introduction

The way that you are assessed will depend on whether you are studying for the AS or the A level qualification. Here are some key differences:

- AS students will sit two 70 mark exam papers, each worth 50% of the AS level. Each 1 hr 30 minute paper could have content from modules 1 to 4.
- A level students will sit 3 exam papers, the first with content from modules 1, 2, 3 and 5, the second with content from modules 1, 2, 4 and 6, and the final paper covering content from all the modules studied.
- A level students will also have their competence in key practical skills assessed by their teacher in order to gain the Science Practical Endorsement. The endorsement will not alter the overall grade but if you pass the endorsement this will be recorded on your certificate and can be used by universities as part of their conditional offer of a place on a course.

The tables below give details of the exam papers for each qualification:

AS exam papers

AS Chemistry					
ASSESSMENT OVERVIEW					
Paper			Marks	Duration	Weighting
Paper 1	Breadth in chemistry		70	1 hr 30 mins	50%
	Section A	Multiple choice	20		
	Section B	Structured questions and extended response questions covering theory and practical skills	50		
Paper 2	Depth in chemistry		70	1 hr 30 mins	50%
	Structured questions and extended response questions, covering theory and practical skills		70		

A level exam papers

A level Chemistry					
ASSESSMENT OVERVIEW					
Paper			Marks	Duration	Weighting
Paper 1	Periodic table, elements and physical chemistry		100	2 hr 15 mins	37%
	Section A	Multiple choice	15		
	Section B	Structured questions and extended response questions covering theory and practical skills	85		
Paper 2	Synthesis and analytical techniques		100	2 hr 15 mins	37%
	Section A	Multiple choice	15		
	Section B	Structured questions and extended response questions covering theory and practical skills	85		
Paper 3	Unified chemistry		70	1 hr 30 mins	26%
	Structured questions and extended response questions covering theory and practical skills		70		
Non-exam assessment	Practical Endorsement for chemistry		Pass/Fail	Non-exam assessment	Reported separately
	Teacher-assessed component common to Chemistry A and Chemistry B (Salters). Candidates complete a minimum of 12 practical activities to demonstrate practical competence. Performance reported separately to the A Level grade. Moderation details still to be confirmed by Ofqual at the time of going to press		0		

Exam strategy

Arrive equipped

Make sure that you have all of the correct equipment needed for your exam. In a transparent bag or pencil case you should have at least:

- pen (black, ink or ball-point)
- pencil (HB)
- 30 cm rule
- eraser which will not smudge
- scientific calculator

Ensure your answers can be read

Your handwriting needs to be clear so that the examiner can easily read it. Even when you are in a hurry make sure that the key words are easy to read.

Plan your time

Think about how many marks are available on the paper and how many minutes you have to complete it. This will give you an idea of how long to spend on each question. Be sure to leave some time at the end of the exam for checking answers. Use a rough guide of a minute per mark, but short answers and multiple choice questions may be quicker. Longer answers may require more time.

Understand the question

Always read the question carefully and spend a few moments working out what you are being asked to do. The command word used will give you an indication of what is required in your answer.

Be scientific and accurate, even when writing longer answers. Use technical terms you've been taught. Always show your working for any calculations. Marks may be available for individual steps, units, or recalling a key formula, not just for the final answer. Remember that for a calculation you are often given marks for the correct technique, even if you have a numerical error in your calculations.

Plan your answer

For the extended response questions, marks will be awarded for your ability to structure your answer logically. You should show how the points are related or follow on from each other. Read the question carefully and even plan your answer with a brief flow chart before starting the full answer.

Make the most of graphs and diagrams

Diagrams and sketch graphs can be used to gain marks. Often they are quicker and can give the same marking points as a prose explanation. However, they must be carefully drawn.

If you are asked to read a graph, pay attention to the labels and numbers on the x and y axis. Remember that each axis is a number line.

If asked to draw or sketch a graph, always ensure you use a sensible scale and label both axes with quantities and units. If plotting a graph, use a pencil and draw small crosses for the points.

Diagrams must always be neat, clear and fully labelled.

Check your answers

For extended response questions, check that you have written the same number of separate points as the number of marks available. For calculations, read through each stage of your working. Substituting your final answer into the original question can be a simple way of checking that the final answer is correct. Another simple strategy is to consider whether the answer seems sensible. Always check the units and the degree of accuracy that you report your answer in.

Sample AS questions – multiple choice

1. How many electrons are there in a $^{24}_{12}Mg^{2+}$ ion?

A 10

B 12

C 14

D 22

[Q1, H156/02 sample paper 2014]

Notice that magnesium is a positive ion. This means there are fewer electrons than in the neutral element. Look carefully at the symbol and make sure that you know which is the mass number and which is the atomic number. You can use the key on the data sheet to help you classify the mass and atomic number.

There is only one mark, so only one answer should be given. Determine which number is the atomic number and then take two electrons from this to calculate the electrons in the ion.

Question analysis

- Multiple choice questions look easy to answer. You need to think carefully about your choice. You may find it helpful to jot down some working out.
- Timing is very important. For calculation questions, you do not need to ensure you have calculated all the way to the final answer, you just need to choose the response which is closest to your estimation, with the correct sign and units.
- The correct answer will be given alongside some other answers known as distractors. They could be the same numerical answer but different units, and you must be careful to select the correct one.
- If you change your mind you should follow the rubrics to make your correction.

Sample student answer

B

Verdict

This is a weak answer. The student has correctly recognised that 12 is the atomic number which is the number of protons in an element. In a neutral atom, this is the same as the number of electrons. However, the student has not taken in to account that the magnesium is a 2+ ion. The correct answer is A.

If you have time at the end of the paper, do check your multiple choice answers. Even if you are not sure, choose one answer and you have a 25% chance of gaining a mark. Do not look for patterns in the answers, there might be a few questions all with the answer A.

Sample AS questions – structured

2. Carbon monoxide can be made in the laboratory by heating a mixture of zinc metal and calcium carbonate. An equation for this reaction is shown below.

$Zn(s) + CaCO_3(s) \rightarrow ZnO(s) + CaO(s) + CO(g)$

(a) A student carried out the reaction of zinc (Zn) and calcium carbonate ($CaCO_3$) in a fume cupboard. The student measured the volume of gas produced.

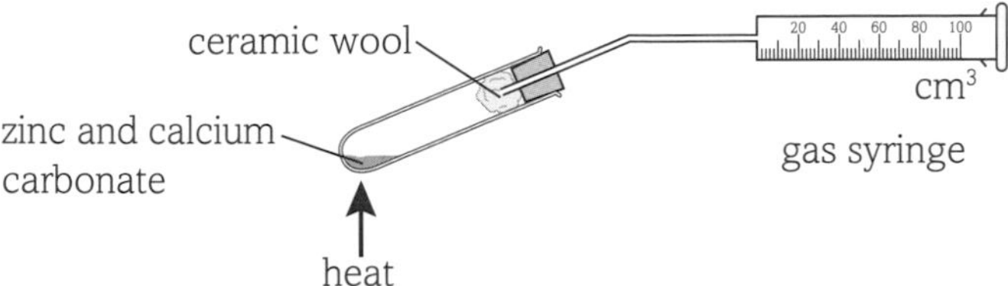

A mixture containing 0.27g of powdered zinc and 0.38 g of powdered $CaCO_3$ was heated strongly for two minutes. The volume of gas collected in the 100 cm^3 syringe was measured. The experiment was then repeated.

(i) Calculate the maximum volume of carbon monoxide, measured at room temperature and pressure, that could be produced by heating this mixture of Zn and $CaCO_3$.

Show **all** your working.

volume of carbon monoxide = ... cm^3

(ii) The student did **not** obtain the volume of gas predicted in **(i)** using this procedure.

Apart from further repeats, suggest **two** improvements to the practical procedure that would allow the student to obtain a more accurate result. [2]

[Q2, H156/02 sample paper 2014]

Notice how the examiner has written some key words in bold. This is to draw your attention to the fact that all the working should be shown, not just the answer that you wish to report, and that two improvements, not more or less, should be detailed.

Question analysis

- Focus on the command word of the question, and ensure you have given the appropriate level of detail. In section i, you are asked to calculate and to show all your working, the units are also given.
- Usually there is a mark for each piece of information. So, if you know a relevant formula or the units an answer should be measured in, state these even if you cannot begin the calculation.
- You answers should be brief and contain key chemistry vocabulary. There is no need for full sentences, brief bullet points or a well labelled diagram can be used to attain full marks.

Sample student answer

(i) The number of moles of $CaCO_3$ is 0.0038 and the number of moles of Zn is 0.0041. So, the limiting reagent is calcium carbonate. From the balanced symbol equation you can see that there would be 0.0038 moles of CO produced.
So, $24{,}000 \times 0.0038 = 91.3\ cm^3$.

(ii) Repeat the experiment ensuring all the apparatus is fully sealed to prevent any loss of gas. Use a gas syringe with more graduations.

Verdict

This is a good answer. In section (i), the student did not show the mole calculations in full, but the reasoning is logical.

In section (ii), the student could have added more detail. This could have been discussing adjustments to the apparatus to ensure the collected gas is being measured at room temperature and pressure. Alternatively the student could have expanded on improving accuracy by eliminating anomalous data and averaging repeated measurements.

Sample AS questions – extended response

3. Students work together in groups to identify four different solutions.

   Each solution contains one of the following compounds:

   - Ammonium sulfate, $(NH_4)_2SO_4$
   - Sodium sulfate, Na_2SO_4
   - Sodium chloride, NaCl
   - Potassium bromide, KBr

   Your group has been provided with a universal indicator paper and the following test reagents:

   - Barium chloride solution
   - Silver nitrate solution
   - Dilute ammonia solution
   - Sodium hydroxide solution

   (a) A student ion your group suggests the following plan:

   - Add about 1 cm depth of each solution into separate test-tubes
   - Add a few drops of barium chloride solution to each test-tube
   - A white precipitate will show which solutions contain sulfate ions
   - Two of the solutions will form a white precipitate.

Describe how you would expand this plan so that all four solutions could be identified using a positive test result.

You should provide all the observations and conclusions that would enable your group to identify all four solutions.

[Q3, H156/02 sample paper 2014]

The command word is describe. This means that a detailed account is needed to show how the reagents listed can be used to identify each chemical.

Question analysis

With longer questions you should plan your answer. Maybe write a list of key points that you want to use. Then number them to make a logical order before writing your answer.

Ensure that you answer the question that is posed. You will use valuable time in expanding your answer outside the question brief and this will not gain you any marks.

Your answer could include well-labelled diagrams.

Sample student answer

The negative ion can be identified by sequencing two tests on the solutions one after the other. Two solutions have already had their negative ions identified as sulfate using the barium chloride solution to each sample solution.

Take a fresh sample of the two samples that did not turn cloudy with barium chloride. Add silver nitrate solution to each test tube. One sample will produce a white precipitate showing that it contains chloride ions – this is the sodium chloride solution. The other sample will produce a cream precipitate showing that it contains bromide ions – this is the potassium bromide solution.

Return to the two sulfate-containing solutions. Take a fresh sample of each into a test tube and add the equivalent amount of dilute sodium hydroxide. Gently heat the solutions in a fume cupboard. The sample that contains ammonium ions will produce ammonia gas, which can be identified as it will turn damp universal indicator paper blue. This sample must be the ammonium sulfate. The other solution will not produce ammonia, so can be identified as the sodium sulfate solution.

Verdict

This is a strong answer as the student has sequenced the tests so that each sample is identified with a positive test result. The answer assumes that the four different solutions were known, but just needed to be allocated to the four samples. If the four samples were completely unknown then further tests would be needed to identify the other metal ions. A flame test would differentiate between sodium and potassium but this would require the solid salt rather than the solution.

Glossary

accuracy: how close to the true value a measurement is.
acid: a chemical that is a proton donor and releases $H^+(aq)$ in solution.
activation energy: the minimum energy required to start a reaction by breaking bonds in the reactants.
aim: identifies the purpose of the investigation. It is a straightforward expression of what the researcher is trying to find out from conducting an investigation. Compare with **hypothesis.**
alcohol: a group of organic compound where the parent functional group is –OH.
alicyclic hydrocarbon: a hydrocarbon with carbon atoms joined together in a ring structure.
alkali: a chemical that reacts with an acid and is soluble in water releasing hydroxide ions, e.g. $OH^-(aq)$ ions.
alkanes: a homologous series of saturated hydrocarbons.
alkenes: a homologous series of unsaturated hydrocarbon.
amount of substance: the quantity whose unit is the mole. Chemists use 'amount of substance' as a means of counting atoms.
amphoteric substances: substances that can react as both acids and bases.
aliphatic hydrocarbon: a hydrocarbon where carbon atoms are joined together in straight or branched chains.
anhydrous: a substance that contains no water molecules.
anion: a negatively charged ion.
anomalies: data points that do not fit the overall trend in the data.
anomalous: results that do not follow the general pattern of the data.
aqueous: a solution in which the solvent is water. It is usually shown in chemical equations by appending (aq) to the relevant formula. For example, a solution of ordinary table salt, or sodium chloride (NaCl), in water would be represented as NaCl(aq).
aromatic hydrocarbons: contain at least one benzene ring.
atom: the smallest object that retains properties of an element. Composed of electrons and a nucleus (containing protons and neutrons).
atomic (proton) number: the number of protons in the nucleus of an atom.
atomic orbital: a region of space where it is likely that you will find electrons. Each orbital can hold up to two electrons, with opposite spins.
atomic radius: a measure of the size of its atoms, usually the mean or typical distance from the nucleus to the boundary of the surrounding cloud of electrons.
atomic structure: theoretically consists of a positively charged nucleus surrounded by negatively charged electrons revolving in orbits at varying distances from the nucleus, the constitution of the nucleus and the arrangement of the electrons differing with various chemical elements. An atom contains equal numbers of negative electrons and positive neutrons and therefore has zero charge overall.
average bond enthalpy: the mean energy needed for 1 mole of a given type of gaseous bonds to undergo homolytic fission.
Avogadro constant, N_A: the number of atoms per mole of the carbon-12 isotope ($6.02 \times 10^{23}\,mol^{-1}$).
base: a chemical that can react with acids and is a proton acceptor.
biodegradable materials: are affected by the action of microorganisms and environmental conditions leading to decomposition.
biological catalysts: catalysts found in nature are called enzymes. Enzymes are of huge importance, for example they help to catalyse reactions in the body.
bioplastics: materials made from a renewable source that are biodegradable.
Boltzmann distribution: the distribution of energies of molecules at a particular temperature, often shown as a graph.
bond angle: the angle that is formed between two adjacent bonds on the same atom.
bonded pair: a pair of electrons that have been shared between two chemically bonded atoms.
bonding region: the space where an electron can be found in a bond.
carbonyls: An organic molecule which contains the C=O functional group.
carboxylic acids: an organic acid containing a carboxyl functional group –COOH.
calorimetry: the quantitative study of energy change in a chemical reaction.
catalyst: a substance that increases the rate of a reaction without being used up during the process.
categoric variable: a qualitative description of a variable.
cation: a positively charged ion.
CFCs: chlorofluorocarbons are a class of organic compounds which contain chlorine and fluorine. They are inert and also non-toxic to humans.
chemical energy: a special form of potential energy stored in chemical bonds.
***cis–trans* isomerism:** a type of E/Z isomerism where two substituent groups attached to each carbon atom of the C=C is the same.
collision theory: a model to help understand and make predictions about how changing temperature, pressure, concentration, surface area or catalyst may change the rate of reaction of a chemical change.
complete combustion: the oxidising of a fuel in a plentiful supply of air.
compound: a substance formed from two or more chemically bonded elements in a fixed ratio, usually shown by a chemical formula.
concentration of a solution: the amount of solute, in mol, dissolved per 1 dm^3 (1000 cm^3) of solution.
concordant results: values that are close to each other and therefore represent reliable quantitative data.
continuous variable: a measured value which could be any number.
control variable: a factor that you must keep constant between experimental runs so that you can compare results.

coordinate bond: see **dative covalent bond**.
covalent bond: a bond formed by a shared pair of electrons between nuclei.
curly arrows: model the flow of electron pairs during reaction mechanisms.
d-orbital: a region within an atom that can hold up to two electrons, with opposite spins.
dative covalent bond: a bond formed by a shared pair of electrons that has been provided by one of the bonding atoms only. Also known as a coordinate bond.
dehydration: a chemical reaction where a water molecules is eliminated from an organic compound.
delocalised electrons: electrons that are shared between more than two atoms.
delta negative: δ^- means 'slight negative charge'. The symbol δ is known as delta, so sometimes you may hear or see these slight charges referred to as delta negative.
delta positive: δ^+ means 'slight positive charge'. The symbol δ is known as delta, so sometimes you may hear or see these slight charges referred to as delta positive.
dependent variable: the factor that you observe in an experiment.
diatomic molecules: molecules composed only of two atoms, of either the same or different chemical elements.
dilute: the process of decreasing the concentration of a solute in solution, usually simply by mixing with more solvent. To dilute a solution means to add more solvent without the addition of more solute.
discrete variables: variables that can only be particular defined numbers.
displacement reaction: a reaction in which a more-reactive element takes the place of a less-reactive element in a compound.
displayed formula: a formula which shows the relative positioning of atoms and the bonds between them.
disproportionation: the oxidation and reduction of the same element in a redox reaction.
distillation: a technique used to separate miscible liquids or solutions.
dot-and-cross diagrams: used to model the electrons in chemical bonding.
dot formulae: gives the ratio between the number of compound molecules and the number of water molecules within the crystalline structure.
dynamic equilibrium: the equilibrium that exists in a closed system when the rate of the forward reaction is equal to the rate of the reverse reaction and all the chemicals have their concentrations maintained.
***E*/*Z* isomerism:** a type of stereoisomerism that is caused by the restricted of rotation around the double bond – two different groups are attached to each carbon atom of the C=C double bond.
electrical conductivity: the degree to which a specified material allows charge to be carried. It is calculated as the ratio of the current density in the material to the electric field which causes the flow of current.
electron: the smallest of the particles that make up an atom, and they carry a negative charge. The number of protons and electrons is equal in each atom.
electron configuration: the arrangement of electrons in an atom or ion.
electron shielding: the repulsion between electrons in different inner shells. Shielding reduces the net attractive force from the positive nucleus on the outer-shell electrons.
electronegativity: a measure of the attraction of an electron in a covalent bond.
electrophile: an electron-pair acceptor.
element: a substance that cannot be broken down into simpler substances by chemical means. An element is composed of atoms that have the same atomic number, that is, each atom has the same number of protons in its nucleus as all other atoms of that element.
elimination: an organic chemical reaction in which one reactant forms two products. Usually a small molecule like water is released.
empirical formula: the simplest whole-number ratio of atoms of each element present in a compound.
endothermic: a reaction in which the enthalpy of the products is greater than the enthalpy of the reactants, resulting in heat being taken in from the surroundings (ΔH is positive).
enthalpy, *H*: the heat content that is stored in a chemical system.
enthalpy change of combustion, $\Delta_c H^\circ$: the energy change that takes place when 1 mole of a substance is completely combusted.
Enthalpy change of formation, $\Delta_f H^\circ$: the energy change that takes place when 1 mole of a compound is formed from its constituent elements in their standard state under standard conditions.
enthalpy change of neutralisation, $\Delta_{neut} H^\circ$: the energy change associated with the formation of 1 mole of water from a neutralisation reaction, under standard conditions.
enthalpy change of reaction, $\Delta_r H^\circ$: the energy change associated with a given reaction.
enthalpy cycle: a pictorial representation showing alternative routes between reactants and products.
enthalpy profile diagram: a diagram of a reaction that allows you to compare the enthalpy of the reactants with the enthalpy of the products.
equilibrium constant, K_c: Chemists consider the position of the equilibrium using the equilibrium constant, K_c. K_c gives a measure of where the equilibrium lays, essentially by giving the ratio between products and reactants
equilibrium law: a law that states that for the equilibrium:

$$aA + bB \leftrightarrow cC + dD$$

$$K_c = \frac{C^c D^d}{A^a B^b}$$

ester: a functional group of COO found in some organic molecules.
esterification: the chemical reaction which forms an ester.
exothermic: a reaction in which the enthalpy of the products is smaller than the enthalpy of the reactants, resulting in heat loss to the surroundings (ΔH is negative).
experiment: an ordered set of practical steps, which are used to test the hypothesis.
extraneous variable: a factor that is not controlled or measured in an experiment but may introduce error into the results.
false positive: a chemical test is when a positive result is produced but not due to the desired product being formed.
first ionisation energy: the energy required to remove one electron from each atom in one mole of gaseous atoms to form one mole of gaseous 1+ ions.

fragmentation: the process in mass spectrometry that causes a positive ion to split into pieces, one of which is a positive fragment ion.
functional group: a group of atoms that is responsible for the characteristic chemical reactions of a compound.
general formula: the simplest algebraic formula for a homologous series.
giant covalent lattice: a three-dimensional structure of atoms that are all bonded together by strong covalent bonds.
giant ionic lattice: a three-dimensional structure of oppositely charged ions, held together by strong electrostatic forces of attraction (ionic) bonds.
giant metallic lattice: a three-dimensional structure of positive ions and delocalised electrons, bonded together by strong metallic bonds.
global warming: the increased average temperature of the planet and atmosphere, thought to be caused by increased concentrations of CO_2. If unchecked it is thought that global warming will lead to climate change.
gradient: the gradient is the measure of a slope in a graph. For a straight line, choose two points on the graph. Note the change in the *y*-axis and divide this by the change in the *x*-axis.
group: a vertical column in the periodic table. Elements in a group have similar chemical properties and their atoms have the same number of outer-shell electrons.
halogenation: an addition reaction where a halogen is added across the C=C.
haloalkanes: a group of chemical compounds derived from alkanes containing one or more halogens. They are a subset of the general class of halocarbons, although the distinction is not often made.
Hess' law: states that the enthalpy change in a chemical reaction is independent of the route it takes.
heterogeneous catalysts: a catalyst used in a reaction which is in a different phase from the reactant. A typical example for this would be a solid catalyst used in liquid reactants, or gaseous reactants passed over a solid catalyst.
heterolytic fission: happens when each bonding atom receives one electron from the bonded pair, forming two radicals.
homogeneous catalysts: a catalyst used in a reaction which is in the same phase as the reactant. This could include for example a liquid catalyst that is mixed with liquid reactants. It could also include a gaseous catalyst with gaseous reactants.
homologous series: a series of organic compounds that have the same functional group with successive members differing by CH_2.
homolytic fission: happens when one bonding atom receives both electrons from the bonding pair.
hydrated: a crystalline compound containing water molecules.
hydration: a reaction where water is a reactant in a chemical reaction.
hydrocarbons: compounds that contain only hydrogen atoms and carbon atoms.
hydrogen bond: a strong permanent dipole–permanent dipole attraction between an electron-deficient hydrogen atom ($O–H^{\delta+}$, $N–H^{\delta+}$, $F–H^{\delta+}$) on one molecule, and a lone pair of electrons on a highly electronegative atom (O, N or F) on a different molecule.
hydrogenation: an addition reaction where hydrogen is added across the C=C.
hydrolysis: a chemical reaction in which water is a reactant. There are alkali hydrolysis reactions where $–OH^-$ is the reacting species.
hypothesis: a prediction and explanation of the chemistry behind the prediction.
incomplete combustion: oxidising a fuel in a limited supply of air.
independent variable: the factor that you are interested in changing to see the effect it has on one other factor.
infrared: refers to the region of the electromagnetic spectrum with wavelengths between 700 nm and 300 μm. It is the region between red in the visible spectrum and microwaves.
infrared spectroscopy: an analytical technique that utilises a substances behaviour on absorbing infrared radiation and produces a spectrum that can be used to identify certain functional groups, such as the OH group.
intermolecular force: an attractive force between neighbouring molecules or atoms.
ion: a positively or negatively charged atom or (covalently bonded) group of atoms (a molecular ion).
ionic bond: the electrostatic attraction between oppositely charged ions.
ionisation: the process of an atom becoming an ion.
isotopes: atoms of the same element with different numbers of neutrons.
kilojoules: a measurement of energy, where 1 kJ = 1000 J.
law of conservation of energy: energy cannot be created or destroyed, only moved from one place to another.
Le Chatelier's principle: when a system in dynamic equilibrium is subjected to a change, the position of equilibrium will shift to minimise the change.
line of best fit: shows the trend in the plotted points – this could be a straight line, curve or any 's' shape.
London (dispersion) forces: attractive forces between induced dipoles in neighbouring molecules.
lone pair: an outer-shell pair of electrons that is not involved in chemical bonding.
margin of error: shows the range that a value lies within.
mass (nucleon) number: the number of particles (protons and neutrons) in the nucleus.
mass spectrometry: an analytical chemistry technique that helps identify the amount and type of chemicals present in a sample by measuring the mass-to-charge ratio and abundance of gas-phase ions.
metallic bonding: the electrostatic attraction between positive metal ions (cations) and delocalised electrons.
meta study and meta-analysis: a type of secondary research. This is using the raw data from a variety of studies and then using it to try to answer the new aim. This is a mathematical approach using statistics and is often used by social scientists.
method: a step-by-step detailed explanation of how to complete an experiment.
molar gas volume: the volume per mole of a gas. The units of molar volume are $dm^3\,mol^{-1}$. At room temperature and pressure, the molar volume is approximately $24.0\,dm^3\,mol^{-1}$.
molar mass, *M*: the mass per mole of a substance. The units of molar mass are $g\,mol^{-1}$.
molar volume: the volume per mole of a gas. The units of molar volume are $dm^3\,mol^{-1}$. At room temperature and pressure, the molar volume is approximately $24.0\,dm^3\,mol^{-1}$.

mole: the amount of any substance containing as many particles as there are carbon atoms in exactly 12 g of the carbon-12 isotope.
molecular formula: shows the numbers and type of the atoms of each element in a compound.
molecular ion, M^+: the positive ion formed in mass spectrometry when a molecule loses an electron.
molecule: a group of atoms held together by covalent bonds.
monomers: small molecules that are used to make polymers.
multiple covalent bond: some non-metallic atoms can share more than one pair of electrons with another atom to form a multiple bond , e.g. a double bond or a triple bond.
neutralisation: the reaction of an acid with a base.
neutron: a particle found in the nucleus of an atom. It is almost identical in mass to a proton, but carries no electric charge.
nomenclature: the naming system for compounds.
nucleons: particles found in the nucleus of an atom (protons and neutrons).
nucleophiles: electron pair-donors.
nucleophilic substitution: a chemical reaction in which an atom or group of atoms is exchanged for a nucleophile.
organic synthesis: A branch of chemistry which designs reactions to make a target molecule.
outline: a summary of the experiment.
oxidation: The loss of electrons, loss of hydrogen, gain of oxygen or an increase in oxidation number.
oxidising agent: a reagent that oxidises (takes electrons from) another species.
ozone layer: an area of high concentration of ozone, O_3, in the stratosphere.
p-orbital: a region within an atom that can hold up to two electrons, with opposite spins.
particle model: a simplification of matter which can be used to explain observations and make predictions.
percentage error: a mathematical way of comparing the experimental value with the actual value.
percentage (%) yield: actual amount, in mol, of product/ theoretical amount, in mol, of product × 100
period: a horizontal row of elements in the Periodic Table. Elements show trends in properties across a period.
periodic table: a tabular arrangement of the chemical elements, organized on the basis of their atomic number (number of protons in the nucleus), electron configurations, and recurring chemical properties.
periodicity: a regular periodic variation of properties of elements with atomic number and position in the Periodic Table
permanent dipole: a small charge difference across a bond that results from a difference in the electronegativities of the bonded atoms.
permanent dipole–dipole interaction: a weak attractive force between permanent dipoles and permanent dipoles or induced dipoles in neighbouring polar molecules.
photodegradable: capable of being decomposed by the action of light, especially sunlight.
pi (π) bonds: sideways overlap of adjacent p-orbitals above and below the bonding C atoms.
plan: a summary of the experiment that you wish to complete.
polar covalent bond: a covalent bond that has a permanent dipole.
polymers: a macromolecules made from small repeating units.
precipitation reaction: the formation of a solid from a solution during a chemical reaction. Precipitates are often formed when two aqueous solutions are mixed together.
precision: the degree to which repeated values, collected under the same conditions in an experiment, show the same results.
pressure: a measure of the force applied over a unit area
primary alcohol: has the functional group attached to a carbon atom with no more than one alkyl group.
primary haloalkane: has the halogen on the end of the parent chain.
principal quantum number, n: a number representing the relative overall energy of each orbital, which increases with distance from the nucleus. The sets of orbitals with the same n value are referred to as electron shells or energy levels.
product: substances made/formed in a chemical reaction.
proton: particle found in a nucleus with a positive charge. The number of these gives the atomic number.
qualitative data: a description of what is being observed.
qualitative test: a simple test where an observation can be used to identify a species.
quantative data: a quantity (number) of what is being observed.
quickfit: a selection of heat resistant glassware with connectors that can be easily put in a variety of arrangements.
radical: a highly reactive species with one or more unpaired electrons.
radiation: the emission of energy as electromagnetic waves or as moving subatomic particles, especially high-energy particles which cause ionization.
radical substitutions: an organic mechanism involving radicals where one or more atoms get exchanged.
rate of reaction: the change in concentration of a reactant or a product in a given time.
reactants: a substance that takes part in and undergoes change during a reaction.
reaction mechanisms: models which show clearly the movement of electron pairs in a reaction.
recycling: collecting, sorting and processing of waste for a different use.
redistillation: the purification of a liquid using multiple distillations.
redox reaction: a reaction in which both reduction and oxidation take place.
reducing agent: a reagent that reduces (adds electron to) another species.
reduction: the gain of electrons, gain of hydrogen, loss of oxygen or a decrease in oxidation number.
reflux: a technique used to stop reaction mixtures boiling away into the air.
relative atomic mass, A_r: the weighted mean mass of an atom of an element compared with one-twelfth of the mass of an atom of carbon-12.
relative isotopic mass: the mass of an atom of an isotope compared with one-twelfth of the mass of an atom of carbon-12.
reliable: results that are similar when they are repeated.
repeating unit: a specific arrangement of atoms that occurs in the structure over and over again.
resolution: the smallest change in quantity being measured that can be observed.

s-orbital: A spherical shaped region within an atom that can hold up to two electrons with opposite spins. Found at the n=1 level and above.

salt: any chemical compound formed from an acid when an H^+ ion from the acid has been replaced by a metal ion or another positive ion, often a metal or ammonium ion, NH_4^+.

saturated compounds: compounds that have only single bonds.

saturated organic chemicals (compounds): organic chemicals that have only single covalent bonds.

saturation: the degree or extent to which a species is dissolved or absorbed compared with the maximum possible, usually expressed as a percentage.

scatter graphs: a method for expressing quantative data, where the independent variable must be continuous and noted on the *x*-axis. The dependent variable must also be continuous but displayed on the *y*-axis.

secondary alcohol: has the functional group attached to a carbon atom with two alkyl groups.

separating funnel: a piece of equipment used to separate immiscible liquids.

shell: a group of atomic orbitals with the same principal quantum number, *n*. Also known as a main energy level.

sigma bonds: made by direct overlap between orbitals on the bonding atoms.

significant figures: the numbers used to represent a quantity that have meaning.

single covalent bonds: where atoms are bonded by one shared pair of electrons between nuclei.

simple molecular lattice: a three-dimensional structure of molecules held together by weak intermolecular forces.

skeletal formula: a simplified structural formula drawn by removing hydrogen atoms from alkyl chains.

solubility: the property of a solid, liquid, or gaseous chemical substance called solute to dissolve in a solid, liquid, or gaseous solvent to form a homogeneous solution of the solute in the solvent.

species: any type of particle that takes place in a chemical reaction.

specific heat capacity: the energy required to raise 1 g of a substance by 1 K.

standard condition: standard sets of conditions (temperature and pressure) for experimental measurements established to allow comparisons to be made between different sets of data. This is usually set at 100 kPa, 298 K and all solutions have a concentration of $1\,mol\,dm^{-3}$.

standard form: method for writing very small or very large numbers. Standard form is always written as:

$A \times 10^n$

where *A* is the first significant figure and n is a description of the direction and number of places that the decimal point has moved.

standard solution: a solution of known concentration. Standard solutions are normally used in titrations to determine unknown information about another substance.

stereoisomers: organic compounds with the same molecular formula and structural formula but having different arrangements of atoms in space.

stoichiometry: the molar relationship between the relative quantities of substances taking part in a reaction.

structural formula: provides the minimum detail to show the arrangement of atoms in a molecule.

structural isomers: compounds with the same molecular formula but different structural formula.

sub-shell: a group of the same type of atomic orbitals (s, p, d or f) within a shell. An 's' subshell can hold a maximum of 2 electrons, a 'p' subshell can hold a maximum of 6 electrons and a 'd' subshell can hold a maximum of 10 electrons.

successive ionisation: values that are a measure of the energy required to remove each electron in turn.

surveys: a type of primary research. It is setting out limits to observe something that is already happening.

sustainability: chemical processes which minimise the use and generation of hazardous materials.

synthetic polymers: human-made polymers.

table: a clear and structured way of recording information about an experiment.

tertiary alcohol: has the functional group attached to a carbon atom with three alkyl groups.

thermodynamics: a branch of physical chemistry concerned with changes in the energy of chemical systems and surroundings.

titrations: the slow addition of one solution of a known concentration (called a titrant) to a known volume of another solution of unknown concentration until the reaction reaches an end point, which is often indicated by a colour change.

unsaturated organic chemicals (compounds): organic chemicals that contain at least one carbon–carbon double covalent bond.

valid experiment: provides information to test the aim of the experiment.

Van der Waals' forces: a type of intermolecular bonding that includes **permanent dipole-dipole** bonding and **induced dipole-dipole interactions (London forces)**

variables: factors that can affect the outcome of an experiment.

volatility: the tendency of a substance to vaporize.

volume: the quantity of three-dimensional space occupied by a liquid, solid, or gas.

water of crystallisation: water molecules that form an essential part of the crystalline structure of a compound.

weighing by difference: a method used to accurately weigh the amount of material transferred. The mass of a container before and after transferring the material is taken and the difference between these values is the mass of material transferred.

yield: the amount of product obtained in a chemical reaction.

Key

atomic (proton) number
atomic symbol
name
relative atomic mass

Period	1 (1)	2 (2)	3	4	5	6	7	8	9	10	11	12	13 (3)	14 (4)	15 (5)	16 (6)	17 (7)	18 (0)
1	1 **H** hydrogen 1.0																	2 **He** helium 4.0
2	3 **Li** lithium 6.9	4 **Be** beryllium 9.0											5 **B** boron 10.8	6 **C** carbon 12.0	7 **N** nitrogen 14.0	8 **O** oxygen 16.0	9 **F** fluorine 19.0	10 **Ne** neon 20.2
3	11 **Na** sodium 23.0	12 **Mg** magnesium 24.3											13 **Al** aluminium 27.0	14 **Si** silicon 28.1	15 **P** phosphorus 31.0	16 **S** sulfur 32.1	17 **Cl** chlorine 35.5	18 **Ar** argon 39.9
4	19 **K** potassium 39.1	20 **Ca** calcium 40.1	21 **Sc** scandium 45.0	22 **Ti** titanium 47.9	23 **V** vanadium 50.9	24 **Cr** chromium 52.0	25 **Mn** manganese 54.9	26 **Fe** iron 55.8	27 **Co** cobalt 58.9	28 **Ni** nickel 58.7	29 **Cu** copper 63.5	30 **Zn** zinc 65.4	31 **Ga** gallium 69.7	32 **Ge** germanium 72.6	33 **As** arsenic 74.9	34 **Se** selenium 79.0	35 **Br** bromine 79.9	36 **Kr** krypton 83.8
5	37 **Rb** rubidium 85.5	38 **Sr** strontium 87.6	39 **Y** yttrium 88.9	40 **Zr** zirconium 91.2	41 **Nb** niobium 92.9	42 **Mo** molybdenum 95.9	43 **Tc** technetium (98)	44 **Ru** ruthenium 101.1	45 **Rh** rhodium 102.9	46 **Pd** palladium 106.4	47 **Ag** silver 107.9	48 **Cd** cadmium 112.4	49 **In** indium 114.8	50 **Sn** tin 118.7	51 **Sb** antimony 121.8	52 **Te** tellurium 127.6	53 **I** iodine 126.9	54 **Xe** xenon 131.3
6	55 **Cs** caesium 132.9	56 **Ba** barium 137.3	57 **La*** lathanum 138.9	72 **Hf** hafnium 178.5	73 **Ta** tantalum 180.9	74 **W** tungsten 183.8	75 **Re** rhenium 186.2	76 **Os** osmium 190.2	77 **Ir** iridium 192.2	78 **Pt** platinum 195.1	79 **Au** gold 197.0	80 **Hg** mercury 200.6	81 **Tl** thallium 204.4	82 **Pb** lead 207.2	83 **Bi** bismuth 209.0	84 **Po** polonium (209)	85 **At** astatine (210)	86 **Rn** radon (222)
7	87 **Fr** francium (223)	88 **Ra** radium (226)	89 **Ac*** actinium (227)	104 **Rf** rutherfordium (267)	105 **Db** dubnium (268)	106 **Sg** seaborgium (269)	107 **Bh** bohrium (270)	108 **Hs** hassium (269)	109 **Mt** meitnerium (278)	110 **Ds** darmstadtium (281)	111 **Rg** roentgenium (280)	112 **Cn** copernicium (285)		114 **Fl** flerovium (289)		116 **Lv** livermorium (293)		

58 **Ce** cerium 140.1	59 **Pr** praseodymium 140.9	60 **Nd** neodymium 144.2	61 **Pm** promethium (145)	62 **Sm** samarium 150.4	63 **Eu** europium 152.0	64 **Gd** gadolinium 157.3	65 **Tb** terbium 158.9	66 **Dy** dysprosium 162.5	67 **Ho** holmium 164.9	68 **Er** erbium 167.3	69 **Tm** thulium 168.9	70 **Yb** ytterbium 173.1	71 **Lu** lutetium 175.0
90 **Th** thorium 232.0	91 **Pa** protactinium 231.0	92 **U** uranium 238.0	93 **Np** neptunium (237)	94 **Pu** plutonium (244)	95 **Am** americium (243)	96 **Cm** curium (247)	97 **Bk** berkelium (247)	98 **Cf** californium (251)	99 **Es** einsteinium (252)	100 **Fm** fermium (257)	101 **Md** mendelevium (256)	102 **No** nobelium (259)	103 **Lr** lawrencium (262)

Index